ACK.

Rec

BY WHOM

D1458502

MCCANDLESS & BARTON

Electric Lifts

MODERN HIGH SPEED GEARLESS
VARIABLE VOLTAGE LIFT
(*Express Lift Co.*)

Electric Lifts

A MANUAL ON THE CURRENT PRACTICE
IN THE DESIGN, INSTALLATION, WORKING
AND MAINTENANCE OF LIFTS

by

R. S. Phillips

M.I.E.E., Staff Engineer
Post Office Engineering Department

FIFTH EDITION

Pitman Publishing

Fifth edition 1966
Reprinted 1970

SIR ISAAC PITMAN AND SONS LTD.
Pitman House, Parker Street, Kingsway, London, W.C.2
P.O. Box 6038, Portal Street, Nairobi, Kenya

SIR ISAAC PITMAN (AUST.) PTY. LTD.
Pitman House, Bouverie Street, Carlton, Victoria 3053, Australia

PITMAN PUBLISHING COMPANY S.A. LTD.
P.O. Box 9898, Johannesburg, S. Africa

PITMAN PUBLISHING CORPORATION
6 East 43rd Street, New York, N.Y. 10017, U.S.A.

SIR ISAAC PITMAN (CANADA) LTD.
Pitman House, 381-383 Church Street, Toronto, 3, Canada

THE COPP CLARK PUBLISHING COMPANY
517 Wellington Street, Toronto, 2B, Canada

©

R. S. Phillips
1966

SBN: 273 41044 X

MADE IN GREAT BRITAIN AT THE PITMAN PRESS, BATH
G0—(T.116)

PREFACE TO FIFTH EDITION

THE fourth edition has been completely revised, all out-of-date matter having been replaced by newer material, and much additional material is also included. Many of the figures have been replaced by illustrations of more recent equipment and additional illustrations of modern equipment have been inserted.

The erection of large blocks of multi-storey flats has required extensive lift installations and details of the practices and equipments for this purpose are now included. Fire precautions in these buildings have also become important and there is now information on this aspect as it affects the lift installations. Considerable progress has been made in the automatic control of groups of high-speed lifts for large high buildings erected in recent years, and details of these control systems are now given. The use of static switching devices in lift controllers has been investigated by several manufacturers, and progress in this field is now described in some detail in the chapter on Controllers. Other aspects which particularly have received fuller treatment are capital and maintenance costs, motor-speed control, gearing, testing, and radio and television interference suppression.

R. S. P.

LONDON
1966

PREFACE TO FIRST EDITION

MANY years have elapsed since the publication of a book dealing with British lift practice, and in view of the rapid development of the industry during recent years the Author feels that the present time is opportune for the production of a work describing modern lift equipment and methods.

The idea originated during the reading of the Author's paper entitled "Modern Electric Passenger Lifts" before the Institution of Post Office Electrical Engineers in 1935. Mention was made, during the discussion, of the lack of published information on the subject, whilst added weight was given to these remarks by subsequent talks with members of some of the lift manufacturing firms.

One of the difficulties encountered in preparing the material was the use, by various authorities interested in lift installation, of different terms for the same items of equipment and methods of control. The publication of the *Code of Practice for the Installation of Lifts and Escalators* by the Building Industries National Council, however, did much to remove these difficulties, and the Author decided to adopt, as far as possible, the terms recommended for general use by this Council. In addition, most of the safety measures embodied in the Code have been carried into this work.

A great deal of the information has been obtained from notes made by the Author during the past few years, whilst the origin of other details has been duly acknowledged in the text.

The value of the book has been considerably increased by the generous assistance rendered by the leading British lift and motor manufacturers who, in addition to supplying information regarding their equipment, were kind enough to loan blocks or photographs from which many of the illustrations have been prepared. In this connexion the Author desires to take this opportunity of thanking the following firms: British Thomson-Houston Co., Etchells Congdon & Muir, Express Lift Co., J. & E. Hall, Marryat & Scott, Metropolitan Vickers Electrical Co., R. J. Shaw, Waygood-Otis, Wm. Wadsworth, and in particular, Mr. W. Wood, Chief Electrical Engineer of the last-named firm, who checked a great deal of the material in Chapter XV and offered valuable suggestions. In preparing Chapter IV the Author gratefully acknowledges help received from British Ropes, Ltd., which firm loaned blocks for many of the illustrations in this chapter.

In conclusion, the Author wishes to state that, although employed by the Post Office, the practices described in this book are not necessarily those adopted by the Post Office Engineering Department. Obviously, however, many of the details are the result of experience gained during the erection and maintenance of Post Office lifts.

R. S. P.

LEEDS
1938

CONTENTS

CONTENTS

Chapter VII

Chapter VIII

Chapter IX

Chapter X

Chapter XI

Chapter XII

Chapter XIX

APPENDICES

INSETS

CHAPTER I

GENERAL

To provide the best possible lift service in a building, consistent with cost and any building restrictions, it is necessary that full consideration be given to the problem, and to do this all the available data regarding the type of building and its occupants should be closely studied. There are three main types of lift, namely *passenger*, *goods*, and *service*. Passenger lifts are those designed primarily for passenger service, goods lifts are mainly for the transport of materials, but may be required occasionally to carry passengers, whilst a service lift is so constructed as regards size, or otherwise, that it is impossible for passengers to enter the car, and hence goods only are carried. In any particular building it will soon be evident which of these types will be necessary, and some consideration will now be given to the selection of lifts for these various purposes.

PASSENGER LIFTS

Capacity. The total capacity of passenger lifts required in any building to give a certain grade of service is determined by the number of occupants and the number of visitors who will be expected to utilize a lift service, and both vary considerably with the type of building. The density of the visitors and occupants will be great for restaurants and theatres and less for offices, hotels, and flats. In this country a large number of buildings are of such size and nature that very little difficulty will be experienced in deciding that one passenger lift will give adequate service. It should be remembered, however, that it is often preferable to install two small lifts if, on the grounds of capacity alone, it is thought that a single lift will have to be of a fairly large capacity, say 3 000 lb. Two small lifts of 1 500 lb. capacity each will give better service than one large lift, the latter probably having been designed to cater for the maximum number of passengers at the periods of peak loading. Consequently, this lift will be running comparatively lightly loaded for most of the day, with a resulting decrease in

efficiency and increase in running costs. On the other hand, with
two small lifts, adequate service could probably be maintained
during the greater part of the day with one lift, the second being
brought into service during peak loads. The advantages to be
gained by installing two smaller lifts often outweigh the extra
initial outlay. Large buildings require a group of lifts and the
best service is obtained if these are installed adjacent to each
other.

Specification. Having decided that the number of occupants
and floors (usually not less than three floors for passenger
service) justify the provision of a lift, we then necessarily have
to prepare a specification detailing the various features
required. Although it is desirable from the safety point of
view that certain items should be specified, the general type of
passenger lift in this country is now so well established that
most of the reputable lift firms will design and supply a perfectly satisfactory and safe lift if provided with only the main
requirements such as well sizes, power supply characteristics,
contract speed, type of control, and contract load. The contract
speed and the contract load are the maximum values as specified
in the contract of purchase. The essential details can confidently be left to any British firm of repute if the buyer is not
in a position to furnish a comprehensive specification. Details
of specification requirements are contained in B.S. 2655 and
B.S. Code of Practice C.P. 407.101.

Speed. When determining a suitable car speed, the height
of the building, distance between stops and quality of service
desired must all be considered. The higher the car speed the
better the resulting service, but it must be remembered that
the cost of a lift increases as the contract speed is raised.
If a large amount of interfloor traffic is anticipated, the car
speed should not exceed about 200 ft. per min., otherwise
most of the running time will consist of acceleration and
retardation and the motor will not have sufficient time in
which to travel for an appreciable distance at its full speed. In
buildings of seven or eight storeys, where a fair amount of traffic
from the ground to the upper floors is expected, speeds up to
400 ft. per min. are now common. A number of the larger
buildings in this country having upwards of about twenty
storeys employ car speeds of up to 1 000 ft. per min. In

America, however, several of the "skyscrapers" have upwards of fifty storeys, and speeds of 1 000 ft. per min. are quite common, whilst the larger of these buildings are equipped with lifts travelling at 1 600 ft. per min. These high speeds are used for express service to the upper floors.

Car speeds are generally selected in relation to the number of floors served, and the usual practice in this country, for general office buildings, is as follows—

No. of Floors	Car Speed in ft per min.
2	100
3–4	100–150
5–6	200–300
7–9	300–400
10–12	400–500
12–15	500–600
over 15	600–1000

When installing a lift in an existing building, however, other factors such as the available top and bottom clearances may limit the lift speed.

Size and Capacity of Car. In deciding the approximate car size it is usual to allow 2 ft.² for each passenger and 3 ft.² for the attendant for cars up to 1 500 lb. capacity. The internal height should be not less than 6 ft. 6 in. To calculate the size of the lift machine, the car loading must be known, and in arriving at this the average weight of each passenger is taken as 150 lb., but frequently $1\frac{1}{2}$ cwt. is used as a rough approximation.

Type of Control. When considering the method of control to be adopted, the type of building and of its occupants and the lift capital cost are the determining factors. For intermittent traffic an attendant is not justified, and an automatic form of control should be installed. On the other hand, when the traffic is likely to be fairly regular, better service may be maintained by employing a car attendant and adopting car switch control. If the traffic is intermittent for the greater part of the day, but definite peak periods are expected, such as in the early morning, at midday, and in the evening, it is advisable to install a dual form of control. With this control, the car is normally worked automatically by the passengers, but by an attendant and car switch control during the periods of heavy traffic. Where the extra cost is warranted, automatic collective control or one of the automatic controls for banks of lifts is adopted. Automatic control is now rapidly replacing attendant control on modern lifts.

Capital Costs. The cost of a lift depends upon the details of the specification on which the tender is based and also varies appreciably with different contractors. As with many other forms of engineering equipment the price quoted is governed to some extent by the nearness of the specification to the contractor's standard items and also just where the customer's particular requirement falls in his standard ranges. A specification may suit one contractor but not another and this will result in a considerable difference in prices. The tendered price will also depend upon the state of the contractor's order book at the time of tendering.

In 1962* The Building Research Station of the Department of Scientific and Industrial Research examined in detail, data on the prices of lift installations from eighty-six successful tenders and sixty-one comparative estimates supplied by lift manufacturers. Much useful information was obtained on how the tender price varies with the height of building, number of floors served, lift speed and type of control. It must be borne in mind, however, that the figures quoted were based on average experience and they may not apply to any particular set of conditions. For example, the costs of alternative designs for a pair of lifts serving eighteen floors might differ by up to £6,000.

As a result of this investigation the following average costs were obtained for different types of lift.

1. *Contract speed of 100 ft. per min.* The cost per lift with simple automatic control varied from £2,800 for lifts serving five floors to £3,800 for lifts serving ten floors. The cost per lift with full collective control and interconnected lifts varied from £4,800 for lifts serving ten floors to £5,900 for lifts serving fourteen floors.

2. *Contract speed of 150 ft. per min.* The cost per lift with

* "The Costs of Lifts in Multi-storey Flats for Local Authorities," by T. L. Knight and A. E. Duck, *Chartered Surveyor*, Feb. 1962.

full collective control and interconnected lifts varied from £5,900 for lifts serving twelve floors to £7,500 for lifts serving eighteen floors. The costs of identical lifts, each serving only half the number of floors were £4,400 and £5,400 respectively.

3. *Contract speed of* 200 *ft. per min.* The cost per lift with full collective control and interconnected lifts varied from £7,500 for lifts serving sixteen floors to £9,800 for lifts serving twenty-four floors. The costs of identical lifts, each serving only half of the floors, were £5,600 and £6,900 respectively.

4. *Contract speed of* 300 *ft per min.* The cost per lift with full collective control and interconnected lifts varied from £9,000 for lifts serving eighteen floors to £10,800 for lifts serving twenty-four floors. The costs of identical lifts each serving only half of the floors were £6,900 and £7,900 respectively.

5. *Contract speed of* 500 *ft. per min.* Lifts running at this speed are usually gearless machines installed in banks with full collective group control. As an example of the 1965 costs of these lifts, a group with cars of 15-passenger capacity serving twelve floors would cost about £13,000 per lift.

From the costs studied it was possible to ascertain the importance of the various design factors in determining the capital cost of the installation.

(a) *Stops at all or alternate floors.* In blocks served by two lifts an economy can be achieved at the expense of quality of service by arranging for each lift to stop at alternate floors, i.e. each floor being served by one lift only. The omission of half the number of landing doors, trims to openings and control panels, resulted in a saving of £150 to £250 per floor in the cost of a pair of lifts.

(b) *Speeds of travel.* In blocks of flats up to about twelve floors high it is usual to install lifts of 100 ft. per min. with single-speed motors. In blocks of fourteen to eighteen floors, speeds of 150, 200 and 250 ft. per min. have been employed, whilst eighteen to twenty-four floors require a speed of 300 ft. per min. to provide a reasonable waiting interval. A 200 ft. per min. lift costs about £1,000 more than a 100 ft. per min. lift, and a 300 ft. per min. lift costs about a further £1,000.

(c) *Control.* Down collective control costs about £25 per lift per floor more than simple automatic control and full collective control adds about another £25 per lift per floor. When a bank of lifts serve the same floors they may be interconnected so that the landing call is answered by the lift which is nearest to the landing and travelling in the required direction. This adds about £900 to the cost of the installation for ten floors and £2,000 for twenty-four floors.

(*d*) *Type of car*. The car normally used for multi-storey flats carries eight passengers or a contract load of 1 200 lb. This will take large perambulators, most furniture or a stretcher in an inclined position. The extra cost of providing ten-person cars is about £150 per lift. For speeds above 200 ft. per min. gradual wedge clamp safety gear is fitted to the car and this will increase the price of the car by about £200.

(*e*) *Height of the building*. This affects the cost because of the increased number of doors, amounts of rope and guides, higher speed and more complex control, and the cost of each extra floor for a pair of lifts is between £400 and £600.

GOODS LIFTS

In many buildings it is necessary to install one or more goods lifts, these being designed primarily for the transport of goods, but occasionally to carry passengers. The design of a goods lift is similar in principle to that of a passenger lift, the main differences being that the car is of rougher construction, the entrances wider and the contract speed rarely exceeds 200 ft. per min. Accurate levelling, however, may be essential to facilitate the loading and unloading of trolleys filled with fragile goods, and in these cases one of the available schemes of corrective levelling is often incorporated. The service is usually intermittent, and the control is therefore either automatic or semi-automatic.

In determining the minimum contract load for a goods lift, careful consideration must be given to the type and size of load to be carried and the method of loading and unloading. For general goods which are distributed over the car floor, the contract load should be four times the weight of the heaviest single item or 70 lb. per square foot of car floor area, whichever is the greater.

If the goods are in a power-operated truck and the truck, together with its load, must be carried in the lift, the minimum contract load must be the total weight of the truck and its

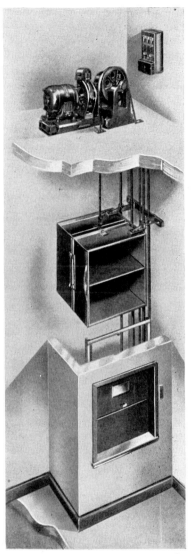

Fig. 1.1. Goods Lift with
Bi-parting Doors

(*Otis Elevator Co.*)

Fig. 1.2.
Typical Service Lift

(*Otis Elevator Co.*)

maximum load. If the goods lift is to be used to transport motor vehicles, the contract load must be equal to the weight of the heaviest vehicles to be carried or 35 lb. per square foot of car floor area, whichever is the greater. Special consideration must also be given to the car guides and their fixing because of the side loads that may be imposed when a heavy truck enters or leaves the car. Heavy goods loads demand well-designed brakes and adequate rope traction to avoid slipping in the sheave grooves. A typical goods lift is shown in Fig. 1.1 This is fitted with bi-parting landing doors which is a feature of large modern goods lifts.

SERVICE LIFTS

A service lift is one which is designed and constructed so that it is possible for goods only to be carried, and the factor of human safety does not therefore enter so largely into the design

FIG. 1.3. A BANK OF SERVICE LIFTS
(*Marryat & Scott*)

of these lifts. Usually the floor area does not exceed 9 sq. ft., and the car height does not exceed 4 ft. The capacity does not exceed 5 cwt. Fig. 1.2 shows a typical service lift car and winding machine. The type is chiefly used in hotels and restaurants for service from the kitchens to the dining-rooms, in banks for the transport of bullion, and in libraries for transport of books. The general principles governing the design are similar to those for a goods lift, but the machine and car are much smaller in size and safety gear is generally omitted, although buffers are fitted under the car and counterweight. The speed of travel is invariably between 50 and 150 ft. per min., whilst the control is either automatic or semi-automatic. Fig. 1.3 shows a bank of automatic service lifts in a large public restaurant. They can be called or sent to all floors, and a loudspeaker-microphone telephone system allows conversation between floors. The cars are lined with Formica and are fitted with rise and fall hatches. The fittings are of stainless steel.

CHAPTER II

TRAFFIC ANALYSIS

THE larger buildings require several passenger lifts to provide adequate service, and the problems associated with the design of such installations are much more complex than when a single lift will fulfil the requirements. Variations of speed, capacity, number and position of lifts are the main factors which effect the quality of service and the cost, and from the large number of possible combinations one must be selected which will best satisfy the requirements of the particular building with due regard to economics.

Preliminary Considerations. A lift installation is the result of joint efforts by the lift designer, architect, builder, and lift contractor, although in many cases the lift contractor combines the functions of lift designer, constructor and erector. The lift designer knows what he requires of the building, the architect is aware of the methods of best fulfilling these requirements, and the builder carries out the necessary structural work. Close liaison between these three during all sketch plan and working drawing stages is often the difference between a costly and inferior installation, and one that is satisfactory in all respects. It is most unsatisfactory for an architect to reach the working drawing stages without having been made aware of the number and positions of the lifts and of the sizes of the wells and machine rooms or, on the other hand, for the lift designer to be asked for this information at so late a stage and perhaps be requested that his lifts should suit some space on the drawings which has been left for this purpose.

The first matter for consideration is the positions to be occupied by the lift wells, the main requirements being that users will be able to pass quickly from the building entrance to the lifts and that the lift exits on the higher floors will be as near as possible to the centres of population of the floors. If the building has one main street entrance, the passenger lifts should be arranged adjacent to each other in a single bank and conveniently situated with regard to the entrance. From the lift service aspect the bank should not be arranged in two sections, one on each side of a central staircase, which practice is sometimes adopted for appearance or to satisfy architectural requirements. With a single bank, a common machine room can be used and so simplify maintenance, it is easier to arrange lift interconnexion facilities and the resultant service is better than if the lifts are separated. Figs. 2.1 and 2.2 show the winding machines and controllers respectively

FIG. 2.1. WINDING MACHINES FOR TWO GEARLESS LIFTS
(*Marryat and Scott*)

for two interconnected duplex-controlled gearless lifts in a common machine room. The contract load is 1 500 lb. and the contract speed 500 ft. per min. If the building has two main entrances, two banks of lifts are necessary, the number in each bank being governed by the number of passengers that will be expected to use each entrance. Although the lift entrances are usually near the stairway, care should be taken in the design to ensure that persons intending to use the lifts are kept clear of those who wish to use the stairs, which are for floor to floor foot traffic and for use in an emergency.

Grade of Service. Consideration must next be given to the quality of service it is desired to provide, and this depends

FIG. 2.2. CONTROLLERS FOR TWO GEARLESS LIFTS

Marryat and Scott

upon the type of building, its rental value, or on the importance
of its occupants. This quality of service is a measure of the speed
with which passengers can be transported to their destinations
and hence is determined by the sum of the time which the
average passenger has to wait for the arrival of a lift and the
time for the lift to reach the desired floor. The shorter these
times the better will be the service provided. The maximum
time a person may have to wait for a lift is termed the Waiting
Interval (W.I.) and is the interval between the arrival of
successive cars. This depends upon the round trip time
(R.T.T.) of each lift and on the number of lifts in the bank.
The R.T.T. is the time which elapses between a lift leaving the
ground floor and again arriving at that floor after making an
average number of stops at the upper floors with an average
number of passengers. It will be appreciated that it is possible
to obtain the same grade of service with a large number of
slow speed lifts as with a small number of high speed lifts, and
economics will determine the best arrangement between these
two extremes. A lift service which has a small W.I. and a
large travelling time always appears to the user to be better
than an equivalent service with a larger W.I. and a smaller
travelling time, as a long wait tends to make a person impatient.
For this reason it is usual to design for a waiting interval of
from 20–60 seconds, the time selected depending upon the
class of building.

No definite standards of service have yet been defined, but
the qualities or grades of the lift services in different buildings
may be assessed by comparing for each building the sum of
the average time a person has to wait for a lift and the average
travelling time. Hence the grade of service is determined by
$\dfrac{\text{W.I.}}{2} + \dfrac{\text{R.T.T.}}{4}$. If there are N lifts in the bank then W.I.
$= \dfrac{\text{R.T.T.}}{N}$ and hence $\dfrac{\text{W.I.}}{2} + \dfrac{N \times \text{W.I.}}{4}$, or $\dfrac{\text{W.I.}}{4}(2 + N)$, is a
measure of the grade of service. It may be noted that in a
single lift installation $N - 1$ and W.I. = R.T.T. In this case
the grade of service is measured by $\dfrac{3\,\text{W.I.}}{4}$. As an example
consider a bank of 4 lifts with an interval of 30 seconds between

the arrival of each lift and in which the R.T.T. is 120 seconds. The average passenger would have to wait 15 seconds at the ground floor and his travelling time would be $\dfrac{120}{4} = 30$ seconds. The total waiting and travelling time would therefore be 45 seconds. From observations that have been made on many existing installations, it is reasonable to classify the grade of service as excellent, good, fair or casual if the value of $\dfrac{\text{W.I.}}{4} (2 + N)$ is not more than 45 seconds, between 45 and 55 seconds, between 55 and 65 seconds, or more than 65 seconds respectively. For any particular building a suitable time will be selected by the lift designer after he has considered closely the building and its occupants.

Size of Car. Usual car sizes for these large buildings are such that the car will accommodate 8, 10, 15 or 20 persons, the car floor area being determined as described in Chapter X. In choosing the car shape it is desirable that the width should be not less than the depth so as to facilitate the rapid entry and egress of passengers. A car which is approximately square in shape is good practice.

Sometimes the car is required to be large enough to accommodate occasionally some special item of equipment and this will therefore indicate the size of car. In the absence of any such special requirement in designing for a bank of lifts, it is necessary to commence by choosing some arbitrary car size as a basis for consideration. For this purpose a 10-passenger car is suitable and subsequent investigation will then show whether a smaller or larger car would be preferable.

Probable Number of Stops. A round trip is composed of a number of factors and of these it is best to consider first the probable number of stops that will be made at the upper floors during the round trip. In the design of a lift service the probable conditions during the periods of heaviest traffic must be closely investigated. In office buildings the peak is generally in the morning when the building is being filled or during the evening when it is being emptied. If the lifts are designed so that they will satisfactorily clear the traffic during the peak period, the service will generally be adequate at all other periods.

Consequently it is necessary to estimate the probable number of stops that will be made during a trip in the peak period. The approximate numbers of persons employed on each floor must be known or estimated as well as the number of visitors likely to travel to the floors. A detailed study of these figures, together with a knowledge of the uses to which the building will be put, will enable the lift designer to estimate the probable peak periods and also the number of persons likely to travel to each floor during these periods. In many cases this information is sufficient for an experienced designer to estimate closely the probable number of stops during a trip in the peak period and where these are likely to be made, without more detailed study or calculation.

In the very large or high buildings an estimate by such means becomes more difficult and in these cases the formula developed below is used—

Let N = the total number of passengers entering the car at the ground floor each trip during the peak period,

P = the total population on all floors served during the peak period,

$P_a P_b P_c$ = population on the 1st, 2nd, 3rd, etc., floor served by the lift,

n = number of floors served above the ground floor.

The probability that any passenger will leave the car at the first floor is $\dfrac{P_a}{P}$, at the second floor $\dfrac{P_b}{P}$, and so on.

The probability that the passenger will not leave the car at the first floor is $1 - \dfrac{P_a}{P} = \dfrac{P - P_a}{P}$

The probability that none of the passengers will leave the car at the first floor is $\left(\dfrac{P - P_a}{P}\right)^N$

Then for any particular floor the probability that no passenger will wish to leave the car is

$$\frac{1}{n}\left[\left(\frac{P - P_a}{P}\right)^N + \left(\frac{P - P_b}{P}\right)^N + \cdots \cdots + \left(\frac{P - P_n}{P}\right)^N \right]$$

Hence the probability that there will be a stop at any particular floor is

$$1 - \frac{1}{n}\left[\left(\frac{P-P_a}{P}\right)^N + \left(\frac{P-P_b}{P}\right)^N + \cdots + \left(\frac{P-P_n}{P}\right)^N\right]$$

Therefore the average probability that there will be stops at n floors is

$$n\left\{1 - \frac{1}{n}\left[\left(\frac{P-P_a}{P}\right)^N + \left(\frac{P-P_b}{P}\right)^N \right.\right.$$
$$\left.\left. + \cdots + \left(\frac{P-P_n}{P}\right)^N\right]\right\}$$

or the probable total number of stops S_n is

$$n - \left[\left(\frac{P-P_a}{P}\right)^N + \left(\frac{P-P_b}{P}\right)^N + \cdots + \left(\frac{P-P_n}{P}\right)^N\right]$$

It follows that the expression within the brackets $\left[\right]$ represents the probable number of landings at which no stop will be made.

If stops are always made at certain floors, e.g. the third and sixth, the corresponding terms are omitted from the formula which, in this example, then becomes

$$S_{n-2} = (n-2) - \left[\left(\frac{P-P_a}{P}\right)^N + \left(\frac{P-P_b}{P}\right)^N + \left(\frac{P-P_d}{P}\right)^N\right.$$
$$\left. + \left(\frac{P-P_e}{P}\right)^N + \left(\frac{P-P_g}{P}\right)^N + \cdots + \left(\frac{P-P_n}{P}\right)^N\right]$$

and $S_n = S_{n-2} + 2$.

If each floor has the same population i.e.

$$P_a = P_b = P_c \cdots = P_n = \frac{P}{n}$$

then the probable number of stops

$$S_n = n - n\left(1 - \frac{1}{n}\right)^N$$
$$= n\left[1 - \left(\frac{n-1}{n}\right)^N\right]$$

Round Trip Time. Having selected what might be a suitable car size (10 persons) and determined the probable number of stops, detailed consideration can next be given to the various factors comprising the R.T.T. during the busy period. The R.T.T. will be composed of times for opening and closing of doors, passengers to leave and enter the car, acceleration and retardation of the car near landings, running at contract speed, and for operations of the car switch, or push buttons.

Door Operation. The time required for opening and closing the doors depends upon the method of operation and on the type and width of the doors. With a 4 ft. entrance and two-speed power operated doors, in which the landing and car doors move simultaneously, the minimum time for opening and closing is about 4 seconds. If centre opening doors are fitted, these move a distance of only half the entrance width and hence the corresponding time is approximately 3 seconds. The doors should be as light as possible to facilitate quick operation.

Manually operated doors or gates are seldom fitted to-day on large passenger lift installations as the time wasted in operation is so great. Further, this time can vary considerably depending upon the particular person performing the opening and closing. The average time for manual opening and closing is about 6 seconds if the car and landing gates are coupled together and double this period if each has to be operated separately.

The width of the entrances is important as a wide entrance enables rapid movement of passengers into or out of the car but it also results in longer door operating times. Entrances are usually 3 ft. 6 in. to 4 ft. 6 in. wide.

Time for Passengers to Enter the Car. This depends upon the number of persons entering, the width of the entrance, to some extent on the business of the occupants of the building, and whether or not an attendant is employed. The loading time is shortened if the entrance is wide and also if a trained attendant drives the car instead of the control being automatic. At the ground floor the entrance should be as large as practicable to enable two or more passengers to enter or leave abreast. At the upper floors, however, the need for wide entrances is not so great, whilst any advantage that they will give by enabling rapid movement of passengers is offset by the time lost in door operation. The time required for passengers to enter the car at the upper floors depends also on the number of passengers already in the car.

When loading the car at the ground floor the average time required for each passenger will vary with the number of passengers entering the car. If the car is partially loaded with, say, 4 or 5 persons, the average time per person will be about one second, but if the number of passengers comprise a full load of, say, 15 persons, the average time for each to enter may be about three-quarters of a second. In this case the earlier passengers instinctively move quicker if there are people behind whilst the later ones may, in turn, be urged by those who have to wait for the next trip. In filling the car an average time of one second for each passenger is reasonable for estimating, but a good attendant will improve on this time. A typical relation between the loading time and the number of passengers in an efficient attendant controlled lift service is shown in Fig. 2.3, in which the contract load is 15 persons and the lift entrances are 4 ft. 6 in. wide. A lift of this size would not normally be operated on automatic control by the passengers but if this were done it is likely that the time required would be at least double those indicated by the graph.

Although it is unusual to estimate for passengers entering the car at the upper floors during the peak period any such passengers would require rather longer time than that needed by one person entering at the ground floor. If a half filled car with 7 passengers stops at an upper floor to take on another 4 persons, these will probably require $1\frac{1}{2}$ seconds each as those already in the car will have to move inwards to make room for them. An average of $1\frac{1}{2}$ seconds for each of these passengers is reasonable for estimating.

Time for Passengers to Leave the Car. During the peak period passengers will leave the car at several of the upper floors and the time required to empty the car at the various landings will depend upon the number of passengers entering at the ground floor, and the number of stops made. At any particular floor the average time for a passenger to leave will be greater than that for entering as several of the passengers may have to move aside to enable him to leave. The time for three passengers to leave a full car would be about 6 seconds, but if these three were the last to leave they would probably do so in half this time. The average time per passenger to empty a full car at one landing would be about $\frac{3}{4}$ second.

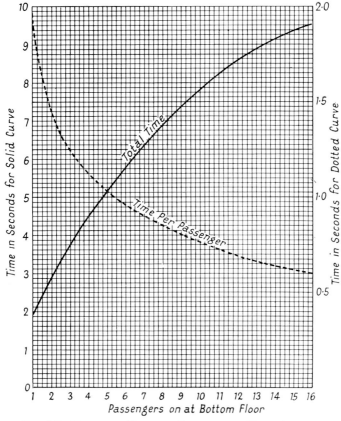

FIG. 2.3. PASSENGER LOADING TIME DURING PEAK ARRIVAL PERIOD IN OFFICE BUILDING

Typical times required for passengers to leave a 15-person car stopping at various numbers of floors are shown in Fig. 2.4 These graphs relate to an efficient attendant controlled installation, but the times will be greater if the car is operated on automatic control.

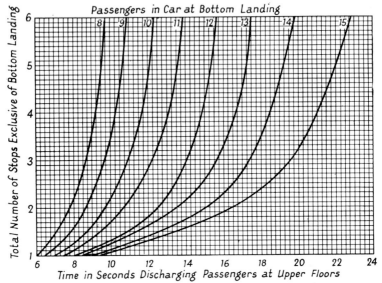

FIG. 2.4. PASSENGERS' UNLOADING TIME DURING PEAK ARRIVAL
PERIOD IN OFFICE BUILDING

Push Button Operation. If automatic control is employed in which the car is operated by push buttons, some time will be required for the selection and operation of the appropriate buttons after the doors have closed. An average of two seconds is reasonable for this purpose.

Car Speeds. Before consideration can be given to the travelling time between floors, a suitable car speed must be chosen. This depends largely on the height of the building, and if of 4 to 6 storeys a contract speed of 200 ft. per minute is quite common. For 6 to 8 storeys this can usually be increased with advantage to 300 ft. per minute. If the building has 8 to 12 storeys a speed of 400 ft. per minute should be considered. In the highest buildings in this country speeds up to 1 000 ft. per minute are in use whilst in America speeds up to 1 600 ft. per minute are used.

For a speed of 200 ft. per minute the drive will usually be a two-speed squirrel-cage induction motor, or a tandem motor, the induction motor being employed for light duty where the number of starts per hour does not exceed about 100. For a heavy duty of about 200 starts per hour at this speed, a geared variable voltage drive or a ventilated tandem motor should be considered. At a speed of 300 ft. per minute the drive will invariably be geared variable voltage and for speeds of 400 ft. per minute and above, will be gearless variable voltage.

Travelling Time. The time to travel between successive stops is composed of periods of acceleration, running at contract speed and retardation. Rates of acceleration depend upon the type of motor used and may be as low as 1 ft. per second per second with a slow speed geared lift or as high as 6 ft. per second per second with gearless variable voltage control. In practice the higher rates are usually associated with high contract speeds. So far as the passenger is concerned there is no limit to the speed at which the car may travel as a velocity of 100 ft. per minute is the same as 1 000 ft. per minute to a passenger travelling in an enclosed lift car. There is, however, the effect of rapidly varying air pressures on the human system and particularly on persons with weak hearts. It is not yet clearly established how fast a man can be dropped through a given change of air pressure without causing serious discomfort or harm. Apart from this aspect the limit is not in velocity but the manner in which it is attained and how standstill is achieved.

The effect of acceleration and retardation on the human body is worthy of some consideration at this stage. Physical discomfort is caused by the movement of the internal organs and the accompanying pressure of these organs against other parts of the body. During acceleration downwards and retardation upwards the muscles of the viscera are partly relieved of their load and the resulting reaction of these muscles tends to lift the visceral mass against the pleura which has not the resisting power of the bony pelvis and thus the stress is transmitted upwards to the lungs and heart. For this reason a sudden acceleration down or a sudden retardation up induces more discomfort than the reverse. A sudden increase in acceleration up or retardation down will affect the ankles of a stout person and the back of a thin person. These effects are worse the more abrupt the change in velocity, particularly if the muscles are not trained to quick response. Physical and mental preparedness will greatly minimize the effects. This may result in acceleration being much more uncomfortable than retardation, and explains why a lift car can be brought to rest much quicker than it can be accelerated without causing serious discomfort. To avoid discomfort the rate of change of acceleration or retardation, i.e. $\dfrac{d^3S}{dt^3}$ must be constant. Since therefore the acceleration must increase at a constant rate, the accelerating force must also increase at a constant rate if upward, and decrease at a constant rate if downward. Consequently during the accelerating period the acceleration and its applied force must gradually increase up to half the contract speed and then both must decrease at a constant rate from this point until maximum velocity has been reached, when the acceleration becomes zero. The retardation period is similar but is reversed. This is the ideal theoretical form of acceleration and retardation, but is rarely approached in practice and is, of course, very different from constant acceleration and retardation.

For the purpose of estimating the travelling time, distance/time curves for the accelerating periods for various car speeds are shown in Fig. 2.5, the points X indicating the ends of the acceleration periods and the commencement of uniform contract speeds. The curves are drawn for constant acceleration which, except at low values, will result in uncomfortable travelling. This fact, however, does not materially affect the distance travelled during the accelerating period. The shorter time required to reach contract speed with a higher acceleration is indicated by the curves C and D, which are for the same contract speed but D has the higher acceleration. The various distances travelled during acceleration show that if the higher speeds are employed for short journeys the car may not reach the contract speed before slowing commences.

The duration of the retardation period depends upon the method of slowing and on the adjustment of the brake but the rate should be as high as possible consistent with comfortable travelling. For the purpose of estimating the travelling time it is usual to assume that the rates of acceleration and retardation are equal. Consequently the retardation curves will be similar to those for acceleration shown in Fig. 2.5 but will be reversed. Curves showing the total travelling times for the same speeds as in Fig. 2.5, when stops are made between floors 30 ft. apart, are shown in Fig. 2.6. The constant speed periods are between points X and Y and from curves similar to these the total travelling time between stops may be estimated. The curves are drawn for values of accelerations and retardations likely to be obtained in practice with the contract speeds

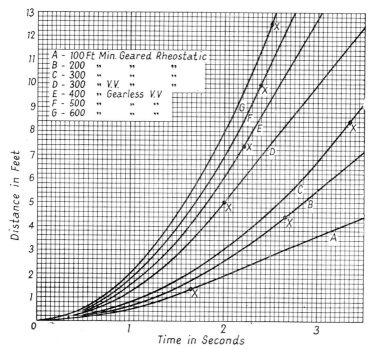

FIG. 2.5. TYPICAL ACCELERATION CURVES FOR VARIOUS LIFTS

quoted. These represent average uniform values but instantaneous accelerations greater than these may be experienced depending upon the method of control employed.

It is possible to calculate the actual travelling time of a round trip during the busy period after the probable number of stops have been ascertained and the distance travelled during an acceleration or retardation period is known or can be estimated.

This may be done as follows—

Let S = the number of stops made between the ground and the uppermost floor at which passengers are unloaded.

        D = the distance in feet between the ground and this top floor.

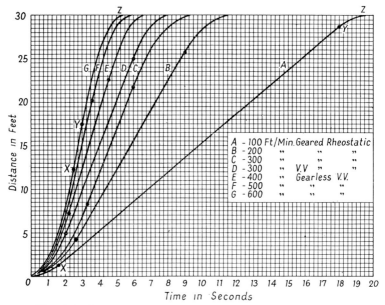

Fig. 2.6. Typical Time/Distance Curves for Various Lifts

V = the contract speed in feet per second.

d = the distance in feet required for acceleration from rest to contract speed which is assumed to be equal to the retardation distance.

The average round trip during the morning busy period in an office building consists of loading at the ground floor, unloading during S stops at the upper floors and a non-stop return to the ground floor.

The distance travelled on the upward journey during the acceleration and retardation period = $2dS$ ft.

And the distance during the acceleration and retardation periods when returning to the ground floor $= 2d$ ft.

Therefore the total acceleration and retardation periods $= 2d(S + 1)$ ft.

The total distance travelled at contract speed $= 2D - 2d$ $(S + 1)$ ft.

Therefore the total time during acceleration and retardation periods $= \dfrac{2d(S + 1)}{\dfrac{V}{2}} = \dfrac{4d(S + 1)}{V}$ sec.

This assumes that the average speed during these periods is half the contract speed.

Time for running at contract speed $= \dfrac{2D - 2d(S + 1)}{V}$ sec.

Therefore, the total travelling time $= \dfrac{2}{V}(dS + D + d)$ sec.

This formula holds good only if the distance between stops is large enough to enable the car to reach the contract speed before slowing commences.

If the distance d_1 between any two stops is less than the distance needed to reach contract speed, the travelling time between these stops will be $2\sqrt{\dfrac{d_1}{f}}$ sec. where f is the average acceleration. The total time must be modified accordingly.

Miscellaneous Times. The lifts in a bank are sometimes dispatched automatically or automatic signals are sent to the operators to ensure that, as far as practicable, the regular schedule is maintained. If this feature is not provided, an additional 10 per cent should be allowed on the R.T.T. to cover possible running off schedule.

Typical Example of Lift Traffic Analysis. After estimating or calculating values for the various factors affecting lift service as described above, consideration can be given to the possible number and types of lifts to meet the particular requirements. For the purpose of demonstrating the principles involved in the traffic analysis of a typical large office building the following example will be considered.

There are 8 floors above the ground or entrance level together with a sub-ground and basement and the occupants and rental value are such that a high grade of service is necessary. One main entrance in the centre of the frontage indicates that a bank of lifts is required. In estimating the probable number of stops during the busy period, traffic to the basement, sub-ground and 1st floors will be ignored. This will either be pro-hibited during these periods or the number of persons desiring service to these floors will be so small that it can be neglected at this stage. The busiest traffic period of the day in this building is considered to be during a particular half-hour in the morning when the building is being filled and it is estimated that 75 per cent of the normal population on the second floor and above will require lift service during this period. Hence the actual number of people to be taken to the upper floors can be assessed from data on the population of each floor. The number travelling to each of these floors is found to be sufficient to warrant making service available to all these floors during the busy half-hour. The total number requiring such service is 662 made up as shown in the table below—

Floor	2	3	4	5	6	7	8
No. of persons requiring service	36	93	160	85	120	105	63

The maximum capacity of the car will be assumed to be 10 persons and it is now possible to calculate the probable number of stops in unloading 10 persons during the busy period.

The distance between the ground and 8th floor is 110 ft. and this justifies considering speeds of 300 and 400 ft. per minute. To provide a high grade of service the door operating time must be small and therefore power operated doors should be provided, and, in addition, the lifts should be attendant operated. The round trip time can now be considered and will comprise door opening and closing, passengers entering and leaving, travelling time and any extras. During each round trip 10 passengers will be carried and after determining the R.T.T. as already outlined it is possible to estimate the number of 10-person lifts required to deal with 662 persons in 30 minutes. The grade of the resulting service, i.e. $\dfrac{\text{W.I.}}{4}(2 + N)$, can also be

LIFT TRAFFIC ANALYSIS

Contract Load (persons)	10	15	20	10	15	20
Contract Speed (ft./min.)	300	300	300	400	400	400
Probable No. of Stops per Trip	5·21	5·97	6·37	5·21	5·97	6·37
Door Operating Time (seconds)	21	24	27	21	24	27
Passengers Entering and Leaving (seconds)	20	33	47	20	33	47
Travelling Time (seconds)	56	58	59	46	48	49
Total Time (seconds)	97	115	133	87	105	123
10 per cent for off Schedule	10	12	13	9	10	12
R.T.T. (seconds)	107	127	146	96	115	135
No. of Lifts	4	3	3	4	3	3
Persons Carried in 30 minutes	672	640	740	750	705	800
Waiting Interval (seconds)	27	42	49	24	38	45
$\dfrac{\text{W.I.}}{4}\,(2+N)$	40	53	61	36	48	57
Grade of Service	Excellent	Good	Fair	Excellent	Good	Fair

assessed. It is necessary to consider other possible combinations of lift sizes and speed, and the table above shows the traffic details for this building with cars of 10, 15, and 20 passengers capacity and contract speeds of 300 and 400 ft. per minute all calculated as described above.

A selection must now be made from these possible installations depending upon the quality of service required and the price that the building owners are prepared to pay for lift service. It will be seen in this case that four lifts of 10 persons capacity each at 300 ft. per minute will handle the traffic, that W.I. is low and the grade of service excellent. Three 15-person lifts at 300 ft. per minute will deal with the traffic sufficiently closely for practical purposes, but the waiting interval is rather high and the grade of service inferior. The improvement in service by running these three 15-passenger lifts at 400 ft. per minute is not sufficient in this case to justify the extra cost. The four 10-passenger lifts at 400 ft. per minute are unnecessarily extravagant, whilst both the 20-passenger installations are too large and the grade of service not good enough for a high-class building. The best arrangement is therefore four 10-passenger lifts at 300 ft. per minute, and the next, and a cheaper one, is three 15-passenger lifts at 300 ft. per minute. For a building of this type, lifts at 500 ft. per minute might also be included in the analysis. In a similar manner, possible arrangements for any particular building can be studied.

The traffic in buildings used for other purposes might not have such pronounced short period peaks as an office building. For example, in a large departmental store the peak load is spread over a much longer period and furthermore is two-way traffic. In considering the design of lift installations some important facts are that high speeds are uneconomical if there is a great deal of floor to floor traffic and that greater improvements in the grade of service can be made by speeding up door operation or time for loading and unloading passengers than by increasing the contract speed. Control by car switch is usually more efficient than by push buttons operated by passengers and the latter should not generally be used for passenger lifts larger than 10-persons capacity as it is uneconomical to allow a large lift to answer single calls. If several lifts are required to handle heavy traffic from a number of landings it is advisable to install signal control in which the attendant only controls starting and the lift stops automatically at each landing where a call has been registered. Dual control should be fitted on one or more lifts in a bank if periods of light traffic such as at night are anticipated when these lifts can be operated on automatic control and the others shut down. It should be borne in mind, too, that in such circumstances automatic power operated doors or at least door closers (although these have the disadvantage of being more difficult to open) are a great advantage in ensuring that a lift is not put out of service by a passenger leaving and failing to close the door.

CHAPTER III

ACCOMMODATION

Well. In most buildings the lift well is placed in or adjacent to the main staircase, but, whilst in some buildings these positions have advantages, lift engineers nowadays frequently avoid the stairs when designing the well and, in fact, in America some building authorities prohibit the erection of the well in the stairway, the latter being regarded solely as a means of exit during emergency. Stairway wells often present difficulty in regard to guide fixings. The advantages of placing the well in a separate portion of the building are that there is not the same tendency for dirt to accumulate, the well can be totally enclosed by walls which will support the guides rigidly, and no special methods of guarding and screening are necessary. A typical guarded stairway well with power-operated three-leaf landing doors is shown in Fig. 3.1. Furthermore, most stairway wells are unsightly, and so the architect who avoids using them for his lifts is better able to contrive a more pleasing appearance for his lift entrances. An example of a totally enclosed non-stairway well with an effective entrance is shown in Fig. 3.2. This is a bank of two passenger lifts in one well with one overhead machine room. The control is duplex collective variable-voltage, geared motor with load of 2 000 lb and speed 300 ft. per min. and the doors are two-speed power-operated. The main point for consideration is that the lift should afford a quick and easy means of access to an exit from the upper floors.

The area of the well is governed by the size and number of the cars, and by the disposition of the car and landing entrances; these entrances also determine the necessary clearances for the car and counterweight. The most common arrangement is with the landing entrances all on the same side of the car, i.e. vertically under each other at the various landings. This results in a simple guide arrangement and a cheaper car construction. Occasionally car entrances on opposite sides are required. Entrances on adjacent sides can be provided but they result in corner guides which is an expensive arrangement,

29

and should be avoided if at all possible. Sufficient pit depth must be allowed to enable the car to come to rest without excessive buffer shock after operating the final switch in the event of the normal stopping switch failing to operate. The

FIG. 3.1. STAIRCASE LIFT
(Hammond and Champness)

depth required depends upon the contract speed and the type of buffer employed and is usually between 3 ft. and 5 ft. with spring buffers and between 5 ft. and 11 ft. with oil buffers and speeds up to 700 ft. per min. Where the pit depth is inadequate for the contract speed, an emergency terminal slow-down switch must be provided to reduce the terminal speed of the lift to that appropriate to the pit depth available. The minimum overhead clearance which should be given is also a function of the contract speed. These clearances are given in Chapter XIII. Provision must be made in the pit bottom for the accommodation of buffers, usually two for the car and two for the counterweight, if spring buffers, and for the fixing of the

FIG. 3.2. TWO PASSENGER LIFT ENTRANCES
(*Hammond and Champness*)

guide ends. Counterweight buffers, whilst not installed on all lifts, are invariably provided when the well does not extend to the basement. In these circumstances a special framing is erected for the buffers, and this must be capable of supporting the weights of the car and counterweight together with a possible impact strain. The lift machine is generally either in the basement or at the top of the well, but in both cases a suitable structure is required at the top of the well. In the former case the top joists must be capable of supporting the diverting

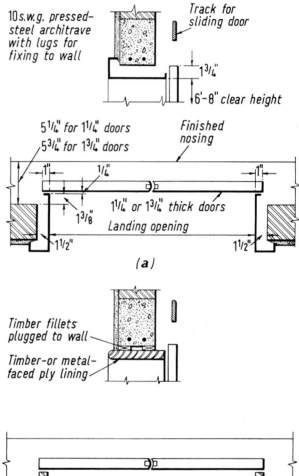

10 s.w.g. pressed-steel architrave with lugs for fixing to wall

Track for sliding door

1³/₄"

6'-8" clear height

5¹/₄" for 1¹/₄" doors
5³/₄" for 1³/₄" doors

Finished nosing

1"

¹/₄"

1"

1¹/₄" or 1³/₄" thick doors

Landing opening

1³/₈"

1¹/₂"

1¹/₂"

(a)

Timber fillets plugged to wall

Timber-or metal-faced ply lining

(b)

FIG. 3.3. TYPICAL LANDING
(a) Pressed-steel architrave, (b) timber or metal-faced ply lining,
Dimensions and labels which are identical in (a),
(Hammond and

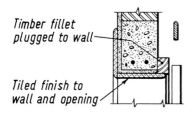

Timber fillet
plugged to wall

Tiled finish to
wall and opening

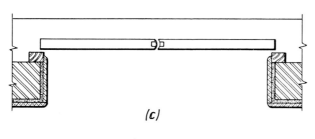

(c)

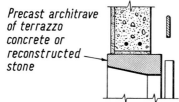

Precast architrave
of terrazzo
concrete or
reconstructed
stone

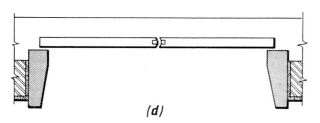

(d)

OPENING FINISHES

(*c*) tiled finish to walls and opening, (*d*) precast architrave of terrazzo.
(*b*), (*c*) and (*d*) are not repeated in (*b*), (*c*) and (*d*).
Champness)

33

pulleys and their resultant loads, whilst in the latter case the top structure is subjected to the weights of the machine, car, counterweight, and control gear. With the basement position, the machine should never be immediately under the well, and it is therefore necessary to provide joists for supporting the bottom pulleys which lead the ropes into the well. It is frequently possible to fix the main supporting joists to the building framework, but in some cases this is impracticable and it then becomes necessary to erect an independent steel structure from the basement and so relieve the building fabric of all lift loads. The well is usually of brick construction faced either with plaster or some form of ornamental glazed brick, except when a steel structure has to be erected.

No equipment, except that forming part of the lift or necessary for its maintenance, should be installed in the well. The inside surface of the well should, as far as practicable, form a continuous flush surface, but if projections which are opposite a car entrance and extending inwards from the general well surface cannot be rendered flush, they should be bevelled on the underside to an angle of 60 deg. from the horizontal.

The finishes required to landing openings usually depend on the architect's requirements, some typical examples being shown in Fig. 3.3 for two-panel centre-opening doors.

Machine Room. The overhead location of the lift machine shows a saving in first cost and has several engineering advantages over the basement or any intermediate position. The machine room should therefore, whenever possible, be situated directly over the well and should have sufficient windows to afford good natural lighting and ventilation. The former is necessary for efficient maintenance, and adequate artificial lighting supplemented by hand lamp points should be provided for carrying out any work necessary during hours of darkness. Efficient ventilation is provided mainly to reduce the temperature and resistance of the motor windings, and thereby increase the electrical efficiency of the plant. An overhead machine room for two passenger lifts with two-speed a.c. motors is shown in Fig. 3.4. In large machine rooms where a number of lift machines is installed it is sometimes the practice to install a Plenum ventilating plant. The incoming air is drawn through filters and thence by ductwork to the generators and motors, a separate exhaust system being provided. The clear room height should not be less than 7 ft. The entrance door should preferably open outwards, and when it is essential to locate the machine below, access doors must be provided at the top of the well to enable the necessary periodical attention

FIG. 3.4. MACHINE ROOM FOR TWO PASSENGER LIFTS
(*Marryat & Scott*)

to be given to the overhead pulleys. The machine room should be restricted to the housing of lift equipment. It is advisable that the walls, floor and ceiling of the machine room be properly finished to reduce to a minimum dust emanating from brick or concrete surfaces. This can have a serious effect on controller contacts, and to prevent dust settling on the oiled guides the well should be properly surfaced.

The area of the room depends upon the lay-out and type of the lift equipment, the size being governed by the car capacity and speed. Ample space must be provided to facilitate maintenance, and means, e.g. a trap door in the floor, whereby any items of equipment may be removed for repair or replacement. In the lay-out care must be taken to ensure that the controller position conforms to the Home Office Regulations, which demand that there should be a clear width, measured from the bare conductors, of not less than 3 ft. on all sides. When the controller is fixed near a wall the 3 ft. clearance must be provided between the controller and the wall, unless the connexions and all parts requiring attention when the controller is working, are placed on the front of the panel when a minimum clearance of 1 ft. 9 in. should be provided. Bare conductors must not be exposed on both sides of a controller passage-way unless the clear width of the passage-way is not less than 4 ft. 6 in. between bare conductors. The floor should be capable of supporting the heaviest lift unit and is usually designed to withstand a concentrated load of 300 lb. on any 4 sq. in. of the floor. This does not require that it shall be capable of supporting 300 lb. on every 4 sq. in. of its surface simultaneously. The sizes of well-designed machine rooms for various capacity single lifts should be approximately those quoted in the tables later in this chapter, which also give information relating to other accommodation requirements.

The figures in the table on page 43 for light traffic lifts relate to single speed lifts with the counterweight at the side of the car. They have single entrances with single sliding panel doors either hand- or power-operated.

The table on page 44 gives minimum dimensions for geared lifts for general office type buildings. Each lift has a single entrance with two-panel power-operated doors, side opening or centre opening. The counterweight is opposite the landing entrances and the machine room directly above the well. The total headroom is the distance from the top landing to the machine room ceiling, and the pit depth the distance from the bottom landing to the pit floor. Fig. 3.5 gives details of the layout of a typical lift for this type of duty and, it is also suitable for the light traffic duty referred to in the preceding paragraph, except that in this case the counterweight position and the doors are different from those mentioned above. Fig. 3.6 shows details of a larger lift with centre opening doors.

The figures in the table on page 45 for heavy traffic passenger lifts relate to gearless lifts with single entrances and the counterweights opposite the landing entrances. The landing doors are power-operated, two-panel, centre opening. For these gearless machines, it will be noted, very large overhead clearances and pit depths are required. The reason for the large top of the well clearances is to accommodate the special roping arrangements needed with gearless machines. Usually a double wrap drive with two to one roping is required. The deep

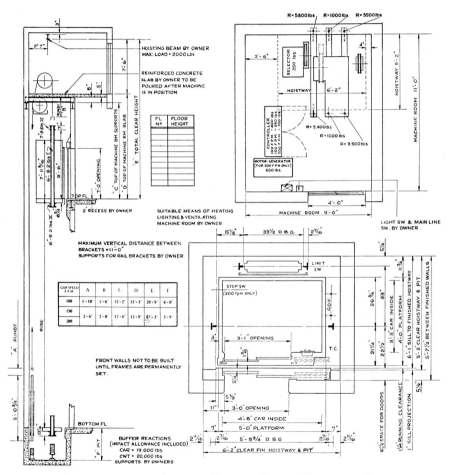

FIG. 3.5. TYPICAL GENERAL PURPOSE GEARED PASSENGER LIFT
1 200 lb. at 100 ft. per min., 150 ft. per min. or 200 ft. per min.
(*Otis Elevator Co.*)

pit permits the use of compensating ropes, generally desirable with long travel gearless lifts. Details of a particular lift for this duty are shown in Fig. 3.7.

The details for general duty goods lifts on page 46 are for full-width single entrances fitted with collapsible shutter gates.

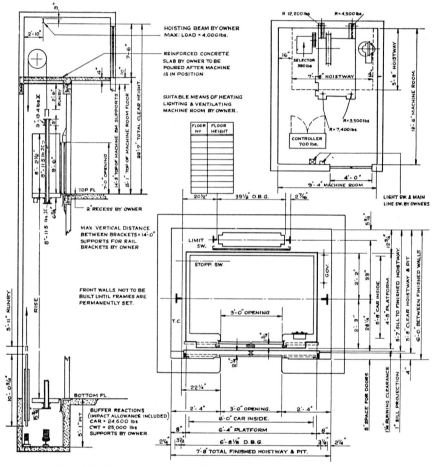

FIG. 3.6. TYPICAL GENERAL PURPOSE GEARED PASSENGER LIFT
2 000 lb. at 300 ft. per min.
(*Otis Elevator Co.*)

The counterweight is at one side of the well. If entrances at front and back are required, the lift well depth must be increased by two inches. The layout of a lift of this type is shown in Fig. 3.8.

For heavy duty goods lifts the figures on page 47 are for single entrances fitted with vertical bi-parting power-operated doors.

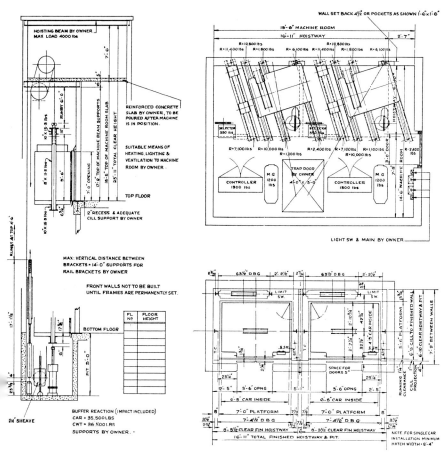

FIG. 3.7. TYPICAL HEAVY TRAFFIC GEARLESS PASSENGER LIFT
3 000 lb. at 500 ft. per min.
(*Otis Elevator Co.*)

To accommodate these doors a minimum floor to floor height of
12 ft. is required. If entrances at the front and back are
necessary, the depth of the well must be increased by 7 in.
Details of the layout of a heavy duty goods lift of this type are
shown in Fig. 3.9.

Minimum dimensions for service lifts from 1 to 5 cwt.

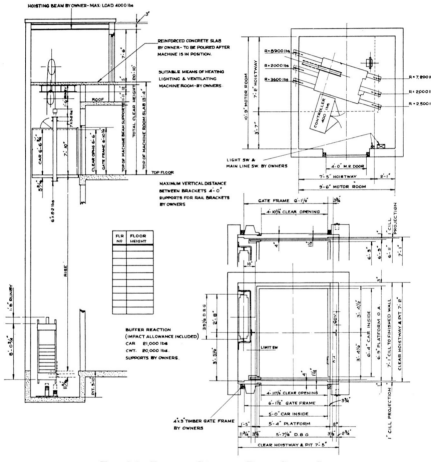

FIG. 3.8. TYPICAL GENERAL DUTY GOODS LIFT
20 cwt. at 100 ft. per min.
(*Otis Elevator Co.*)

capacity are shown in the table on page 43. The speed of these
small lifts is either 50 or 100 ft. per minute. The general
layout of a service lift is shown in Fig. 3.10.

When, for any reason, an overhead machine room is not

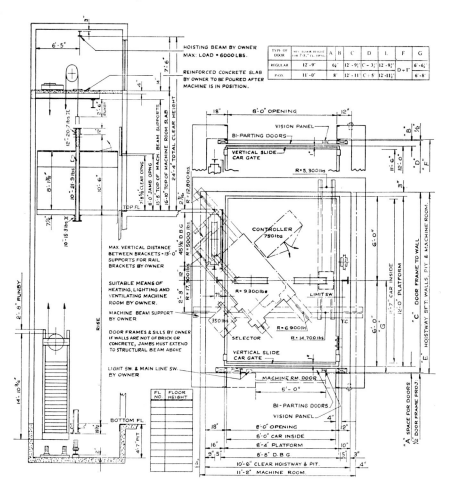

FIG. 3.9. TYPICAL HEAVY-DUTY GOODS LIFT

80 cwt. at 100 ft. per min.

(*Otis Elevator Co.*)

possible, the equipment is usually installed in the basement and for such installations the data in the table on page 48 are representative of good practice.

Substantial rolled steel joists or reinforced concrete beams will be required in the machine room floor to carry the lift loads. If the machine is above the well, the total load to be supported will be the sum of the weights of the machine, car,

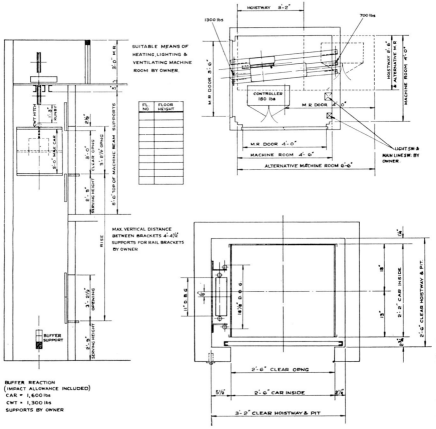

FIG. 3.10. TYPICAL SERVICE LIFT
2 cwt. at 100 ft. per min.
(*Otis Elevator Co.*)

MINIMUM DIMENSIONS FOR SMALL SINGLE SPEED GEARED LIFTS FOR LIGHT PASSENGER TRAFFIC

Load Persons	Load lb.	Speed ft. per min.	Lift Well Width ft. in.	Lift Well Depth ft. in.	Machine Room Width ft. in.	Machine Room Length ft. in.	Top Landing to Mach. Room Floor ft. in.	Total Head-room ft. in.	Pit Depth ft. in.	Mach. Room Height ft. in.	Landing En-trance Height ft. in.	Landing En-trance Width ft. in.	Car Platform Width ft. in.	Car Platform Depth ft. in.	Top of Well Load cwt.
4	600	100	5 8	3 10	7 0	9 0	13 6	21 0	3 3	7 6	6 8	2 3	3 8	3 1	74
6	900	100	6 3	4 5	7 0	10 0	13 6	21 0	3 3	7 6	6 8	2 6	4 2	3 8	99
8	1 200	100, 150	6 9	4 10	8 0	10 0	13 6 / 14 0	21 6	3 3 / 3 9	7 6	6 8	2 9	4 8	4 1	112

MINIMUM DIMENSIONS FOR OVERHEAD SERVICE LIFTS

Load cwt.	Speed ft. per min.	Lift Well Width ft. in.	Lift Well Depth ft. in.	Machine Room Width ft. in.	Machine Room Depth ft. in.	Top Landing to Mach. Room Floor ft. in.	Total Head-room ft. in.	Pit Depth ft. in.	Mach. Room Height ft. in.	Landing En-trance Height ft. in.	Landing En-trance Width ft. in.	Car Platform Width ft. in.	Car Platform Depth ft. in.	Top of Well Load cwt.
1	50 100	3 6	2 7	5 0	6 0	9 0	13 0	3 0	4 0	—	—	2 3	2 3	20
2	50 100	3 9	2 10	6 0	7 0	9 3	13 3	3 0	4 0	—	—	2 6	3 6	27
3	50 100	4 3	3 4	6 0	7 0	9 6	13 6	3 0	4 0	—	—	3 0	3 0	37
4	50 100	4 9	3 9	6 6	7 6	10 0	14 0	3 0	4 0	—	—	3 3	3 3	52
5	50 100	5 0	4 0	6 6	7 6	10 6	14 6	3 0	4 0	—	—	3 6	3 6	70

43

MINIMUM DIMENSIONS FOR GENERAL PURPOSE OVERHEAD GEARED PASSENGER LIFTS

| Load | | Speed ft. per min. | Lift Well | | Machine Room | | Top Landing to Mach. Room Floor | Total Head-room | Pit Depth | Mach. Room Height | Landing En-trance Height | Landing En-trance Width | Car Platform | | Top of Well Load cwt. |
Persons	lb.		Width	Depth	Width	Length							Width	Depth	
			ft. in.	ft. in.	ft. in.	ft. in.	ft. in.	ft. in.	ft. in.	ft. in.	ft. in.	ft. in.	ft. in.	ft. in.	
8	1 200	200	6 4	5 10	9 0	13 0	14 9	22 9	5 0	8 0	7 0	3 0	5 0	4 0	145
10	1 500	150	7 4	5 8	9 0	13 0	14 9	22 9	4 0	8 0	7 0	3 0	6 0	4 0	147
10	1 500	200	7 4	5 8	10 0	13 0	14 9	22 9	5 0	8 0	7 0	3 0	6 0	4 0	153
10	1 500	300	7 4	5 8	10 0	14 0	15 0	23 0	5 0	8 0	7 0	3 0	6 0	4 0	192
13	2 000	150	7 8	6 4	10 0	13 0	14 9	22 9	4 0	8 0	7 0	3 0	6 4	4 8	160
13	2 000	200	7 8	6 4	10 0	14 0	14 9	22 9	5 0	8 0	7 0	3 0	6 4	4 8	170
13	2 000	300	7 8	6 4	10 0	15 0	15 0	23 0	5 3	8 0	7 0	3 0	6 4	4 8	220
16	2 500	150	8 4	6 8	10 0	14 0	14 9	22 9	4 0	8 0	7 0	3 6	7 0	5 0	195
16	2 500	200	8 4	6 8	10 0	15 0	14 9	22 9	5 3	8 0	7 0	3 6	7 0	5 0	210
16	2 500	300	8 4	6 8	11 0	15 0	15 0	23 0	5 3	8 0	7 0	3 6	7 0	5 0	275
20	3 000	150	8 4	7 2	11 0	13 0	14 9	22 9	4 3	8 0	7 0	3 6	7 0	5 6	230
20	3 000	200	8 4	7 2	11 0	14 0	14 9	22 9	5 3	8 0	7 0	3 6	7 0	5 6	246
20	3 000	300	8 4	7 2	12 0	16 0	15 0	23 9	5 3	8 0	7 0	3 6	7 0	5 6	323

MINIMUM DIMENSIONS FOR GEARLESS OVERHEAD HEAVY TRAFFIC PASSENGER LIFTS

Load		Speed ft. per min.	Lift Well				Machine Room				Top Landing to Mach. Room Floor		Total Head-room		Pit Depth		Mach. Room Height		Landing Entrance Height		Landing Entrance Width		Car Platform				Top of Well Load cwt.
Per-sons	lb.		Width		Depth		Width		Length														Width		Depth		
			ft.	in.	ft.	in.	ft.	in.	ft.	in.	ft.	in.	ft.	in.	ft.	in.	ft.	in.	ft.	in.	ft.	in.	ft.	in.	ft.	in.	
10	1 500	350	7	4	5	10	12	0	26	0	18	0	27	6	8	3	9	6	7	0	3	0	6	0	4	0	475
10	1 500	500	7	4	5	10	12	0	26	0	19	0	28	6	9	3	9	6	7	0	3	0	6	0	4	0	475
13	2 000	350	7	8	6	6	13	0	28	0	18	0	27	6	8	3	9	6	7	0	3	0	6	4	4	8	560
13	2 000	500	7	8	6	3	13	0	28	0	19	0	28	6	9	3	9	6	7	0	3	0	6	4	4	5	560
16	2 500	350	8	4	6	10	14	0	29	0	18	0	27	6	8	3	9	6	7	0	3	6	7	0	5	0	660
16	2 500	500	8	4	6	10	14	0	29	0	19	0	28	6	9	3	9	6	7	0	3	6	7	0	5	0	660
20	3 000	350	8	4	7	4	15	0	30	0	18	0	27	6	8	3	9	6	7	0	3	6	7	0	5	6	760
20	3 000	500	8	4	7	4	15	0	30	0	19	0	28	6	9	3	9	6	7	0	3	6	7	0	5	6	760
20	3 000	600	8	4	7	4	15	0	30	0	20	0	29	6	10	0	9	6	7	0	3	6	7	0	5	6	880
20	3 000	700	8	4	7	4	15	0	30	0	21	0	30	6	11	0	9	6	7	0	3	6	7	0	5	6	880

MINIMUM DIMENSIONS FOR OVERHEAD GEARED GENERAL DUTY GOODS LIFTS

Load cwt.	Speed ft. per min.	Lift Well Width ft. in.	Lift Well Depth ft. in.	Machine Room Width ft. in.	Machine Room Length ft. in.	Top Landing to Mach. Room Floor ft. in.	Total Head-room ft. in.	Pit Depth ft. in.	Mach. Room Height ft. in.	Landing Entrance Height ft. in.	Landing Entrance Width ft. in.	Car Platform Width ft. in.	Car Platform Depth ft. in.	Top of Well Load cwt.
10	50	6 4	4 11	8 0	10 0	13 9	21 3	3 0	7 6	6 6	4 0	4 4	4 4	112
10	100	6 4	4 11	8 0	10 0	13 9	21 3	3 3	7 6	6 6	4 0	4 4	4 4	112
15	50	6 10	5 11	9 0	11 0	14 0	21 6	3 3	7 0	7 0	4 6	4 10	5 4	156
15	100	6 10	5 11	9 0	11 0	14 0	21 6	4 0	7 6	7 0	4 6	4 10	5 4	162
20	50	7 6	6 11	9 0	13 0	14 0	21 6	4 0	7 6	7 0	5 0	5 7	6 4	200
20	100	7 6	6 11	9 0	13 0	14 0	21 6	4 0	7 6	7 0	5 3	5 7	6 4	218
20	150	7 6	6 11	9 0	13 0	14 6	22 0	4 6	7 6	7 0	5 3	5 7	6 4	230
30	50	8 5	8 5	10 0	13 0	14 9	23 6	4 3	8 9	7 6	6 0	6 4	7 10	280
30	100	8 5	8 5	10 0	13 0	14 9	23 6	4 3	8 9	7 6	6 0	6 4	7 10	300
30	150	8 5	8 5	10 0	13 0	15 3	24 0	5 0	8 9	7 6	6 0	6 4	7 10	312
40	50	9 8	9 5	10 0	14 0	16 0	25 3	4 6	9 3	7 6	7 0	7 4	8 10	370
40	100	9 8	9 5	10 0	14 0	16 0	25 3	4 6	9 3	7 6	7 0	7 4	8 10	400
40	150	9 8	9 5	10 0	14 0	16 6	25 9	5 3	9 3	7 6	7 0	7 4	8 10	410

The above dimensions are for single entrance lifts.
For two opposite entrances, add 2 inches to the well depth.

MINIMUM DIMENSIONS FOR OVERHEAD GEARED HEAVY DUTY GOODS LIFTS

Load cwt.	Speed ft. per min.	Lift Well Width	Lift Well Depth	Machine Room Width	Machine Room Depth	Top Landing to Mach. Room Floor	Total Head-room	Pit Depth	Mach. Room Height	Land-ing En-trance Height	Land-ing En-trance Width	Car Platform Width	Car Platform Depth	Top of Well Load cwt.
		ft. in.	ft. in.	ft. in.	ft. in.	ft. in.	ft. in.	ft. in.	ft. in.	ft. in.	ft. in.	ft. in.	ft. in.	
30	50, 100	8 10	8 10	9 6	14 0	16 0	24 6	4 6	8 6	7 6	6 0	6 4	8 0	282
30	150	8 10	8 10	9 6	14 0	16 0	24 6	4 8	8 6	7 6	6 0	6 4	8 0	290
40	50, 100	9 10	9 10	10 6	14 0	16 6	24 6	4 6	8 6	7 6	7 0	7 4	9 0	368
40	150	9 10	9 10	10 6	14 0	16 6	25 6	5 1	9 0	7 6	7 0	7 4	9 0	404
60	50, 100	11 0	12 4	11 6	15 0	17 0	26 0	4 8	9 0	7 6	8 0	8 4	11 6	539
60	150	11 0	12 4	11 6	15 0	17 0	26 0	5 4	9 0	7 6	8 0	8 4	11 6	558
80	50, 75	11 0	15 4	12 0	16 0	17 6	26 6	5 4	9 0	8 0	8 0	8 4	14 6	713
80	100	11 0	15 4	12 0	16 0	17 6	26 6	5 4	9 0	8 0	8 0	8 4	14 6	752
100	50, 75	13 2	15 4	13 0	17 0	17 6	27 0	5 6	9 6	8 0	10 0	10 4	14 6	884
100	100	13 2	15 4	13 0	17 0	17 6	27 0	5 6	9 6	8 0	10 0	10 4	14 6	935

The above dimensions are for single entrance lifts. For two opposite entrances, add 7 in. to the well depth.

TYPICAL DIMENSIONS FOR BASEMENT MACHINE ROOMS

Load		Machine Room		Mach. Room Height	Pent-house Height	Top of Well Load
	cwt.	Width	Depth			
		ft. in.	ft. in.	ft. in.	ft. in.	cwt.
	1	5 0	7 0	7 0	4 0	31
	2	6 0	8 0	7 0	4 0	46
	3	6 0	8 0	7 0	4 0	60
	4	6 6	8 6	7 0	4 0	83
	5	6 6	8 6	7 0	4 0	102
Persons	lb.					
4	600	7 0	11 0	8 0	4 6	107
6	900	7 6	12 0	8 0	4 6	159
8	1 200	8 0	14 0	8 0	5 0	213
10	1 500	8 6	15 0	8 0	5 0	246
15	2 250	9 0	16 0	8 0	5 6	330
20	3 000	9 6	18 6	9 0	5 6	410

load, counterweight, controller, and sundry other small gear. The greater part of this load is "live," and the supporting joists must therefore be designed to carry double the "dead" load, and, further, as the load is unequally distributed, comparatively large section joists are necessary. In calculating the size of the overhead supporting beams, the total load is taken to be the weight of all the equipment resting on the beams plus twice the maximum static load suspended from the beams. The load resting on the beams includes the complete weights of the winding machine, sheaves, controller and all auxiliary equipment and the load suspended from the beams includes the sum of the tensions of all ropes suspended from the beams. The latter is the "live load" which must be doubled to allow for impact and acceleration stresses. Factors of safety of 5 and

7 are allowed for steel and concrete respectively in designing the beams, the deflection of which, with loads as stated above, should not exceed $\frac{1}{2500}$ of the span. The actual arrangement of these josist depends upon the lay-out of the machine, typical supporting joists being shown in Figs. 3.5–3.10.

In the tables on pages 43–48 the load quoted is the equivalent dead load at the top of the well. This is the load due to the tensions in the suspending ropes and to the weight of the winding machine and does not include the weights of the controller, motor generator set or any other auxiliary equipment. For an overhead geared 2-speed motor drive, the weight of the controller and other gear amounts to about 10 per cent of the overhead well load quoted. If a variable voltage drive is used, the weight of the overhead control gear, motor generator set and other equipment, will be approximately 20 per cent of the overhead loads stated in the tables.

With the basement location, overhead joists will be required to support a total load equal to twice the sum of the weights of car, load, and counterweight, and as this is a "live" load, the equivalent "dead" load will be twice this figure. Overhead joist arrangements for basement machines are shown in Figs. 3.11 and 3.12.

The overhead well loads shown on page 48 for basement machines are also the total top loads as all the equipment, except the overhead load supporting pulleys, is located in the basement. A basement winding machine with the diverting pulley mounted on the machine is shown in Fig. 3.13, and this arrangement eliminates the necessity of supporting the pulley by structural steelwork fixed to the building. Lifting joists must also be fixed in the motor room to enable the machine to be placed in position and to allow of ready removal of any part for repair or renewal. A permanent danger notice should be displayed on the outside of the door and one near the machinery and the room should be kept locked.

Noise. In some buildings, notably hospitals, silence is of paramount importance, and various measures must be adopted if it is desired to reduce the lift noise to the lowest possible level. This is generally a difficult problem, as with modern steel-framed buildings noise is readily transmitted to rooms which may be at a considerable distance from the motor room.

In dealing with the problem of noise it is necessary first to reduce the noise generated in the motor room itself to a

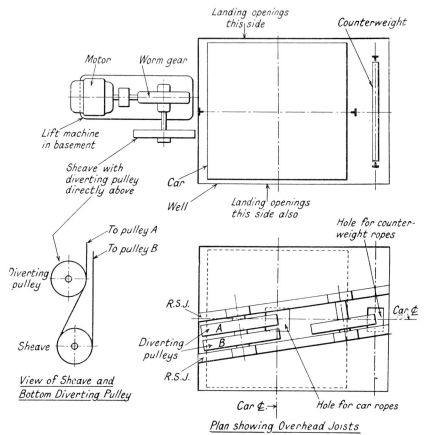

FIG. 3.11. DETAILS OF TRACTION DRIVE WITH MACHINE BELOW

minimum. A reasonably silent motor may be obtained by carefully selecting the type, paying due regard to its design, and by employing plain bearings. Motors fitted with commutators should have the micas undercut and the brushes carefully bedded. Motor hum may be considerably reduced by using a motor frame size larger than that required to give the necessary

FIG. 3.12. OVERHEAD JOISTS AND PULLEYS
(*Marryat & Scott*)

FIG. 3.13. BASEMENT MACHINE
(*Otis Elevator Co.*)

horse-power output and thus working the magnetic circuit below its saturation point. Brake noises will be diminished if the brake is immersed in oil, and often by incorporating torque motor brakes instead of the usual type, whilst hum due to a.c. brakes can be reduced by using shaded poles, i.e. inserting copper loops in the face of each magnet core. The gearing should be silent in operation if the gear teeth are accurately cut, and the same applies to gearing on floor selectors. Considerable noise emanates from the controller switches during the starting, acceleration and stopping periods, and only by giving attention to design or totally enclosing the controller is it possible to diminish these noises to a negligible amount.

Having eliminated the noises generated in the machine room as much as possible, our next step is to prevent that which remains from being transmitted to other parts of the building. Direct transmission of noise by the air can be reduced by lining the machine room walls with some sound insulating material such as Cabot's quilting or "Absorbit." Noise transmitted via the well may be greatly diminished by providing a free circulation of air between the machine room floor and the top of the well. This entails building a false floor about 3 ft. below the machine room floor and ensuring a free circulation of air through this space. Vibration transmitted by the building structure can be reduced by inserting slabs of compressed cork or similar material between the concrete bed on which the machine baseplate rests and the supporting joists. As an additional precaution, the ends of the joists may be surrounded by insulating material in boxes before bedding in the walls.

Other lift noises are due to the motion of the car and counterweight in the well, and to the operation of gates and doors. The former may be diminished by fitting a retiring lock release cam to the car and thus preventing the cam from hitting the gate lock striker arm when passing landings. The noise due to the motion of the car and counterweight on their guides may be prevented from being transmitted to adjacent rooms by inserting a felt sleeve round each guide fixing bolt. A more complete but also more expensive method of effecting this is to construct double well walls, the inner wall carrying the guides, the intervening space forming a sound absorbing chamber.

Gate noises may be reduced by the insertion of rubber buffers between the pickets, rubber bumpers at the extremes of travel, and by fitting hardwood bottom gate tracks instead of steel tracks.

Fire Precautions. The *British Standard Code of Practice on Fire Precautions in Flats and Maisonettes over 80 ft. in Height* contains recommendations regarding lifts and the following are extracts from this Code which also represents good practice for lifts in other types of high building.

A lift introduces a negligible fire hazard provided that landing doors are fire-resisting, the machine room is at the top, and permanent ventilation is provided at the top of the well. There is, therefore, no objection to arranging lifts within a staircase enclosure on any floor, including the entrance floor.

Lift wells. The walls enclosing lift wells should have a fire resistance of not less than one hour except where the lifts are in a staircase enclosure. Wells should have permanent vents at the top not less than 1 sq. ft. in area. Lift machine rooms should be separated from lift wells by the enclosing wall of the well or by a floor of the machine room.

Doors in Lift Wells. The doors in the enclosing walls of lift wells should, in conjunction with their frames, have a fire resistance of half an hour. They should be fitted with an automatic closing device which will ensure that the doors are kept closed at all times except when persons are entering or leaving the lift. The doors, when closed, must be effectively smoke-tight.

Doors to Lift Machine Rooms. The doors to lift machine rooms, where located within the building, should in conjunction with their frames, have a fire resistance of half an hour, and be made self-closing.

Lift Machine Rooms. Lift machine rooms should preferably be sited at the top of the well and should always be so sited if the lifts open out of a common approach route or staircase that provides the only means of escape from a dwelling.

Fire Lifts. A passenger lift or lifts should be arranged so as to be available for the exclusive use of firemen in any emergency by providing at ground level a switch in a glass-fronted box marked "FIRE SWITCH" which operates a control whereby firemen can obtain the use of the lift without interference from the landing call points. Alternatively the fire switch may be protected by a metal cover and unlocked by a key which would pass the dry riser box and any other locks which would require to be opened by the fire brigade.

A sufficient number of lifts should be arranged as fire lifts to ensure that in flats, every floor (except, under the circumstances described below, the top floor) and in maisonettes every entrance floor, has direct access to at least one such lift. A fire lift should not be more than about 15 ft. from a main staircase if that is the only staircase to which there is access, or about 50 ft. if there is another staircase on the same floor to which there is access. In addition, if a fire lift is not in a main staircase enclosure or within 15 ft. of a door in a main staircase enclosure it should be within 15 ft. of a smokestop door that leads to a main staircase.

In order to ease the difficulty of accommodating the space necessary for over-run at the top of the well for a high-speed lift it is considered that a fire lift need not serve the top floor of a building provided that the lift is not more than 15 ft. from a main staircase on the floor below, the hydrant outlet on the top floor is within the staircase enclosure or in a ventilated lobby adjoining the staircase, and the number of flats on the floor does not exceed eight.

A fire lift should have a platform area of not less than $15\frac{1}{2}$ sq. ft. and be capable of carrying a load of 1 200 lb. Its speed should be such as will reach the top floor from ground level (non-stop) in one minute. The electric supply to any fire lift should be provided by a sub-main circuit exclusive to the lift except that where the lift is one of a battery of not more than six lifts (whether fire lifts or not) the other lifts may be fed from the same supply. The cables supplying current to the lift motor should pass through routes of negligible fire risk.

CHAPTER IV

TYPES OF DRIVE

Two types of drive are employed for lift work, namely, the *traction* or *sheave* drive, and the *drum* drive.

TRACTION DRIVE

In this case one set of ropes is used, the ropes passing from the car round a cast-iron or steel grooved sheave and thence to the counterweight. Friction between the ropes and the sheave grooves therefore supplies the force necessary to raise or lower the car. The sheave is secured to a turned mild steel shaft by two sunk keys at right angles to each other. Alternatively, keys may be eliminated by bolting the sheave rim to a gear and sheave centre. This centre is a casting with two flanges, one of which carries the sheave, the other carrying the worm-wheel rim, the shaft being pressed into the bore of the centre. The outer end of the driving shaft is carried in a pedestal bearing fitted with readily renewable bearings of gunmetal or white metal, preferably of the split pattern, in order to compensate for wear. Sometimes the sheave shaft is supported in tapered roller bearings. In those cases in which no outer shaft bearing is employed, the outer edge of the sheave is fitted with an extended flange to minimize the danger of the ropes leaving the sheave. A lift machine with no outer bearing is shown in Fig. 4.1.

The main advantage of the traction drive is that if either the car or counterweight comes into contact with the buffers, the drive ceases and there is no danger of the car being wound into the overhead structure, as would be possible with a drum machine. With very high rises, however, overwinding of the car may be possible if the counter weight is on its buffers. The weight of the ropes on the counter weight side may be sufficient to provide traction to drive the ropes. In such cases special precautions are taken to prevent this occurring. Other advantages of this method are cheapness, simplicity, and the fact that standard equipment may be used irrespective of the height of travel. A typical gearless traction machine employing

55

FIG. 4.1. LIFT MACHINE WITH NO OUTER SHEAVE BEARING
(*Wm. Wadsworth & Sons*)

FIG. 4.2. GEARLESS TRACTION MACHINE
(*Marryat and Scott*)

double-wrap roping, and of contract load 1 500 lb. and contract speed 500 ft. per min. is shown in Fig. 4.2.

Sheaves. Sheave is the name given to a pulley to which power is applied, and is that part of the lift machine transmitting driving power to the lift ropes. The sheave is of disc

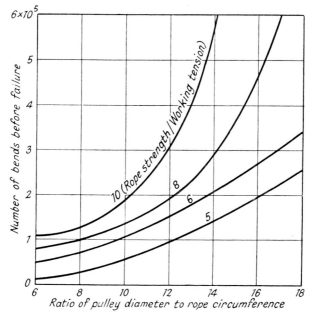

FIG. 4.3. GRAPH SHOWING RELATION BETWEEN NUMBER OF BENDS BEFORE FAILURE AND INCREASING PULLEY DIAMETER
(*Junior Institution of Engineers Journal*)

construction, i.e. without spokes, and to ensure a long life for the ropes the sheave diameter should be as large as practicable, the minimum depending upon the rope diameter, material and construction, and the maximum car speed. A larger diameter sheave should be employed for Seale or Warrington ropes than for ropes of uniform construction, whilst ropes of low tensile strength and consequently high ductility may be used with sheaves of smaller diameter than high tensile strength ropes would require. The ratio of the sheave diameter to the rope diameter should be not less than (0·015S + 37) with a minimum value of 40. This formula, in which S is the rope speed in feet per minute, is applicable to the flattened strand ropes and to all the main eight types of rope described in Chapter V, with the exception of the 6 × 19 (9/9/1) Seale which is stiffer than the other seven constructions.

For this Seale rope the formula should be modified to (0·015S + 44) with a minimum value of 47.

For service lifts, the ratio of sheave diameter to the rope diameter should be not less than 30.

The importance of specifying a large sheave diameter is clearly shown in Fig. 4.3,* which is a graph showing the relation between the number of bends before failure and increasing sheave diameters for different values of the ratio $\dfrac{\text{rope strength}}{\text{working tension}}$. The graph is plotted from results obtained during a test taken on a sample of $\frac{6}{19}$ ordinary lay rope. It also shows that the rope performance is improved if this ratio is large.

The type of sheave groove usually employed is vee-shaped, having an included angle of from 35° to 40°. With a small groove angle the traction is large, but it is necessary to use hard ropes so as to minimize rope wear. Details of a typical vee-grooved traction sheave are shown in Fig. 4.4. On modern high-speed lifts a U-groove, similar to that used on a drum or pulley, is frequently employed, but the traction is so low with this groove as to necessitate the use of the double-wrap method of roping. The U-groove, however, has the advantages of longer rope life and a greater degree of silence, the latter being particularly noticeable at the higher car speeds of 600 ft. per min. and over. Other types of sheave groove sometimes used (Fig. 4.5) are a modified form of vee-groove and an undercut groove, these being compromises between the vee- and U-types. In the round-seat undercut groove the rope seats are the same radii as the pulley grooves, and the width of the undercut should not exceed 0·8 of the diameter of the rope. The rope unit pressure with this groove is greater than with a U-groove, but not as severe as the vee-groove.

* "Wire Ropes," paper by W. A. Scoble read before the Junior Institution of Engineers.

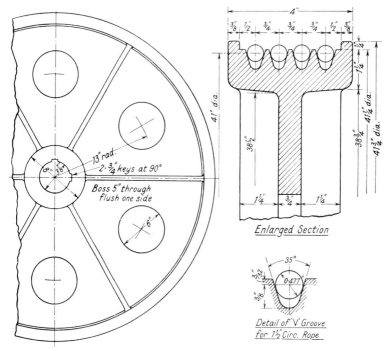

FIG. 4.4. VEE-GROOVED SHEAVE FOR 1¾ IN. CIRCUMFERENCE ROPES
AND 41 IN. ROPE CENTRE DIAMETER
(*R. J. Shaw & Co.*)

The grooves of a traction sheave must be maintained in good
condition as any uneven wear which would cause, say, one rope
to run deeper in its groove than the other ropes, will result in
this rope travelling slower than the others and consequently
slipping in its groove in its efforts to keep pace with the other
ropes.

It will be appreciated that all traction drives rely for their

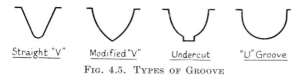

FIG. 4.5. TYPES OF GROOVE

effectiveness upon friction between the ropes and the sheave grooves, the tractive force available depending upon the coefficient of friction between the ropes and the sheave, the groove angle, and the amount of rope wrap. These three factors govern the ratio which can exist between the rope tensions on the two sides of the sheave before slipping occurs. This maximum ratio between the rope tensions on the "tight" and "slack" sides of the sheave may be calculated as follows for both single-wrap and double-wrap drives.

SINGLE-WRAP DRIVE.　In Fig. 4.6 let

T_1 = tension on tight side of rope.

T_2 = 　,,　　,, slack 　,,　　　,,

θ = angle subtended by that portion of the rope in contact with the sheave.

bc = indefinitely small portion of BC.

$d\theta$ = angle subtended by bc

T = tension in rope at c.

$T + dT$ = 　　,,　　　,,　　　,, b.

S = resultant pressure of the sheave on portion bc of rope.

μ = coefficient of friction between rope and sheave.

Then at the moment when slipping occurs

$$(T + dT) - T = \mu S.$$

but 　　　　　　$S = T . d\theta$ .... (Triangle of forces, Fig. 4.6 (b).)

$\therefore$ 　　　　　　$dT = \mu T d\theta.$

$\therefore$ 　　　　　　$dT/T = \mu d\theta.$

$\therefore$ 　　　　$\int_{T_2}^{T_1} \frac{dT}{T} = \mu \int_0^{\theta} d\theta.$

$\therefore$ 　$\log_e (T_1/T_2) = \mu\theta$, or $T_1/T_2 = e^{\mu\theta}$ 　. . . . . . (1)

Since however, the rope lies in a vee-groove as shown in, Fig. 4.6 (c), the effect is to increase the resistance to slipping due to the wedging action between the two surfaces. The resistance to slipping in element bc is $2\mu R$.

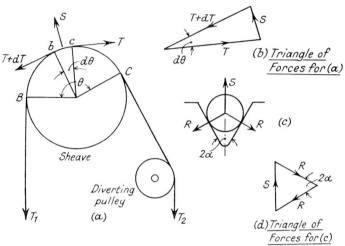

FIG. 4.6. ROPE AND SHEAVE FORCES IN SINGLE-WRAP DRIVE

But from Fig. 4.6 (d), $S = 2R \sin \alpha$, where 2α is the angle of the groove.

$$\therefore \quad 2R = S \operatorname{cosec} \alpha.$$

Hence the resistance to slipping in bc is $\mu S \operatorname{cosec} \alpha$.

In equation (1) above, in which the effect of the vee-groove has been neglected, the resistance to slipping in bc was μS. For this equation to be applicable to a rope in a vee-groove it is, therefore, necessary to substitute $\mu \operatorname{cosec} \alpha$ for μ and the equation then becomes

$$T_1/T_2 = e^{\mu\theta \operatorname{cosec} \alpha} \quad . \quad . \quad . \quad . \quad . \quad (2)$$

It will be observed from equation (2) that when an idle pulley is employed and the angle θ is thereby reduced, the traction is consequently less. To be rigidly accurate, the area of contact should be considered as μ is a function of the unit pressure.

DOUBLE-WRAP DRIVE (see Chapter V). Using symbols as shown in Fig. 4.7

$$T_1/T_3 = e^{\mu \theta_1 \, cosec \, a}$$

and

$$T_3/T_2 = e^{\mu \theta \, cosec \, a}.$$

$$\therefore \qquad T_1/T_2 = (T_1/T_3) \times (T_3/T_2)$$

$$= e^{\mu \theta_1 \, cosec \, a} \times e^{\mu \theta \, cosec \, a}.$$

$$\therefore \qquad T_1/T_2 = e^{\mu \, cosec \, a \, (\theta + \theta_1)}.$$

The traction is therefore increased by employing the double-wrap drive.

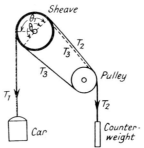

FIG. 4.7. DOUBLE-WRAP DRIVE

Coefficient of Friction. From the above equations it is observed that the maximum ratio of rope tensions is a function of the coefficient of friction between the ropes and sheave. The minimum value of the coefficient of friction necessary to drive a lift car in any particular installation may be calculated as follows.

Assume that the weights of the car and load are 3 000 lb. and 2 000 lb. respectively, and that the maximum speed of travel is 400 ft. per min.

Let counterweight = car + 50% load
                  = 4 000 lb.

If the car reaches its maximum speed in 2 sec., then the average acceleration is 400/120 = 3·3 ft. per. sec. per sec. The maximum rate of acceleration may be taken as 6·6 ft. per sec. per sec.

After acceleration is completed, $T_1/T_2 = 5\ 000/4\ 000$.
During the period of maximum acceleration when the
tendency to slip is greatest, $\dfrac{T_1}{T_2} = \dfrac{\left(5\ 000 + \dfrac{5\ 000 \times 6\cdot6}{32\cdot2}\right)}{4\ 000 - \dfrac{4\ 000 \times 6\cdot6}{32\cdot2}} = 1\cdot9.$

From above $\quad T_1/T_2 = e^{\mu\theta \, \text{cosec } a}.$

If $\theta = 180°$ and $2a = 40°$;

then $\qquad e^{\mu\pi \, \text{cosec } 20°} = 1\cdot9.$

$\therefore \qquad\qquad \mu = (\log_e 1\cdot9)/\pi \, \text{cosec } 20° = 0\cdot07.$

For a steel cable on a cast-iron sheave the value of μ varies
from approximately $0\cdot15$ to $0\cdot40$ as the surface pressure
increases from 100 lb. per in.^{2} to 600 lb. per in.^{2} These figures
are for dry surfaces, but the value of μ may possibly be as low
as $0\cdot06$ if the cables are greasy. Hence the importance of
ensuring that the amount of lubricant used is kept to a
minimum.

Diverting Pulleys. These are idle pulleys used to change the
direction of the ropes. With both drum and traction machines
it is frequently necessary to employ pulleys to divert the ropes
from the sheave or drum to the well. These pulleys are of
similar construction to the sheaves, but the grooves are arcs
of circles of a depth not less than one-third of the diameter
of the rope. The radius of the groove should be larger than the
radius of the rope by one-twentieth to accommodate ropes 5
per cent over nominal size, as most new ropes are oversize.

The pulley diameter is determined in a similar manner to
that described for sheave diameters, and a fact which must
be borne in mind is that if the rope is bent to the curvature of
the pulley, then a small angle of contact has the same effect
on the rope as a large angle. It is, therefore, incorrect to
assume that if the angle of contact is small a smaller pulley
may be employed. A pulley is shown in Fig. 4.8.

DRUM DRIVE

In a drum drive, one end of the car ropes and one end of the
counterweight ropes are securely fastened by clamps on the
inside of a cast-iron or steel drum, the other ends being fastened

to the car and counterweight respectively. One set of ropes is wrapped clockwise round the drum and the other set anticlockwise, hence when one set is unwrapping the other is being wrapped on the drum. There should be not less than one complete turn of the ropes on the drum when the car or counterweight has reached the extreme limit of its travel and clearance. As the car travels, the ropes move along the drum in spiral grooves cut on its periphery, the grooves being similar in radii

FIG. 4.8. DIVERTING PULLEY
(*J. & E. Hall, Ltd.*)

to those described for pulleys, but their depths should be not less than one-third of the diameter of the rope. The pitch of the groove should be such that there is a clearance of not less than $\frac{1}{16}$ in. between parts of the rope when coiled on the drum. The drum diameter should be as large as practicable in order to obtain a satisfactory rope life, the minimum diameter being determined in a similar manner to that described for sheave diameters.

The disadvantage of the drum drive is that as the height of travel increases, the drum becomes unwieldy, and it is therefore seldom used for rises of more than 100 ft. Because of the many advantages of the traction method, the drum is now almost obsolescent.

CHAPTER V

ROPING SYSTEMS AND ROPES

SEVERAL different roping schemes are employed to transmit power from the winding machine to the car. The actual method adopted depends upon local conditions; the situation of the winding machine, and the speed and loading of the car. It is, however, important that particular care should be given to the selection of the roping system, as upon this depends, to a large extent, the life which will be obtained from the lifting ropes. The roping should be as simple as possible and employ the minimum number of pulleys.

The lift machine is usually situated either near the top of the well or as near to the bottom of the well as practicable, but the location which permits of the best and cheapest roping scheme is immediately above the well. Further advantages of the top of the well position are a lower lift capital cost, reduced power consumption and the loads on the overhead structure are usually smaller than with the machine in the basement, the overhead loads for the two positions being as follows—

(a) *Machine Overhead*.
   Load = Lift machine + Control gear + Car + Car load + Counterweight.

(b) *Machine Below*.
   Load = 2 (Car + Car load + Counterweight).

The loadings in the two cases are shown in Fig. 5.1. Hence, in the event of the lift machine (motor, brake, and gearing) together with the control gear weighing more than the combined weights of the car, load, and counterweight, the overhead beams will be subjected to smaller loads with the basement location than if the machine is overhead. An intermediate floor is sometimes utilized for the winding machine when it is not permissible to install in the top or bottom positions.

Traction Drives. The traction drive shown in Fig. 5.2 (*a*) is the simplest of all roping schemes and is employed wherever possible. It is sometimes called the *half-wrap* or *single-wrap*, *one-to-one* system, because the ropes wrap the traction sheave

once only (for approximately 180°) and the peripheral speed of
the sheave is equal to the car speed. On account of the small
arc of contact, a vee-sheave together with comparatively hard
ropes is used. When the diameter of the sheave can be made
equal to the distance between the car and counterweight
supports, a diverting pulley is unnecessary.

A *double-wrap* drive, sometimes referred to as a *full-wrap* or

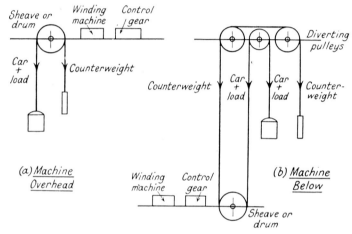

Fig. 5.1. Diagram Showing Loads on Overhead Structures

cross-over drive, is shown in Fig. 5.2 (*b*). The lifting ropes pass
from the car around the sheave, pulley, sheave again, pulley,
and thence to the counterweight if the pulley is not directly
under the sheave. It is thus seen that the ropes wrap the
sheave for approximately 360°. The pulley is placed directly
under the sheave as in Fig. 5.2 (*b*) when the dimensions of the
car and sheave permit. Most modern high-speed gearless lifts
make use of this double-wrap drive which, because of the
increased rope wrap, allows the use of U-shaped grooves on the
sheave. This drive is also employed where the out-of-balance
load is large, and there might be danger of rope slip with a half-
wrap system. With a full-wrap drive the load on the sheave
bearings is double that with a single-wrap drive and, of course,
the frictional resistance is considerably greater. It should be

noted that in these roping sketches only one rope is shown, but the number used may be as many as six, or even eight, whilst the minimum number which should be employed is two. Therefore,

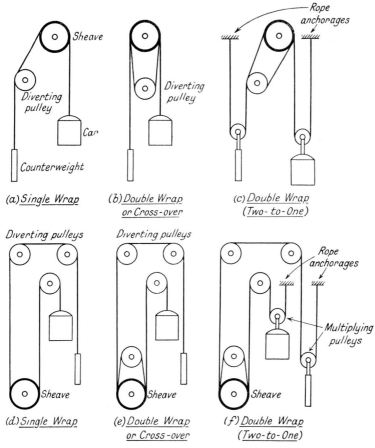

FIG. 5.2. SYSTEMS OF ROPING FOR TRACTION DRIVE

if six ropes are used with a double-wrap drive, then twelve grooves will be required on the sheave and twelve on the pulley when the pulley is not directly under the sheave.

Fig. 5.2(c) shows a *double-wrap, two-to-one* roping system which gives a peripheral sheave speed of twice the car speed. The ropes are securely anchored to one of the structural beams and then pass round a multiplying pulley fixed to the counterweight, pulley, sheave, pulley, sheave, a second multiplying pulley fixed to the car frame, and finally to a second anchorage. This is usually employed on large capacity slow-speed goods lifts with the object of removing a portion of the load from the

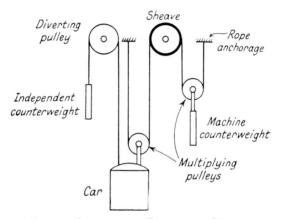

FIG 5.3. SINGLE-WRAP TWO-TO-ONE ROPING
Two counterweights

winding machine and, further, of permitting the use of a smaller number of lifting ropes, since these are not subjected to such stresses as with a one-to-one roping system. The two-to-one roping also enables gearless machines to be used for comparatively low car speeds of about 200 ft. per min. A disadvantage of the method, however, is the comparatively short rope life obtained due to the reverse bends employed.

For very heavy duty lifts, a method which has been employed still further to reduce the load carried by the winding machine makes use of a second counterweight as shown in Fig. 5.3. The addition of this independent counterweight causes some loss of traction, and its mass must be sufficiently low, therefore, to allow enough weight on the machine counterweight to maintain adequate traction. This lightening of the machine load is particularly desirable in the case of large gearless machines where the whole load, plus the weight of the motor, is normally carried by the motor bearings.

Fig. 5.2 (*d*), (*e*), and (*f*) show roping schemes similar to Fig. 5.2 (*a*), (*b*), and (*c*) respectively, but with the winding machine located below the well.

Another drive which has been used in recent years for high-speed gearless machines introduces a 3 : 1 speed ratio between the motor and the car. This is shown diagrammatically in Fig. 5.4. The advantage of this drive is that it permits the gearless machine to be used for low car speeds or for a given car speed; it enables the motor to be run at a higher speed than would be necessary if a 1 : 1 or 2 : 1 drive was employed. This results in a more stable motor control with smoother acceleration and more accurate levelling.

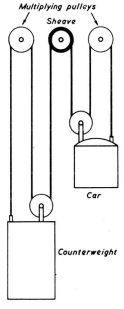

FIG. 5.4. THREE-TO-ONE ROPING

Drum Drives. A simple drum drive is shown in Fig. 5.5 (*a*), in which both car and counterweight ropes are securely anchored to the drum, and then pass to the car and counterweight respectively, via idle pulleys where necessary.

In the drive shown in Fig. 5.5 (*b*) an independent counterweight is employed in addition to the normal machine counterweight. The effect of this independent counterweight is to remove part of the car load from the drum shaft bearings and enable the machine counterweight to be reduced. These two counterweights are sometimes employed on large goods lifts, the independent counterweight being about 400 lb. lighter than the machine counterweight. Both counterweights run in the same guides, with the independent counterweight above, the machine counterweight cables passing through the independent counterweight.

Fig. 5.5 (*c*) and (*d*) show the drum located in the basement, the former employing a machine counterweight only, and the latter both machine and independent counterweights.

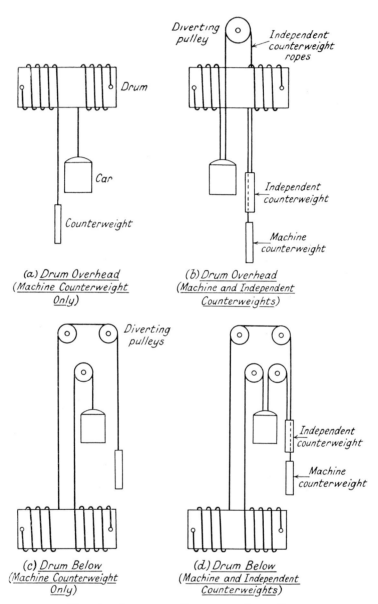

(a) *Drum Overhead*
(*Machine Counterweight Only*)

(b) *Drum Overhead*
(*Machine and Independent Counterweights*)

(c) *Drum Below*
(*Machine Counterweight Only*)

(d) *Drum Below*
(*Machine and Independent Counterweights*)

Fig. 5.5. Systems of Roping for Drum Drive

Compensating Ropes. These are fitted on long-travel lifts having a rise of more than about 100 feet with the objects of making the load on the motor constant during a journey from one end of the well to the other and of eliminating the effect of the rope weight in reducing the traction at the ends of travel. When the car is at the bottom of the well, the load on the motor is increased by the weight of the lifting ropes, which may be appreciable with high rises. Several methods of rope compensation are in use, but probably the best of these is that illustrated in Fig. 5.6. Ropes, equal in size and number to the lifting ropes, are secured to the car and pass to the counterweight via a diverting pulley and rope tension weight fixed in the well bottom.

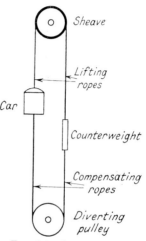

Fig. 5.6. Compensating Ropes

Low-speed lifts are sometimes fitted with compensating chains instead of ropes, cotton sash rope being woven through the links or the chains enclosed in a canvas hose to reduce the noise.

On very high-speed lifts (800 ft. per min. and above) now being installed in this country it is necessary to prevent the compensating rope tension weight from rising, so as to limit the jump of the car or counterweight if either hits the buffers or when the safety gear is applied. This locking-down of the tension weight is achieved by a ratchet mechanism on the compensating weight guides which in turn are fixed to the building foundations.

ROPES

Material. The lifting ropes employed on lifts are of stranded construction, each strand consisting of a number of steel wires. The steel used is either of special acid quality, termed Grade A, in which the phosphorus and sulphur contents are each not more than 0·04 per cent or acid quality, termed Grade B, in which the phosphorus and sulphur contents are each not more than 0·05 per cent. As an alternative, there has been a tendency in recent years to use wire drawn from basic steel made by the basic open-hearth process. Two grades of basic steel are available with similar limits for impurities to those quoted above for acid steel. Improvements in the technique of making basic steel and the fact that it can be made from home-mined ores are the reasons for extending its use to lift ropes. Great strength and flexibility are the most important properties of a lift rope, the former being obtained by using a steel of high carbon content and the latter by using a stranded rope construction. The tensile strength of the steel used in the manufacture of the ropes described later is 70 to 80 or 80 to 90 tons per square inch. A higher tensile strength of 110 to 120 tons per square inch is sometimes used for the inner nine wires of Seale ropes, although the outer nine wires are of 70 to 80 tons per square inch. The high tensile strengths are obtained by using a steel of moderately high carbon content, cold working, by drawing the wires through a series of dies and, in addition, subjecting the steel to a heat treatment known as patenting. To obtain the best type of wire there must be a close adjustment of the carbon content and of the method of drawing.

The fibre cores of all lift ropes consist of jute, hemp, or manilla, impregnated with a special lubricant, the purpose of which is to act as a lubricating medium for the internal parts of the rope and to assist the rope when at work by reducing friction. Furthermore, the lubricant preserves the wires and cores from deterioration from the results of damp, especially when not in actual use. The lubricant must be free from acid, of a penetrating nature and not liable to harden or peel off. For traction drives it is necessary that the insides of the wires be lubricated, but the outside surface of the rope must be dry. This is difficult to achieve in practice, and a new rope usually needs several wipings with paraffin-soaked rag during its first few months of use in order to remove surplus lubricant squeezed out from the impregnated core.

Size. The size of a rope is its nominal diameter and for lift purposes, usually varies between $\frac{1}{4}$ in. and $\frac{15}{16}$ in. The diameter is that of the circumscribed circle and is measured over each pair of opposite strands at each of three places at least 5 ft.

apart, with a rope caliper. The average of these three measurements is taken as the rope diameter. When thus measured before tensioning, the diameter shall not be less than the nominal diameter by more than 2 per cent and shall not exceed the nominal diameter by more than 5 per cent.

Lays. Two methods are employed in laying the wires and strands of a lift rope, namely the *Albert* or *Lang's* lay and the *Ordinary* or *Regular* lay. In a Lang's lay rope the strands are laid up to form the rope in the same direction as its wires were twisted to make the strand as in Fig. 5.7 (*a*). In the ordinary construction, however, the strand wires are twisted in one direction and the completed strands are laid up in the opposite direction to make the rope as illustrated in Fig. 5.7 (*b*). The advantage of Lang's lay rope is that it offers a bigger wearing surface when in use, and therefore can reasonably be expected to give a longer life than ordinary lay rope. Furthermore, a Lang's lay rope is more flexible than an ordinary lay rope. On the other hand, considerable experience is necessary in manipulating Lang's lay ropes as the tendency to kink and untwist is greater than with an ordinary lay. Hence, unless care is exercised in handling the rope, it may be incorrectly installed, with serious results to the length of service. In this country Lang's lay ropes are generally used, whilst in America the ordinary lay is very popular.

The rope is usually of right-hand lay, which means that it is laid up to the right, or clockwise direction when looking at the rope end. Some ropes, however, are of left-hand lay, these being wound in the anti-clockwise direction looking at the end. Therefore, in addition to specifying the rope lay, the direction must also be stated, e.g. "Lang's lay, right hand." Fig. 5.7 (*a*), (*b*), and (*c*) show right-hand Lang, left-hand ordinary, and right-hand ordinary lays respectively. In some sets of lift ropes, right- and left-hand ropes are used in an effort to produce a non-rotating combination. For example, if four lifting ropes are installed, two ropes would be of right-hand lay and the other two of left-hand lay.

The length of lay of a rope is the distance, parallel with the axis of the rope, in which a strand makes one complete turn about the axis of the rope. The lay of the strand, similarly, is

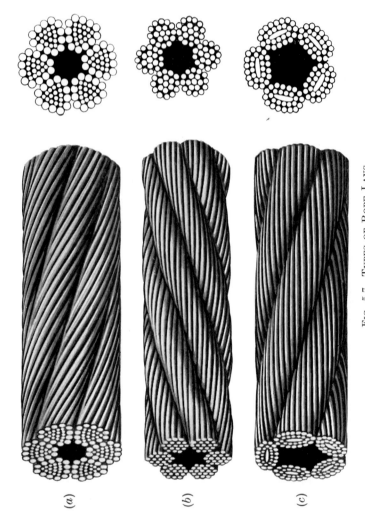

FIG. 5.7. TYPES OF ROPE LAYS

(a) Lang's lay, right hand. (b) Ordinary lay, left hand. (c) Ordinary lay, right hand.

(*British Ropes*)

the distance in which a wire makes one complete turn about the axis of the strand.

Factor of Safety. The minimum factor of safety of the combined lifting ropes should be 10, based on the contract load plus the weight of the car and the accessories. Although the above is usually specified and is termed the "Rope Factor of Safety," it does not give a true indication of the margin of safety because some allowance must also be made for the initial stresses in the wire during manufacture, the additional forces in the rope during acceleration and the bending stresses set up when the rope passes over the sheave and pulleys. With high car speeds acceleration is also higher and the acceleration stresses therefore greater. In order to allow for these stresses a greater factor of safety should be used with high car speeds than with low speeds. Whilst 10 is reasonable for car speeds up to 400 ft. per minute, a factor of 11 should be used for speeds between 400 and 700 ft. per minute, and a factor of 12 for speeds between 700 and 1 400 ft. per minute. With a traction drive, however, the number of ropes needed to obtain adequate traction may exceed the number required by simply applying the above factor of safety. For lifts which do not carry passengers, e.g. service lifts, a factor of safety of 8 is adequate for all speeds.

B.S. 329 defines the factor of safety as—

$$\frac{F \times n \times K}{W}$$

where F = nominal breaking strength of the rope,

   n = number of separate suspension ropes,

   K = roping factor, i.e. 1 for 1:1 roping
           2 for 2:1 roping
           3 for 3:1 roping

   W = load suspended on the "car ropes" at rest in the same units as F.

Round Strand Ropes. The ropes generally employed for lift work are round stranded and usually have six or eight strands. Some of the most popular rope constructions are described below.

(i) 6 × 19 (12/6/1). Each strand consists of nineteen wires

laid with one layer of twelve wires round another layer of six wires round one central wire, all wires being of the same diameter. The six strands are laid round an impregnated core to form the rope.

(ii) 6 × 19 (12/6/1) WITH SIX FILLER WIRES. This is similar to the 6 × 19 rope above, but has in addition six smaller filler wires laid in the interstices between the six-wire and twelve-wire layers. The object of these small wires is to fill the spaces between the two layers and so assist the rope to retain its shape. In assessing the rope strength these small filler wires are considered as not bearing any load.

(iii) 6 × 19 (9/9/1) SEALE. Wires of different sizes are employed in this rope, each strand of nineteen wires being made up of an outer layer of nine large wires, then a layer of nine small wires and finally one central large wire. The six strands are laid round an impregnated core in the usual manner. This rope is stiffer than the 6 × 19 uniform rope and is therefore not as suitable as the uniform rope for small diameter sheaves or pulleys or when the drive has reverse bends in the roping system. Because of its equal lay and more solid rope construction, however, it is a better rope than (i) above, for general use on traction sheaves.

(iv) 8 × 19 (12/6/1) WITH SIX FILLER WIRES. This is similar to (ii) above, except that it has eight strands instead of six and is therefore a little more flexible but has less resistance to abrasion because of its smaller diameter wires. It is also not quite as strong as a 6 × 19 rope of similar diameter.

(v) 8 × 19 (9/9/1) SEALE. This is similar to the 6 × 19 Seale, but has eight strands instead of six and is thus more flexible than a 6 × 19.

Other round strand constructions less frequently employed for lifts than those above are as follows—

(vi) 6 × 12 AND FIBRE. This has six strands each consisting of a single layer of twelve wires laid round a fibre core. The six strands are in turn laid up round a central fibre core. This is not suitable for traction drives as, in these circumstances, the rope does not retain its shape.

(vii) 6 × 24. Each strand of this rope consists of twenty-four wires all of the same size laid in two layers, one of fifteen wires round another of nine wires round a fibre core. The six strands are laid around a central fibre core. Like (vi) it is not used for traction machines.

(viii) 6 $\times$ 37. This rope is similar to the 6 $\times$ 19 uniform rope (i), but has an additional outer layer of eighteen wires resulting in a particularly flexible rope. It is not suitable for traction drives because of the small diameter of the wires.

Another round stranded rope sometimes used in America is the "Warrington" construction which, like Seale ropes, uses wires of different sizes. The large wires resist abrasion, and the small ones have more resistance to bending fatigue. In the 6 $\times$ 19 Warrington rope three sizes of wires are employed. Each strand is formed of an outer layer of twelve wires which are alternately of large and small diameter, then a layer of six wires of medium diameter and, finally, a central medium diameter wire. The six strands are made up to form the rope in the usual manner. Like the Seale rope, the Warrington suffers from lack of flexibility. Some of these round rope sections are shown in Fig. 5.8.

The most popular ropes at the present time for high-speed traction lifts are (ii), (iii) and (v) above, with high-tensile steel, preformed, and with a hard sisal core, these having replaced (i) which, until recent years, was used to a large extent. (ii), (iii) and (v) are generally used with undercut sheave grooves.

Ropes of Special Strand Construction. The quality of the steel is the same as that for round strand ropes. The two constructions of these flattened strand ropes are known as 5/27, 28, or 29 *oval* and 6/25 *F*. The 6/25 *F* (12/12/triangle) construction is the one generally adopted for lifts.

(i) 5/27, 5/28, OR 5/29 OVAL. This rope is made of five oval strands laid around an impregnated fibre core. Each strand consists of twenty-seven, twenty-eight, or twenty-nine wires laid in two layers around a core consisting of a flat wire, an elliptical wire, or a flat strand of a number of round wires.

(ii) 6/25 F. Six flattened strands are laid around an impregnated core to form this rope. The strand construction is twelve large wires laid around twelve smaller wires around a triangular shaped core which may consist of one or more shaped wires or one or more round wires.

The manufacture of flattened strand ropes was the logical development of the desire to produce a rope which, by reason

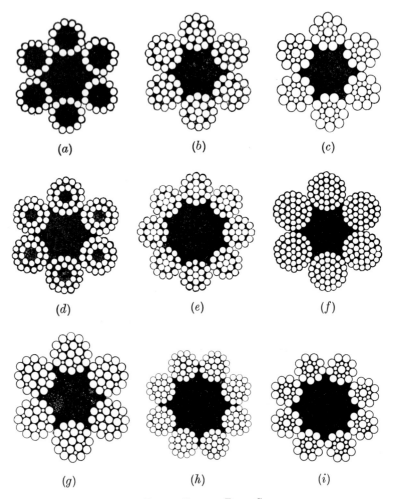

FIG. 5.8.   ROUND STRAND ROPE SECTIONS

(a) 6 × 12 and fibre	(b) 6 × 19	(c) 6 × 19 Seale (9/9/1)
(d) 6 × 24	(e) 8 × 19	(f) 6 × 37
(g) 6 × 19 with fillers (12/6/1)	(h) 8 × 19 with fillers (12/6/1)	(i) 8 × 19 Seale (9/9/1)

(*British Ropes*)

of its power to resist abrasion, would give a longer life than did the round strand rope. It will be readily seen that with ropes of this type, frictional wear is spread over a greater number of the outer wires of the rope. In a round strand rope the wear is taken on one wire in each strand when the rope is

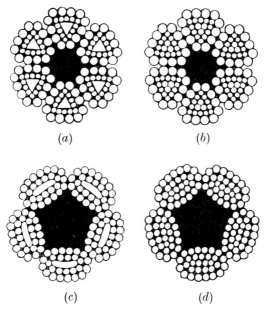

(a)   (b)

(c)   (d)

FIG. 5.9. SPECIAL STRAND ROPE SECTIONS

(a) 6/25 F (12/12/triangle)   (b) 6/25 F (12/12/3)
(c) 5/27 oval (16/11/oval)   (d) 5/27 oval (16 11/4)
(*British Ropes*)

new, and these wires are considerably reduced in sectional area before any appreciable wear is taken on the adjacent wires. In the flattened strand construction, owing to the friction being distributed over a greater external surface, the wear is much more even and the loss of sectional area of the vital outside wires much slower. Owing to the smooth surface and more nearly circular cross-section obtained with a flattened strand rope, the wear on sheaves and pulleys is reduced to a minimum.

Flattened strand ropes show approximately 150 per cent more wearing surface than round strand ropes. Despite this, the amount of these special strand ropes in use on lifts is not nearly so great as that of round strand ropes. Special strand rope sections are shown in Fig. 5.9.

Preformed Ropes. A special process known as *preforming* was introduced into this country from America and is now incorporated in the manufacture of what are termed *preformed* ropes. In a rope of this type the strands are passed through a preforming head which gives to the wires the exact final shape which they will take up in the completed rope. In the manufacture of ordinary rope, the wires are held forcibly in position throughout the life of the rope, as can be seen by cutting such a rope at any point, when the strands and wires will immediately fly apart. Preforming the wires prevents this, as they all lie naturally in their true positions, free from internal stress. Lang's lay or ordinary lay ropes can be preformed; such ropes are sold under various trade names, probably the best known being the "Bluestrand Tru-lay."

Although the initial cost of a preformed rope is greater than that of a corresponding rope of ordinary construction, the manufacturers claim that the many advantages of this rope far outweigh its extra cost. The advantages of this rope which have been borne out by practical experience may be summarized as follows—

(i) The rope is completely "dead" due to the reduction of internal stresses.

(ii) It has longer life, due to the uniform distribution of, and the large reduction in, the internal stresses. This claim has been confirmed both by practical experience and by laboratory tests. Regarding the latter, the Third Report of the Institution of Mechanical Engineers Wire Ropes Research Committee states that on pulleys of corresponding diameters, Tru-lay ropes are distinctly superior in 6×19 ordinary lay ropes, particularly under the severe conditions imposed by the use of small pulleys. In summarizing, the report states that Tru-lay ropes were superior to those made by the usual method of manufacture, particularly on small pulleys.

(iii) The load is evenly balanced on individual strands and wires.

(iv) There is no tendency to high strand even under the severest conditions.

(v) When the outer wires break from long wear there is no tendency for them to fray out from the body of the rope, but they continue to lie in their proper places. This prevents damage to adjacent wires and to sheaves and pulleys.

(vi) It is more easily spliced as there is no need to seize the strands.

(vii) There is less tendency to "kink" than with ordinary ropes.

Preformed ropes are now being used in increasing numbers for lift work.

ROPE FASTENINGS

No car or counterweight rope should be repaired or lengthened by splicing, continuous lengths being invariably employed. Several methods are used for terminating the ropes at the car and counterweight, the best and most generally adopted being by *spliced return loops*, *clipped return loops*, or individual *tapered babbited sockets*. Loops must not bear directly on their fixings but must be lined with proper thimbles. In all cases the fastenings should be capable of sustaining a load of not less than 80 per cent of the ultimate strength of the undisturbed rope.

Spliced Ends. Splicing forms a satisfactory method of terminating a rope end, but care is necessary in forming the splice. and the work should be carried out by an expert, to secure the best results. The rope is passed round a thimble, and sufficient free end of the rope left to form a splice of adequate length; after which the core is cut out from the free end. The free strands are opened out and each strand threaded through those of the main rope according to a definite schedule. Special tools are used to facilitate the operations of opening out the main rope strands and inserting those of the free end. The splice should have at least three tucks with a whole strand of the rope and two tucks with one half of the wires cut out of each strand made, under and over, against the lay of the rope. When the splice has been made, any unevenness may be removed by carefully pounding with a wooden mallet, thus leaving a perfectly uniform exterior. Finally, the splice is carefully bound with either hemp or fine stranded wire as shown in Fig. 5.10. The

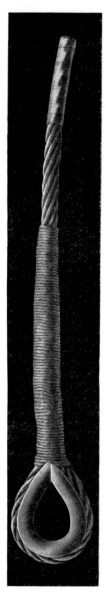

Fig. 5.10. Lift Rope Thimble and Splice
(*British Ropes*)

Fig. 5.11. Correct Method of Fitting Bulldog Grips
(*British Ropes*)

Fig. 5.12. Bullivant's Double Throat Clamp
(*British Ropes*)

lengths of satisfactory splices should be approximately as follows—

Diameter of Rope (in.)	Length of Splice (in.)	Diameter of Rope (in.)	Length of Splice (in.)
$\frac{5}{16}$	7	$\frac{3}{4}$	17
$\frac{3}{8}$	9	$\frac{13}{16}$	19
$\frac{1}{2}$	11	$\frac{7}{8}$	21
$\frac{9}{16}$	13	$\frac{15}{16}$	24
$\frac{5}{8}$	15		

Bulldog Clips. Clips form an effective method of fastening rope ends provided they are of good design, of the proper size, and are correctly fitted. They have an advantage over other methods of fixing in that the rope can be more readily adjusted to take up stretch. The correct method of fixing these clips is shown in Fig. 5.11. For ropes up to $\frac{15}{16}$ in. diameter at least three clips per rope end are recommended, for ropes over $\frac{15}{16}$ in. and up to $1\frac{1}{4}$ in. diameter four clips should be used, and for ropes over $1\frac{1}{4}$ in. diameter five clips per end. After the clips have been fitted it is advisable, the following day, again to tighten up the nuts, as it is often found this is necessary owing to the rope squeezing down.

The clips must always be fitted so that the castings are on the loaded rope, or in other words, the U-bolts must always be on the short end of the rope.

An improved form of clip known as "Bullivant's Double Throat Clamp" is shown in Fig. 5.12. This clip is claimed to be more effective than the ordinary bulldog pattern as it prevents the crushing of the rope which occurs when the ordinary clamps are used.

Sockets. When properly made, the white-metal and socket method of rope capping is probably the strongest known, but great care must be taken to see that the proper alloy is used at the right temperature, and that the wires are perfectly clean and free from grease. An open tapered socket is shown in Fig. 5.13. Several methods are employed for fitting the sockets, one of the best of which is as follows—

The end of the rope to be socketed should be bound with soft iron wire for at least one inch more than the length of the chamber, and again above this length for a further six or eight inches. After placing the socket on the rope, the first binding should be removed, the rope end unlaid, and each individual wire straightened out so that the group of wires resembles a brush. Each wire should then be turned over to form hooks, facing inwards to the centre of the rope, and the fibre centre cut out. Each wire must be properly cleaned, preferably with petrol, which quickly takes away any grease or dirt and leaves the wires dry. After this the wires should be roughened with emery cloth. When all the wires have been cleaned they are

FIG. 5.13. OPEN TAPERED SOCKET
(*British Ropes*)

drawn into position in the socket. The socket is then slightly heated to prevent the too rapid chilling of the white metal, thus ensuring its penetration. After this the socket is fixed in a vice and asbestos yarn or clay wrapped round the rope, under the socket, so as to plug the mouth of the socket and prevent the metal running through. The socket should be at a temperature of about 212°F. immediately before the molten metal is poured in. Some powdered resin should then be dusted amongst the wires. The metal used must have a low melting point so as not to anneal or take the temper out of the wire and, further, must have practically no contraction and set very hard. The melting point should be below 750°F. and excess heat must not be applied to the metal, otherwise the temper of the wires will be damaged. The correct temperature is readily indicated by inserting a chip of soft dry wood, e.g. a match stick, which slightly chars when the metal is ready for use. The proper temperature for pouring is between 635°F. and 685°F. After running in the white-metal, it should be allowed to cool naturally, after which it will be found that the cone has been formed throughout the whole length of the barrel.

Rope Equalizing Gear. Some means are usually provided on lift ropes to equalize the load on the individual suspension

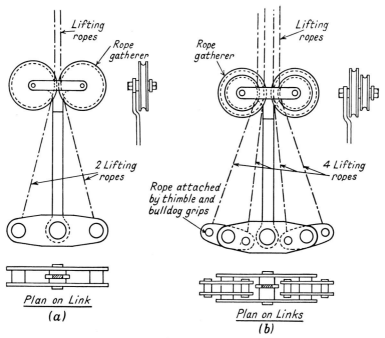

FIG. 5.14. ROPE EQUALIZING GEAR
(*a*) With two ropes   (*b*) With four ropes

ropes. Several different forms of equalizing gear are in use, but the principle of the simplest and probably most widely used is shown in Fig. 5.14 (*a*) and (*b*). With two ropes, each rope passes over a guide pulley or rope gatherer and the ends are attached, one at each side of a lever. This lever is free to rotate and takes up a position inclined to the horizontal, as one rope stretches more than the other. Thus each rope is subjected to the same tension. Three levers and four pulleys are used for a four-rope system as in Fig. 5.14 (*b*), the principle of operation being similar to that for two ropes.

Rope Tension. Some manufacturers do not use rope equalizing links as described above and, particularly for high buildings and speeds, prefer forged screwed individual thimble rods for securing the rope ends. These screwed rods are fitted in a plate on the car crosshead and permit individual rope adjustment to obtain the same tension in each rope. Equalizing rope tension is important, as if most of the strain is on one rope, excessive stretch will occur on it and thus shorten its life. On the other hand, if one rope is relatively loose it might rub in its groove and cause excessive wear. Several methods have been used for checking rope tension, including instruments which measure the rope deflexion from the vertical at about the mid-position in the well, when a known sideways tension is applied to the rope. The most common practice, however, is to lower the car to the bottom landing and for the maintenance man to fit planks in the well mid-position from where he carefully checks the rope tensions by hand. His mate then makes the necessary adjustments to the thimble rod nuts whilst standing on the car crosshead. Rope tension changes during use owing to core compression, wire stretch and creepage due to twisting of the ropes over the sheaves and it should therefore be checked at regular intervals.

CHAPTER VI

MOTORS

General. Several different types of motor suitable for lift work are available, the particular motor chosen depending upon the supply characteristics, car speed, and quality of service to be provided. For many purposes an ordinary commercial motor of speed between 750 r.p.m. and 1 200 r.p.m., and having certain special features, is suitable. Speeds of between 600 r.p.m. and 900 r.p.m. are usually preferred, whilst at speeds above 1 000 r.p.m. there is difficulty in obtaining the necessary degree of silence, even if special precautions are taken in the design of the motor room and the mechanical equipment. Furthermore, the higher the speed the greater the kinetic energy and the more powerful the braking effort required, but on the other hand the price decreases as the speed increases.

The main requirements of a lift motor are, a starting torque equal to at least twice the full load torque, quietness, and low kinetic energy; the last feature is necessary to obtain rapid acceleration and deceleration, together with a minimum amount of brake lining wear. In addition, the armature or rotor shaft must be capable of withstanding the high stresses due to braking and should be extended at the free end and made square, in order that the lift may be operated by hand, with a hand-wheel, in emergency and for effecting adjustments. The motor will be required to run in the same direction, either as a motor or as a generator, it being noted that when the load is a hoisting one, e.g. full load up, the machine functions as a motor, and when the load is overhauling, e.g. empty car up, the machine operates as a generator.

Size. The theoretical horse-power of the motor required to drive any lift is calculated as follows—

Assume that the maximum car load is 10 cwt., the maximum speed 250 ft. per min., and that the counterweight is equal to the car plus 50 per cent maximum load—

The out-of-balance load = 5 cwt.

$$\text{i.e. h.p.} = \frac{5 \times 112 \times 250}{33\ 000} = 4 \cdot 25$$

87

In practice, the horse-power required for satisfactory service will be considerably more than the theoretical horse-power, the actual amount depending upon the overall mechanical efficiency of the lift, which may be anything between about 30 per cent and 60 per cent (this varying with the size of the lift and the drive employed). Suitable motor sizes for various car loads and lifting speeds are shown in Table (*a*) below.

(*a*) H.P. OF MOTORS FOR LOW-EFFICIENCY GEARED LIFTS

Car Speed (ft. per min.)	Contract Load (cwt.)					
	5	10	15	20	30	40
50 . . .	2	3·5	5	6	8·5	11·5
100 . . .	3	5·5	8	10	15	20
150 . . .	4	8	11	15	21	28
200 . . .	5	10	14	18	27	37
250 . . .	6	12	17	23	34	45
300 . . .	7	14	21	27	40	54

The figures quoted in the table are applicable to lifts with low efficiency irreversible worm gearing, and with the counter-weight balanced for half the maximum car load.

In modern lifts various methods are employed to improve the overall mechanical efficiency. High-efficiency worm and worm-wheel gearing is used, and for the sheave shaft tapered roller bearings are sometimes employed. Care is also taken to suspend the car from a point above its centre of gravity instead of from the geometrical centre of the crosshead and so reduce the side thrust on the guide shoes. Roller guide shoes still further increase the mechanical efficiency. The adoption of such measures may result in mechanical efficiencies of geared machines as high as 65 per cent, or 70 per cent for gearless machines. When such steps are taken, the overall efficiency of a geared machine may be as high as 60 per cent and the horse-power of the motors required would then be about two-thirds of those shown in Table (*a*). Suitable motor sizes for modern lifts with high-efficiency gearing are shown in Table (*b*).

(*b*) H.P. OF LIFT MOTORS FOR MODERN HIGH-EFFICIENCY GEARED LIFTS

Contract Load		Car Speed in ft. per min.				
		50	100	150	200	300
Goods Car Load in cwt.	5	1·25	3·5			
	10	2·0	4·5	6·0	8·0	
	15	3·0	6·0	8·0	10·5	
	20	4·5	8·0	10·5	14·0	
	30	6·0	10·5	14·0	20·0	
	40	8·0	14·0	20·0	25·0	
	60	10·5	20·0	30·0		
Passenger Car Load in lb.	600		3·5			
	900		3·5	6·0		
	1 200		4·5	8·0	7·5	
	1 500		6·0	8·0	10·0	12·5
	2 000			10·5	12·5	15·0
	2 500			12·5	15·0	20·0

DIRECT-CURRENT MOTORS

Motors for Car Speeds up to about 100 ft. per min. A single-speed shunt or compound wound motor is employed for these low speeds, the larger sizes being equipped with commutating poles to give sparkless commutation in both directions of rotation. With car switch control, however, single-speed motors are sometimes used for car speeds up to 150 ft. per min. As previously mentioned, the machine must be capable of running as a generator or a motor, and this entails special arrangements being made in the case of a compound motor. The diagrams in Fig. 6.1 show the directions of the currents in the windings of a compound wound motor when "motoring" and "generating." When operating as a motor, both fields are in the same direction, but on change over to a generator, the back e.m.f. becomes greater than the applied voltage, and the two fields oppose each other. This results in a weakened field, reduced dynamic braking power, and an increase in the car speed. It is therefore necessary to arrange that the controller cuts out the series

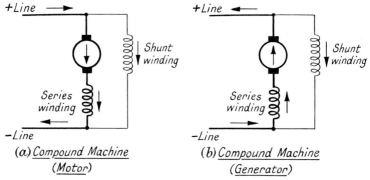

(a) *Compound Machine* (*Motor*)　　　(b) *Compound Machine* (*Generator*)

FIG. 6.1. CURRENTS IN WINDINGS OF A COMPOUND WOUND MACHINE

winding when the full speed has been reached, the object of the series winding being to provide a high starting torque.

When the main contactor closes, the brake shoes are released and the shunt field is energized via the starting resistances. Two methods are employed to cut out the starting resistances and thus obtain acceleration. In the first method, an oil dash-pot or a mechanical time relay is used to control the movement of the accelerating magnet plunger, and by this means the steps of starting resistance are cut out on a definite time basis, determined by the dashpot or relay adjustment. With the second method the accelerating switch coil is connected across the motor armature and the cutting out of the starting resistances is thus dependent upon the motor speed. At the instant of closing the armature circuit, the voltage across the armature is small, but as the motor speed rises the armature voltage increases. Hence, the voltage across the accelerating magnet coil gradually increases and its contact arm cuts out the starting resistance at a rate proportionate to the increase in motor speed. The series field is cut out of circuit immediately the starting resistance has all been cut out.

Deceleration is obtained by re-inserting the starting resistance and by connecting a diverter resistance across the armature, thus producing a slow levelling speed, not subject to such wide variations in speed as with a series resistance. Immediately the main contactor opens, a braking resistance is

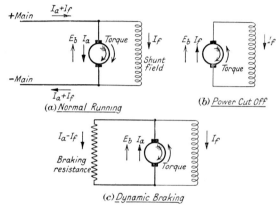

FIG. 6.2. DYNAMIC BRAKING WITH D.C. MOTOR

placed in parallel with the armature. By this means the kinetic energy of the lift is converted to heat, which is dissipated in the resistance, and the resulting dynamic braking assists the mechanical braking action. The effect of the dynamic braking resistance across the armature is shown in Fig. 6.2. In (a), which shows the normal running conditions, the back e.m.f. E_b is in the opposite direction to that of the main voltage and the armature current I_a. When the power is cut off, the conditions are as shown in (b), the field current I_f, which is almost equal to the normal value, being maintained by the armature back e.m.f. The addition of the braking resistance gives the conditions shown in (c), the reversed armature current causing a reverse braking torque, the value of which depends upon the magnitude of the parallel resistance. Hence the rate of deceleration caused by dynamic braking may be low or high, depending upon whether a high or low value of resistance is employed.

Motors of about 15 h.p. and over are sometimes provided with an additional shunt field in parallel with the main shunt field, the auxiliary field being automatically cut out as the speed rises, and re-inserted during the slowing down period. This auxiliary field increases the starting torque and improves the speed regulation.

Motors for Car Speeds between about 100 ft. per min. and 200 ft. per min. Two-speed motors, having speed ratios of 2–1, 3–1, or 4–1, are used in order to obtain a slow speed for efficient landing. The motor is similar to that used for single-speed working except that it is arranged for shunt regulation. In a typical two-speed motor having speeds of 400 r.p.m. and 800 r.p.m., acceleration from zero is obtained by cutting out the series starting resistance, as for a single-speed motor. The field is at full strength (all resistance out) whilst running up to 400 r.p.m. Increase from 400 to 800 r.p.m. is obtained entirely by shunt field weakening, effected by inserting field resistance. Deceleration from 800 to 400 r.p.m. is obtained by short-circuiting the field resistance, and below 400 r.p.m. by introducing the series starting resistance and shunting the armature by means of a diverter resistance, as for a single-speed motor. Dynamic braking is also resorted to, prior to the application of the mechanical brake.

Motors for Car Speeds above 200 ft. per min. During recent years there has been a big demand for car speeds above 200 ft. per min., due to the withdrawal of certain building restrictions, and the consequent erection of higher buildings. In this country the maximum car speed at present is about 1 000 ft. per min., whilst in America speeds up to 1 600 ft. per min. are employed in the "skyscrapers," whose heights in some cases exceed 1 000 feet. With these high speeds it has been possible to make use of a specially designed slow-speed d.c. motor, the shaft of which is coupled directly to the driving sheave, without the use of gearing. This reduction in motor speed and elimination of gearing results in a smaller kinetic energy of moving parts. The motor, brake, and sheave are mounted on a common bedplate to form a single unit, and it is thus seen that the motor bearings carry the load of the car, counter-weight, and the pull due to the ropes around the idle pulley (if a double-wrap drive) in addition to the weight of the motor armature, sheave, and brake drum. The motor is a shunt wound machine having a speed of between 50 r.p.m. and 120 r.p.m., depending on the duty, and is not usually equipped with commutating poles, these being scarcely necessary at such low speeds. On account of its large size it is not practicable to provide a range of speed of more than 1·5 to 1 by field control. In the earlier gearless machines the speed was controlled rheostatically by means of a combination of series and parallel resistances as in the case of the ordinary d.c. motor, but the best and most modern method is the application of the variable voltage or Ward-Leonard principle (see Chapter VII). The major part of the speed variation is accomplished by varying the voltage applied to the motor, the remaining small change being effected by field control.

Although the first cost of a gearless machine is considerably more than that of a geared machine, the gearless motor normally does not need replacement throughout the life of the lift because of its low running speed. For the same reason the maintenance costs are low. Because of the absence of gearing the efficiency is higher and hence the power consumption less. A better service can be obtained because of the higher accelera-tion, and the travelling is much smoother. Until recent years the gearless machine was seldom used for car speeds below 500 ft. per minute, but its superior performance over the geared machine has resulted in its employment for speeds as low as 350 ft. per minute.

Fig. 6.3 shows two gearless motors with brakes, sheaves and governors. These machines are part of an installation arranged with 2 to 1 double-wrap roping, interconnected signal collective control and contract loads and speeds of 2 500 lb. and 500 ft. per minute respectively.

A gearless motor complete with its brake and sheave is shown in Fig. 6.4. In this particular arrangement the motor armature is overhung outside the two main bearings on an extension of the shaft. Only the rope sheave and brake drum are mounted between the bearings. This design reduces the distance between the bearings, resulting in increased strength

Fig. 6.3. Two Gearless Machines with 2 to 1 Roping,
Double-wrap Drive
(*Express Lift Co.*)

Fig. 6.4. Gearless Machine
(*J. and E. Hall*)

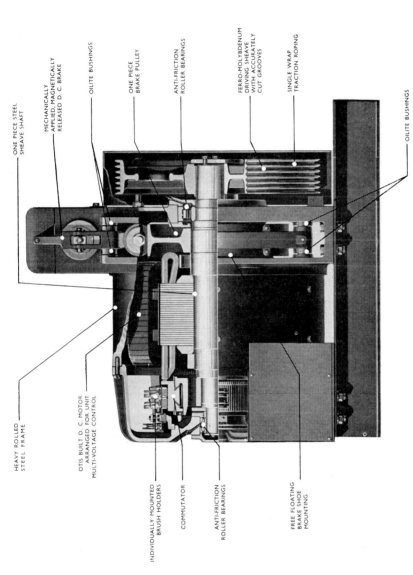

ONE PIECE STEEL
SHEAVE SHAFT

MECHANICALLY
APPLIED, MAGNETICALLY
RELEASED D. C. BRAKE

OILITE BUSHINGS

ONE PIECE
BRAKE PULLEY

ANTI-FRICTION
ROLLER BEARINGS

FERRO-MOLYBDENUM
DRIVING SHEAVE
WITH ACCURATELY
CUT GROOVES

SINGLE WRAP
TRACTION ROPING

OILITE BUSHINGS

HEAVY ROLLED
STEEL FRAME

OTIS BUILT D. C. MOTOR
ARRANGED FOR UNIT
MULTI-VOLTAGE CONTROL

INDIVIDUALLY MOUNTED
BRUSH HOLDERS

COMMUTATOR

ANTI-FRICTION
ROLLER BEARINGS

FREE FLOATING
BRAKE SHOE
MOUNTING

FIG. 6.5. GEARLESS MACHINE FOR SINGLE-WRAP ROPING
(Otis Elevator Co.)

due to lower shaft stresses for the same amount of material.
This arrangement has other advantages over the alternative
method of mounting both electrical and mechanical components between the two bearings. The mechanical portion,
the size of which is determined mainly by the load and sheave
reaction, can be standardized independently of the motor.
The size of the motor is determined by torque and the duty
cycle, so that, if necessary, more than one size of motor can be
used with the same mechanical equipment. The motor can
also be easily withdrawn for maintenance without disturbing
the ropes.

The Otis gearless machine for single-wrap roping, shown in
Fig. 6.5, is of compact design and takes up little floor space.
The sheave is overhung and the main shaft supported on roller
bearings at the commutator end and between the sheave and
brake-drum.

ALTERNATING-CURRENT MOTORS

POLYPHASE SUPPLY

Motors for Car Speeds up to about 100 ft. per min. The single-
speed squirrel-cage motor is suitable for these low speeds,
although it suffers from the disadvantages of high starting
current and a tendency to overheat if the duty is severe. The
slip should be kept within reasonable limits to avoid too great
a speed variation. It is necessary, also, that the motor should
develop a high torque for starting and accelerating, but this
means that it will inherently have a high slip, so that some
compromise must be made in the design. Further, a low slip
is obtained by reducing the rotor resistance, which is an
indication that the starting current will be high. In practice,
the starting kVA should not exceed about 5 kVA per h.p.,
the starting torque should be not less than 225 per cent and
not more than 275 per cent full-load torque. The slip at full
load should not exceed $7\frac{1}{2}$ per cent for motors up to 2 h.p.,
6 per cent for motors between 2 h.p. and $7\frac{1}{2}$ h.p. and 12 per cent
for motors over $7\frac{1}{2}$ h.p. These motors are often switched
directly across the lines, but it is frequently necessary to provide
smoother acceleration by the use of starting resistances in the
stator circuit. The short-circuiting of these resistances is

controlled by an air or oil dashpot. The squirrel-cage motor is not suitable for duties exceeding 90 starts per hour.

The standard rating for a squirrel-cage motor for lifts is 90 starts per hour and the standard duty cycle for 90 starts per hour comprises the following sequence repeated indefinitely—

(*a*) A period of 2 seconds starting and accelerating, developing a static torque of not less than $2\frac{1}{4}$ times rated full-load torque;

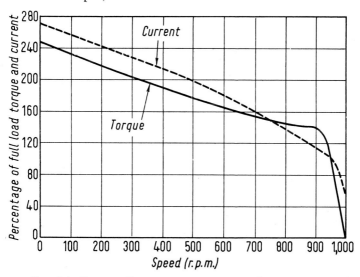

Fig. 6.6. Typical Characteristics of a.c. Squirrel-cage, Single-speed Lift Motor
(*Bull Motors*)

(*b*) A period of 8 seconds running at rated horse-power;
(*c*) A period of 30 seconds "off" (standstill);
(*d*) A period of 1 second starting and accelerating, developing a static torque of not less than $2\frac{1}{4}$ times rated full-load torque;
(*e*) A period of 8 seconds running at no load;
(*f*) A period of 31 seconds "off" (standstill).

The characteristics of Bull Motors' single-speed squirrel-cage induction motor designed for lift duty are shown in Fig. 6.6, from which it will be seen that the starting torque is 250 per cent of full-load torque, the full-load slip is 5 per cent and the starting current 275 per cent of full-load current. The main features are high starting torque, low starting current and slip, smooth acceleration and quiet running. Considerable improvements in design have been made in recent years, particularly in the smoothness of acceleration and the elimination of the starting "howl" usually associated with high-torque squirrel-cage motors. This motor is switched direct on to line without buffer resistance or choke in series, and standard sizes are from 3 h.p. to $12\frac{1}{2}$ h.p. at full-load speed of 940 rev/min. In order to improve the running characteristics still further, a flywheel of approximately equal WR^2 value to the rotor is sometimes fitted. This is 10 in. dia. $\times \frac{1}{2}$ in. for the 3 h.p. size and 10 in. dia. $\times 2$ in. for the $12\frac{1}{2}$ h.p. size.

Better speed regulation, lower starting current and smoother acceleration are obtained by using a wound rotor motor which is accelerated by cutting out the rotor resistances in one or more steps, these being controlled by dashpots. Although the speed can be varied by rotor resistance, this machine is essentially single-speed, as the insertion of rotor resistance results in low efficiency and poor speed regulation. Hence, the rotor resistance is used for the short starting period only, when the loss in performance is unimportant. This type costs about 50 per cent more than the squirrel-cage motor but may be used for duties up to 180 starts per hour. For lift work this type of motor should have a starting kVA not exceeding 3·25 kVA per h.p., a starting torque not less than 225 per cent full-load torque and the full-load slip should not exceed 6 per cent. The connexions for a single-speed slip-ring motor with two steps of starting resistance are shown in Fig. 6.7.

The standard rating for a slip-ring motor is either 90 or 180 starts per hour, the latter being the maximum. The standard 90 starts per hour duty cycle is as defined above for squirrel-cage motors. The standard duty cycle for 180 starts per hour comprises the following sequence repeated indefinitely—

(a) A period of 2 seconds starting and accelerating, developing a static torque of not less than $2\frac{1}{4}$ times rated full-load torque;

(*b*) A period of 8 seconds running at rated horse power;

(*c*) A period of 10 seconds "off" (standstill);

(*d*) A period of 1 second starting and accelerating, developing a static torque of not less than $2\frac{1}{4}$ times rated full-load torque;

(*e*) A period of 8 seconds running at no load;

(*f*) A period of 11 seconds "off" (standstill).

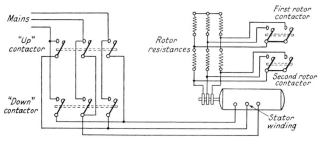

Fig. 6.7. Single-speed Slip-ring Motor with Two Steps of Rotor Resistance

Motors for Car Speeds between about 100 ft. per min. and 200 ft. per min. For these car speeds it is necessary to employ one of the several available types of motor capable of running at more than one speed, in order to obtain a slow landing speed.

Squirrel-cage Induction Motors. The squirrel-cage induction motor can be adapted for pole changing by regrouping the stator winding to give two polar combinations corresponding to the two speeds. For one combination of poles, the coils in any one phase give poles of similar polarity round the machine, the opposite poles being induced in the spaces between the coils. By reversing every alternate coil, the poles produced are alternately north and south and only half the original number of poles is produced. Pole changing can therefore only be used when the two speeds required are in the ratio of two to one. One method of obtaining half the number of poles, i.e. double the speed, by this reversal of current in half the winding, is shown in Fig. 6.8. The winding is connected in star for the larger number of poles, and in two parallel star windings for the smaller number of poles. For the direction of rotation to remain unchanged when the poles are changed, two of the phases must be reversed in relation to the line wires. The motor starting and running connexions are shown in Fig. 6.9. Another method, in which the windings are connected in delta for the larger

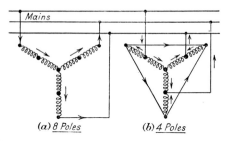

Fig. 6.8. Stator Windings of Two-speed Squirrel-cage Motor
Star—Two Parallel Star

number of poles, and in two parallel star circuits per phase for the smaller number of poles, is shown in Fig. 6.10. This is a better method, as an improved performance is obtained and no

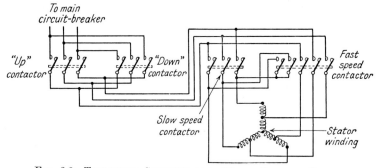

Fig. 6.9. Two-speed Squirrel-cage Motor Control Circuit

reversal of line wires and phases is necessary to maintain the same direction of rotation. A further disadvantage of pole changing is that, if the coil pitch is correct for one of the polar combinations, it is inefficient for the other.

In another form of two-speed squirrel-cage motor, two separate stator windings, wound in the same slots, are employed to give the pole change corresponding to the required two speeds. Ratios up to six to one are possible with this double wound motor, the high and low speed windings being cut in and out of circuit by means of a contactor. The disadvantage of this type is the complexity of the double winding and the difficulty of repair, especially if the damaged winding is the under one of the two.

These two-speed squirrel-cage motors are usually started up with resistance in the high-speed winding, whilst smooth deceleration is obtained by inserting a buffer resistance, either in the slow- or high-speed winding during transition to slow speed. The use of a choke instead of a buffer resistance results in a smoother and less peaked curve of braking torque.

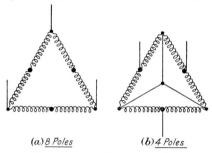

(a) 8 Poles   (b) 4 Poles

FIG. 6.10. STATOR WINDINGS OF TWO-SPEED SQUIRREL-CAGE MOTOR
Delta—Two Parallel Star

Typical torque/speed and current/speed characteristics of a Bull Motors 10 h.p., 4-to-1 speed range "Super Start" motor are shown in Fig. 6.11. When switched direct on line, which the firm recommends, the torque/speed curve is seen to be almost flat up to approximately 60 per cent full speed, when the torque gradually falls to the full-load value at about 7 per cent slip. A "buffer" choke or resistance is necessary in changing to the low-speed winding for deceleration. Ample braking torque is available for deceleration, and the effect of chokes in the circuit, for the adjustment of this torque, is to "flatten" the curve so that smoother acceleration is achieved. It will be noted that the full-load slip at levelling speed is about 10 per cent. The starting current is about 360 per cent of full-load current when developing 250 per cent full-load torque. The standard motor is suitable for most passenger-lift applications up to 200 ft. per min. but for heavy duty it can be provided with a ventilator unit which is readily fitted to the top of the frame in a similar manner to that shown on the heavy-duty tandem motor illustrated later in this chapter. As with the single-speed motor, a small flywheel for adjusting the total inertia of the machine is sometimes fitted. The combination of buffer choke/resistance and flywheel produces very smooth

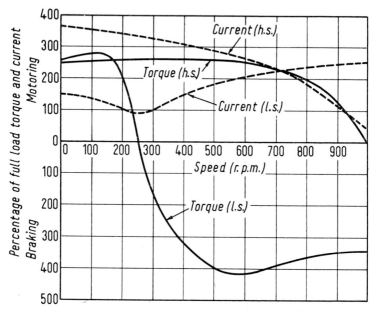

FIG. 6.11. TYPICAL TORQUE/SPEED AND CURRENT/SPEED CURVES OF 4:1 50 ∼ "SUPER START" TWO-SPEED SQUIRREL-CAGE LIFT MOTOR

h.s. = high-speed winding; *l.s.* = low-speed winding
(*Bull Motors*)

operation. This motor is available in standard sizes from 4 to 20 h.p. with 3/1 speed change, 5 to 20 h.p. with 4/1 speed change and 10 to 20 h.p. with 6/1 speed change.

Another method of effecting smooth acceleration with lift motors of this type employs an induction regulator in the motor supply circuit; the regulator applying the starting voltage gradually to the high-speed winding without the use of resistances. The induction regulator is a transformer with a variable ratio of transformation, and is built like an induction motor, the stator forming the primary, and the wound rotor the secondary. The rotor is not free to rotate but its position can be altered through one pole pitch by means of a small torque motor. This turning of the rotor varies the voltage applied to the high-speed winding, and results in very smooth acceleration.

SLIP-RING INDUCTION MOTORS. Speed changes, by the methods employed with squirrel-cage motors, may be obtained with a slip-ring motor, but this usually involves the use of two separate rotor windings arranged similarly to the stator windings. The rotor connexions must be changed at the same time as the stator connexions and this, together with the slip-rings, involves extra complications in the control circuit, although an improved performance is obtained. The use of two rotor windings may be avoided by carrying the rotor currents through internal short-circuited paths during one of the speeds, but this involves the loss of the slip-ring characteristics on one speed.

TANDEM MOTORS. The Tandem, two-speed a.c. motor, which has been employed on a number of lifts, consists of a wound rotor and a squirrel-cage rotor, forming the high- and low-speed sections respectively. These are assembled on the same shaft and the two frames bolted together to form a single two-bearing unit. Various speed ratios are obtainable and in a typical motor having a six-to-one ratio the speed obtainable from the wound rotor portion is 960 r.p.m. whilst the low speed from the squirrel-cage end is 160 r.p.m. In this particular motor, the slip-ring section is wound for 6 poles, and the squirrel-cage section for 36 poles. Performance curves for the sections of this motor are shown in Fig. 6.12, from which it will be seen that if the lifting speed is at its maximum, this can be reduced to one-sixth, by transferring the supply from the slip-ring section to the squirrel-cage section. Both sections are magnetically and electrically independent internally and each is capable of producing twice the full load torque, for starting purposes. The slip-ring portion is of standard design for lift service, whilst the low-speed end should have low slip and high starting torque to ensure a constant speed at all loads. A rotor of the motor is shown in Fig. 6.13 in which the squirrel cage section is of special construction. This consists of alternate

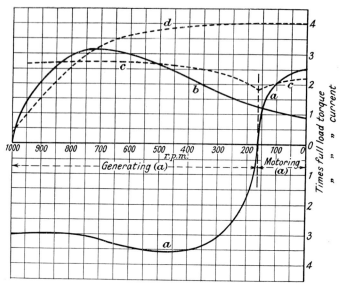

FIG. 6.12. CHARACTERISTIC CURVES OF THE TWO WINDINGS OF
TANDEM MOTOR

(a) Torque of S.C. motor
(b) Torque of S.R. motor with no rotor resistances
(c) Current of S.C. motor
(d) Current of S.R. motor with no rotor resistances

(*Metropolitan Vickers Gazette*)

FIG. 6.13. THE ROTOR OF A TANDEM LIFT MOTOR
(*Bull Motors*)

104

segments of steel and copper secured by steel end rings of inverted "L" sections, the segments being brazed to the steel end rings and the latter welded to the steel hubs. This construction results in a flat speed torque characteristic during deceleration.

It is usual to start up on the high-speed winding with resistance in the rotor circuit, the resistance being gradually cut out until final running speed is attained. To change from high to low speed, a switch disconnects the high-speed winding and another switch energizes the low-speed section. On change over to low speed, dynamic braking occurs, and from the characteristic curves it will be seen that a reverse torque, up to approximately four times full load torque, is obtained. The inertia of the rotating parts helps to smooth out the change, but it is also necessary to insert a buffer resistance or a choke coil in the star point of the squirrel-cage stator winding to prevent the deceleration from being excessive. The resistance is gradually cut out during deceleration, leaving the slow-speed winding fully energized. The starting and running connexions for this type of motor are shown in Fig. 6.14.

The control of the Bull Super Tandem incorporates some patented features which make its characteristics different from the normal type described above. On changeover to the low speed section the high speed stator is not disconnected from the supply but is kept in circuit with the low speed stator whilst a resistance is inserted in the slip-ring rotor. The wiring for the two sections and the necessary resistances are shown in Fig. 6.15, and the speed torque curves for a 6 : 1 speed ratio in Fig. 6.16. The high speed accelerating resistance is between contactors 3 and 2, but in practice this consists of more than the simple one step shown in the diagram. During acceleration contactor 3 is closed and the various steps between 3 and 2 gradually cut out until finally No. 2 is closed and the motor is running at high speed. Contactor No. 3 opens after the closure of No. 2. The high speed section is thus running at contract speed without any rotor resistance and the squirrel-cage section is disconnected from the line by contactor No. 1. Slowing to the final levelling speed is performed by the contactors operating in the sequence shown in Fig. 6.16, which removes the levelling resistance step by step

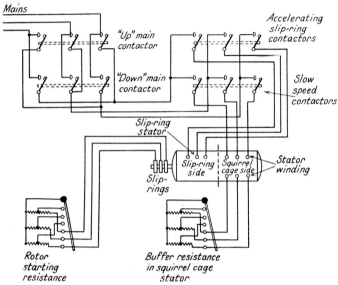

FIG. 6.14. TANDEM MOTOR CONNEXIONS

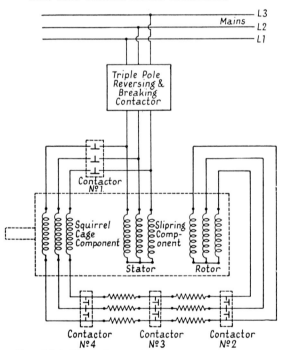

FIG. 6.15. BULL SUPER TANDEM LIFT EQUIPMENT

(*Bull Motors*)

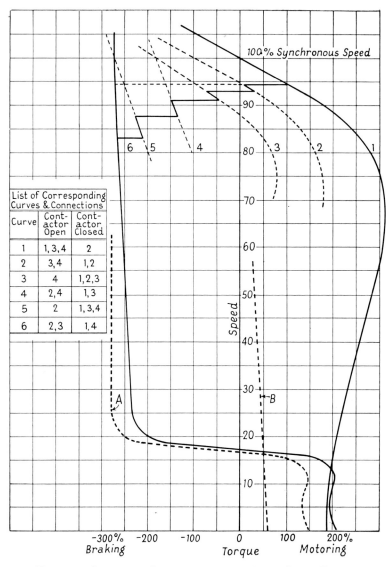

FIG. 6.16. LEVELLING CHARACTERISTICS OF BULL SUPER TANDEM
LIFT EQUIPMENT CONNECTED AS SHOWN IN FIG. 6.15

from the low speed stator and inserts it in the high speed rotor. The transition from the high speed (curve 1) to the low speed (curve 6) is thus effected smoothly.

Curves *A* and *B* illustrate the speed torque characteristics of the low speed and high speed sections acting separately. Curve 6 is the summation of these two and is the torque at the motor shaft during low speed. The torque contributed by the high speed section is additive to that of the low speed section when motoring whilst it does not seriously reduce the net braking torque. The static torque at low speed must, of course, be at least 200 per cent full load torque as the lift may be called upon to re-level. In the arrangement described this requirement can be fulfilled and the low speed section kept to the minimum size since it has to provide only about 140 per cent full load torque, the remaining 60 per cent being supplied by the high speed section. A small reduction of the slip-ring rotor resistance will increase the low speed static torque without appreciably affecting the braking characteristics. Another advantage of this method of control is that the full load slip at levelling speed (less than 6 per cent) is somewhat lower than would otherwise be the case and this results in a smaller variation of speed with varying load.

The price of a tandem motor is nearly double that of the two-speed squirrel-cage motor (this ratio depends on the h.p.), but it has a higher efficiency and can be used for heavier duties. The approximate lift ratings of tandem motors of sizes 10, 20 and 30 h.p. are 140, 120 and 100 starts per hour.

For very frequent starts and stops a heavy duty type of the Bull Super Tandem is available. This consists of the standard motor with a built-in independently running fan unit which provides additional continuous ventilation. Fig. 6.17 shows this heavy duty tandem, the standard machine being similar except that the fan housing is not fitted. The ventilating unit consists of a fractional h.p. vertical spindle motor driving a propeller fan so arranged that it assists the paddle fan in the introduction of air to the interior of the machine. The paddle fan fitted to the standard and heavy duty machines is located in the centre between the two rotor components. Approximate lift ratings for this heavy duty model are 240, 220 and 180 starts per hour for sizes of 10, 20 and 30 h.p. respectively.

A.C. COMMUTATOR MOTORS. Motors of the variable speed, shunt commutator type were popular for lift work, and speed changes of seven to one may be obtained in the smaller sizes, whilst changes as high as fifteen to one are possible in the larger sizes. They are, however, now seldom used for lifts. The arrangement of the windings differs from that of the ordinary induction motor in that the primary winding is located on the

FIG. 6.17. HEAVY DUTY TANDEM MOTOR
(*Bull Motors*)

rotor and the secondary winding on the stator. In addition to the primary winding, which is connected to the supply by means of sliprings and brushes, a regulating winding is placed in the same rotor slots. This latter winding is connected to a commutator in a similar manner to the armature winding of a d.c. machine. The commutator is provided with two brush rockers which can be moved relatively to each other by means of a small pilot motor. One end of each phase of the stator winding is connected to a brush stud of one rocker and the other end of the phase to the corresponding brush stud of the opposite rocker. Hence, the greater the distance the two sets

of brushes are moved apart, the greater will be the amount of
regulating winding connected in series with the secondary. The
pilot motor, by moving the brushes in this manner, varies the
e.m.f. injected into the secondary winding, this e.m.f. being
zero when the brushes connected to the ends of the same
phases of the secondary are in line, i.e. in contact with the same
segments. Under these conditions the motor runs as an ordinary

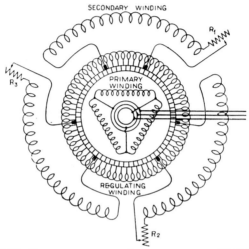

FIG. 6.18. DIAGRAM SHOWING ARRANGEMENT OF WINDINGS OF
COMMUTATOR MOTOR
(*B.T.-H.*)

induction motor at a speed slightly less than synchronous speed.
The e.m.f. which is induced by the primary in the secondary,
however, depends upon the speed of rotation, being zero at
synchronous speed and reaching a maximum value at standstill.
The effect of the injected e.m.f. is to compel the motor to change
from synchronous speed to that at which the induced secondary
e.m.f. balances the injected e.m.f. By rotating the brushes in
one direction or the other, the injected e.m.f. is made negative or
positive, resulting in speeds below or above the synchronous
speed.

Resistances $R1$, $R2$, and $R3$ may be introduced in the secondary winding to obtain creeping speeds as shown in the diagram
of the motor windings in Fig. 6.18. The insertion of resistance,

however, adversely affects the characteristic, the speed drop
from no load to full load being greater than for speed regulation
by brush shifting alone.

Motors for Car Speeds above 200 ft. per min. Geared a.c.
motors are in use for speeds above 200 ft. per min., but these
are exceptions rather than general practice. It is not practicable
to produce a gearless a.c. machine, and the gearless d.c. motor
operating from a variable voltage motor generator set is the
method usually employed. The mains motor is invariably of
the squirrel-cage induction type.

Single-phase Supply

Motors for Car Speeds up to about 100 ft. per min. Repulsion-induction Motors. This motor starts as a repulsion
motor in order to obtain the necessary high starting torque,
after which, the brushes are lifted by a centrifugal governor,
and the commutator short-circuited, the motor finally running
as a squirrel-cage induction motor.

The "Type C.K.S." repulsion-induction motor manufactured
by the British Thomson-Houston Co., however, and to which
the following description refers, has no centrifugally operated
commutator short-circuiting and brush-lifting device, and the
motor is, therefore, not so complicated as the usual type mentioned above. The rotor laminations have two concentric sets
of slots joined by narrow radial slits in which are placed thin
metallic strips. These slots contain two distinct windings; a
commutator winding (similar to a d.c. armature winding) in
the outer slots and a cast aluminium squirrel-cage winding in
the inner slots. The stator is wound with a simple single-phase
winding.

During the starting and accelerating periods, the flux produced by the stator winding links with the outer winding
only, due to the high reactance of the squirrel-cage, and the
motor acts as a repulsion motor giving a high starting torque.
As the motor accelerates, the reactance of the squirrel-cage
decreases, so that more and more of the flux links with this
winding and both windings now assist in the acceleration, thereby producing a large torque. On light loads, i.e. at speeds
above the synchronous speed, the squirrel-cage exerts a
braking torque and prevents the speed from increasing to

more than 2 per cent or 3 per cent above the synchronous speed.

The motor is started by switching direct on the lines, the starting current being approximately $3\frac{1}{2}$ times the full load current, whilst twice the full load torque is developed. The starting current may, if required, be reduced by inserting an external resistance in the line circuit, when the torque developed will be reduced as the square of the starting current. Typical characteristic curves of this motor are shown in Fig. 6.19, and the arrangements of the motor windings in Fig. 6.20.

With high efficiency gearing and an overhauling load, it is possible for the torque developed to be insufficient to effect reversal of rotation of the motor in the event of the car switch being instantly reversed. A time delay is therefore provided in conjunction with the main reversing contactor in order to delay the reversal by approximately 0·5 sec., to enable the brake to operate and reduce the motor speed. The motor will then develop the necessary reverse torque.

CAPACITOR MOTORS. During recent years the cost of condensers has been reduced and their reliability increased, and this has made practicable the use of capacitor motors for single-phase lifts. This motor is quieter than the repulsion-induction motor and may be of either the squirrel-cage or slip-ring type, the latter being employed for the larger sizes.

Electrically, these motors are, in effect, of the two-phase induction type, but operate on a single-phase supply, this being accomplished by the use of condensers which are housed in a separate metal capacitor unit as shown in Fig. 6.21. The stator has two windings; a main and an auxiliary winding displaced electrically by 90°, the condensers being connected to the latter winding. To obtain the high starting torque required for lift service and to ensure quiet running, the value of the capacitance at starting must considerably exceed that at full speed, and the effective capacitance must therefore be reduced as the motor speed increases. This is accomplished automatically by a centrifugal switch which operates at a predetermined speed, and as the motor slows down the switch resets ready for the next start.

The starting current of the squirrel-cage type is $3\frac{1}{2}$ times full-load current and the starting torque twice full-load torque. The

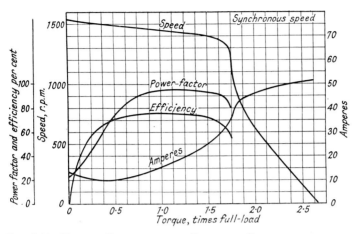

FIG. 6.19. TYPICAL CHARACTERISTIC CURVES OF REPULSION-INDUCTION
MOTOR

Type CKS 3 816, 3 h.p., 1 450 r.p.m., 4-pole 210-volt 50-cycle, single-phase
(*B.T.-H.*)

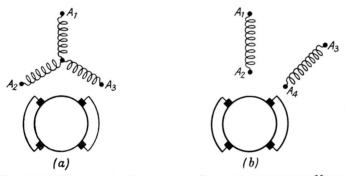

FIG. 6.20. WINDINGS OF SINGLE-PHASE REPULSION-INDUCTION MOTOR

    (*a*) For motors up to 3 h.p.
        For one rotation connect A_1 and A_2 to line
        For opposite rotation connect A_1 and A_3 to line
    (*b*) For motors above 3 h.p.
        For one rotation join A_2 to A_3 and connect A_1 and A_4 to line
        For opposite rotation join A_2 to A_4 and join A_1 and A_3 to line

(*B.T.-H.*)

113

(a)   (b)

FIG. 6.21. CAPACITOR TYPE SINGLE-PHASE LIFT MOTOR
(a) Motor   (b) Condensers
(B.T.-H.)

connexions for starting, acceleration, and running are shown in Fig. 6.22.

The slip-ring type has a lower starting current ($2\frac{1}{4}$ times full load current) for an equivalent torque than the squirrel-cage type, and is used when the supply authority imposes starting

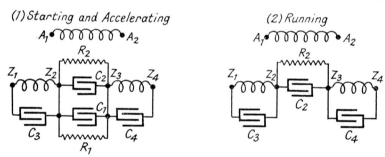

(1) Starting and Accelerating   *(2) Running*

Discharge resistances are connected permanently across condensers as shown

FIG. 6.22. CONNEXIONS OF SINGLE-SPEED SQUIRREL-CAGE CAPACITOR
LIFT MOTOR

A_1, A_2	Main winding	C_1	Starting condenser
Z_1, Z_2 }	Auxiliary winding	C_2, C_3, C_4	Running condensers
Z_3, Z_4		R_1, R_2	Discharge resistances

For one rotation connect Z_1 to A_1, and Z_4 to A_2; and for other rotation connect Z_4
to A_1, and Z_1 to A_2 and connect A_1 to line 1 and A_2 to line 2.
(B.T.-H.)

current limitations. The condenser and winding connexions
are shown in Fig. 6.23.

Another method of obtaining a lift drive from a single-phase

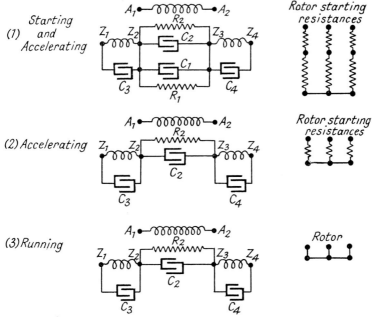

Discharge resistances should be connected
permanently across condensers as shown

FIG. 6.23. CONNEXIONS OF SINGLE-SPEED SLIP-RING CAPACITOR
LIFT MOTOR

A_1, A_2	Main winding	C_1	Starting condenser
Z_1, Z_2	} Auxiliary winding	C_2, C_3, C_4	Running condensers
Z_3, Z_4		R_1, R_2	Discharge resistances

For one rotation connect Z_1 to A_1 and Z_4 to A_2; and for other rotation connect
Z_4 to A_1 and Z_1 to A_2 and connect A_1 to line 1 and A_2 to line 2

(*B.T.-H.*)

supply is to employ a rectifier and use a d.c. motor. Except
when grid-controlled rectifiers are used, however, this method
has the disadvantage that dynamic braking is not possible
unless some other form of rotating machinery is also being
supplied by the rectifier, and will thus allow the lift motor to
function as a generator.

The single-phase to three-phase static converter, manu-factured by Westinghouse Brake & Saxby Signal Co. has been successfully used for lifts by providing a three-phase supply for driving a three-phase lift motor from a single-phase mains supply. The principle of operation is shown in Fig. 6.24 in which the inductance L and capacitance C provide the three voltages V_S, V_L and V_C each equal to the single-phase supply V_S with a fixed load resistance R. The voltage balance is

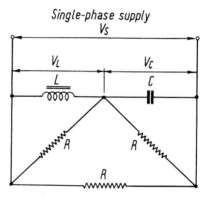

FIG. 6.24. STATIC CONVERTER

upset if the load resistance is varied. The problem of varying loads as with lifts, however, can be solved by connecting a further inductance in parallel with the capacitor and then proportioning the operating flux density in these inductances to bring about a change in their values as the load varies. The capacitor and parallel inductance then behave like a capacitance which varies with change of load. An 8 kVA short-time rated converter has been specially designed for lift work. Because of the high starting torque required it is necessary, for reliable starting, to use phase converter equipment having a full-load rating greater than that of the motor. As the short-time rated phase converter tends to overheat on light load it is essential that a mains isolating switch should be fitted to arrange for the switching to take place on the single-phase side and thus ensure that the phase converters are not left on open circuit. All controls, relays and reversing contactors are arranged across the R and B phases leaving the Y phase for operating the motor. The converter is adjusted so that at full load the phases are balanced. At varying lift loads, phase unbalance does, however, occur, but in spite of this, lifts run satisfactorily from these converters.

Motors for Car Speeds between 100 ft. per min. and 200 ft. per min. A rectifier and a d.c. motor may be used or a motor generator set and variable voltage control.

Motors for Car Speeds above 200 ft. per min. The gearless d.c. machine with a motor generator set, the motor being wound for single-phase working, is the best practice.

CHAPTER VII

VARIABLE-VOLTAGE EQUIPMENT

THE variable-voltage or Ward-Leonard method is the best for varying the speed of gearless motors, but it is also successfully employed with high grade geared d.c. machines. The results obtained are far better than those given by rheostatic methods, in which series and diverting resistances are used.

A motor generator set is employed, the generator of which supplies voltage to the lift d.c. motor; the motor of the set receiving its supply from the mains. The set motor is a standard, constant speed machine, the type used depending upon the characteristics of the power supply available, whilst the generator is usually a shunt-wound d.c. machine, having its armature permanently connected (electrically) to the armature of the lift motor. When the mains supply is a.c., a small exciter is installed in addition to the motor generator set to provide the direct current for the lift motor and generator fields and magnets of the brake and control gear. In recent years it has been the practice to dispense with an exciter and obtain the necessary d.c. supply from a static rectifier using germanium or silicon diodes. Automatic levelling may be performed without the use of an auxiliary machine, and in these cases the generator is provided with an additional shunt field which supplies the low voltage necessary to drive the lift motor at the slow levelling speed.

Since the lift motor receives its supply from the generator it is seen that the speed and direction of the car travel vary with the magnitude and polarity of the generator voltage, these in turn depending upon the strength and direction of the generator field. Acceleration of the car is obtained by strengthening the generator field, the lift being brought up to its final full speed by weakening the lift motor field. Retardation is effected by reversing these operations.

The advantages of this type of control are—

(i) The acceleration and retardation are smoother than that which is obtainable with any other form of control, as the generator voltage is free from any sudden changes.

(ii) It may be used on any supply by employing a suitable driving motor.

(iii) The controller is comparatively simple and its contactors need only be constructed to handle small currents of the order of a few amperes, as opposed to the large contactors handling the full power as in the case of rheostatic control.

(iv) The maintenance and running costs are relatively low since the controller, usually the chief source of trouble, is simple, and there are no rheostatic losses when accelerating, retarding, or running at reduced speeds.

(v) Its ability to re-generate to the supply mains.

When car switch control is employed the motor generator set is started by pressing a button fitted in the car, and the attendant can therefore shut down the set during slack periods. With automatic control, however, an automatic shut-down device is often incorporated, and is adjusted so that if no calls are received for, say, ten minutes, the motor generator set is automatically stopped, and motor generator stand-by losses are therefore eliminated. When a call is received the set is automatically started, but the car will not start until the motor generator full speed has been attained.

Speed Regulation. The speed/load characteristic of the set as a whole is a falling one and some method is invariably adopted to raise the characteristic, i.e. to make the speed independent of car load variations. To provide accurate floor levelling, however, it is desirable in practice to arrange that the motor speed when lifting full load is slightly higher than the no-load speed, and this involves a rising characteristic. This artificial variation is introduced so that the amount of brake slip when lifting full load is practically the same as when lowering full load.

Several methods have been used for improving the regulation, one of which is the over-compounding of the generator field to give an increase in generator voltage when operating against full load. When the load is overhauling, however, and the lift motor functions as a generator, the current in the series winding is reversed and opposes the current in the shunt field. The series winding, being designed primarily for use when the generator is supplying current to the winding motor under full load conditions, is therefore generally of such a strength that it is possible for the series winding to overpower the shunt field. This is guarded against by incorporating some device which, when the lift motor operates as a generator, automatically and gradually shunts the current flowing through the series winding and thus prevents overpowering of the shunt field.

Another method, and one which has proved satisfactory, makes use of a separate small booster, whose armature is connected in series with the generator shunt field, and its field

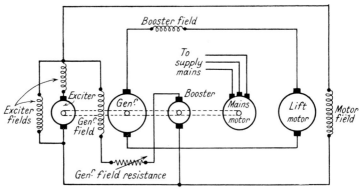

FIG. 7.1. VARIABLE-VOLTAGE SET WITH BOOSTER AND EXCITER

in series with the generator armature, as in the diagram in Fig. 7.1. Hence, as the generator load increases, the speed of the lift motor tends to decrease, but the booster generates a correspondingly increased e.m.f., thus boosting the generator shunt field and holding up the lift motor speed. With an over-hauling load, the generator armature current and booster field current are reversed, and thus the voltage generated by the booster is also reversed. The resultant generator field and output voltage are therefore reduced in values, and this counteracts the tendency of the overhauling load to produce an increase of speed.

A modification of the above method consists of a motor generator set, the generator of which has a common armature winding specially elongated to embrace the two separate (generator and booster) field systems, and one commutator together with its brush gear is therefore eliminated. The principle of this arrangement is shown in Fig. 7.2. The generator field system has shunt and interpole windings but no series windings. Demagnetizing windings are superimposed on the shunt field poles, the former being in use only during stopping and whilst the lift is stationary. The booster field system has three sets of series windings and by this means four combinations of windings can be obtained to give various values of boost. The booster field system also has a demagnetizing winding in series with that on the generator field, so that the two always function simultaneously when brought into action.

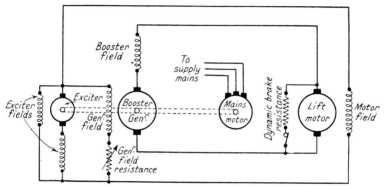

FIG. 7.2. VARIABLE-VOLTAGE SET WITH GENERATOR AND SELF-CONTAINED BOOSTER

During operation the current generated in the armature passes through the booster field on its way to the lift motor winding. If the lift machine is "motoring," then the loop current passing through the booster field winding generates a voltage in the armature which is added to the generator voltage. This ensures a rising voltage on the loop circuit with increase in "motoring" load and is so arranged to give a flat or slightly rising speed characteristic. If the lifting machine is "generating," then the generated current passing through the booster field winding generates a voltage opposed to the generator voltage already present. This ensures a decreasing voltage on the loop circuit with increase in generating load. It will be noted that there is no interaction between the two field systems but simply an

arithmetical addition or subtraction of volts to or from the
generator volts as the load changes in magnitude and characteristic.

A four-unit variable voltage set manufactured by the British
Thomson-Houston Company is shown in Fig. 7.3. This set
comprises a d.c. variable-voltage generator flexibly coupled

FIG. 7.3. VARIABLE-VOLTAGE MOTOR-GENERATOR SET WITH EXCITER
(*B.T.-H.*)

to a skirt-mounted a.c. induction motor and separately
excited from a vee-belt driven exciter mounted on top of the
generator. A small series booster or control exciter is mounted
on top of the set and direct-coupled to the main exciter. For
variable voltage geared drives this small series exciter is not
provided. The set is arranged for three-point support (without
a base plate) on pads of resilient material which also help to
prevent any noise or vibration being transmitted to the
building. The generator frame, commutator endshield and
skirt are of welded steel, and each machine has spring-loaded
ball-bearings. One advantage of the belt-driven exciter is
that the same set can be used on a 50- or 60-cycle supply merely

by changing one pulley. On the smaller powered units requiring
less power for excitation it is sometimes more economical to
use a rectifier instead of an exciter, and the variable voltage
set in these cases is a simplified two-unit arrangement.

The circuits for the speed control of this set are shown in

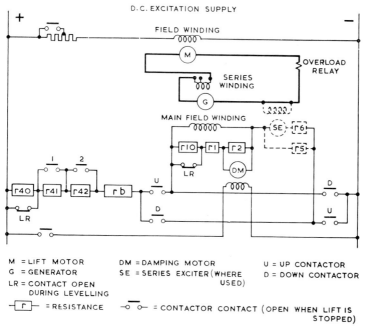

Fig. 7.4. Circuits of Variable-voltage Motor-generator Set
(B.T.-H.)

Fig. 7.4, which includes in addition to the series exciter a small
damping motor for speed control, connected across the generator main field winding. The levelling speed is adjusted to
approximately 20–25 ft. per minute by means of the resistances
$r1$, $r2$, and $r40$. Resistance $r1$ is adjusted to its minimum value
and $r2$ to about three-quarters of the resistance of the generator
field. This levelling speed may be raised by decreasing $r40$,
and vice versa. The intermediate speeds are raised by decreasing $r41$ or $r42$, and vice versa, whilst the full speed is raised by

decreasing rb, and vice versa. Adjustment of the "compounding" (the relation between lift speeds, particularly levelling speeds, during hauling or overhauling conditions) is made by altering the series field strength of the main generator, this being done by adjustment of the tappings of the field coils. Alternatively this may be effected by providing an adjustable diverter resistance across the field coils. Increasing the number of series field turns in circuit or increasing the diverter resistance will increase the "compounding," i.e. with an empty car the lift speed DOWN will exceed the corresponding speed UP. A further adjustment of "compounding" at a full speed is provided by the series exciter whose effect is adjusted by resistances $r5$ and $r6$. These should be adjusted so that the speed DOWN is about 5 per cent less than the UP speed. Increasing $r5$ will increase the "compounding," and vice versa. The first step of speed regulating resistance to be cut, $r41$ determines the value of the first peak current. Decreasing $r41$ reduces this peak current. Decreasing $r2$ also gives a finer adjustment in reducing the starting peak current and so softens the acceleration and makes the starting "surge" less noticeable. To increase the quickness of response when starting, $r1$ is increased in small steps. The quickness of response when levelling is increased by increasing $r10$. Decreasing $r2$, $r1$, or $r10$ also has the effect of reducing the steady speeds.

The adjustment of the "compounding" and levelling speeds may be clearer from a consideration of Fig. 7.5, which shows typical speed curves at low and high speeds for geared and gearless machines. The speeds (ordinates) are plotted against armature current expressed as a percentage reading of a moving-coil ammeter. There are four ordinates drawn, viz.—

At D, -50 per cent F.L. current, or empty car UP, also contract load DOWN; gearless machine. (The actual figure may vary from 50 per cent.)

At d, -5 per cent F.L. current, or empty car UP, also contract load DOWN; geared machine. (The actual figure may differ, or sign may be $+$.)

At O, Zero current in the lift motor.

At U, $+100$ per cent F.L. current, or contract load UP, also empty car DOWN, for both types.

Curve $D_2d_1AU_2$ represents speed curve for gearless machine, apparently level-compounded.

Curve d_1AU_2 represents speed curve for geared machine, apparently under-compounded.

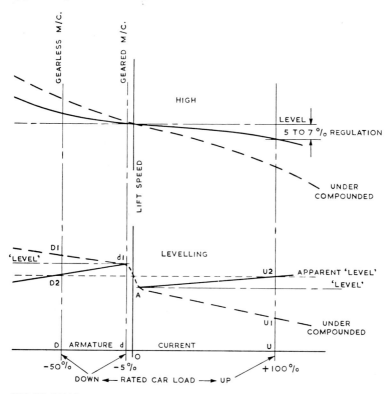

Fig 7.5. Speed Curves for Variable-voltage Sets
(*B.T.-H.*)

Curve $D_1d_1AU_1$ represents speed curve for gearless machine under-compounded.

Curve d_1AU_1 represents speed curve for geared machine under-compounded.

NOTE. Level-compounding is obtained when lines D_1d_1, and

AU_1 are parallel to the X axis. U_1 will then be slightly less than D_1 or d_1. It will be seen that curve $D_2d_1AU_2$ shows actual over-compounding. Curve d_1AU_2 also shows actual over-compounding. The cranked curve prevents these facts from being appreciated when only two readings, Up and Down, are taken.

It will be noted that, with the geared machine, point d_1 will not change appreciably as the compounding is changed, because the current at this point is so small.

At high speed, the compounding is set so that the speed drop

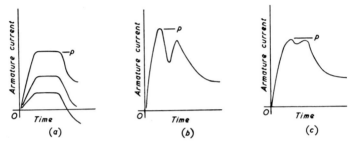

FIG. 7.6.   ACCELERATION CURVES FOR VARIABLE-VOLTAGE SETS
(*B.T.-H.*)

between no-load and rated load Up is from 5 to 7 per cent. As will be seen from the curves, this is more important for the gearless machine, as serious under-compounding may cause the overspeed device to trip.

Fig. 7.6 shows the results of adjustments of acceleration and "peak current" limitation, the maximum peak current being measured by an ammeter in the armature loop circuit when accelerating the car Up with contract load. At steady speed the current will fall to a value representing the out-of-balance load plus friction.

Curves (*a*), (*b*), and (*c*) show typical armature current curves, during acceleration plotted against time. The time taken to accelerate will be from 1·5 to 3 seconds according to the particular lift being considered. These three curves assist in recognition of the type of accelerating curve by observation of the ammeter. Curves (*a*) are three curves for contract load Up (biggest current), balanced load, and contract load Down.

The flat top shown in practice by the ammeter pointer remaining fairly steady during acceleration period, then falling off, shows that the best possible use of the machines is being made and is ideal. Curve (b) shows contract load UP only. The ammeter pointer flicks quickly up and down. This type of curve is associated with serious sparking and a surge sensation at starting. It feels like quick acceleration, but the equipment is badly adjusted and the best use is not being made of the machines. Curve (c) shows a well-adjusted compromise. The ammeter pointer moves more slowly and is steadier. The second peak is nearly equal to the first, and the "dip" between them is small. The machines are being used quite effectively, and the accelerating distance will be little different from (b). The sensation when riding in the car is of smooth acceleration. Curve (c) shows the kind of curve that is to be desired. The improvement shown by (c) over (b) is, generally, due to lower values of resistances $r1$ and $r2$. In each case the maximum peak current is indicated by p. p should not be allowed to exceed the maximum current permitted for the machines in question.

In order to make use of the damping motor, the resistances $r41$, $r42$, etc., must not be cut out all at once; at least two steps, separated by a time interval, are essential. Acceleration in one step renders the damping motor practically useless, and results in a single very large peak of current with corresponding severe effects.

A variable-voltage motor-generator set with the a.c. squirrel-cage induction motor below and the d.c. generator above, combined in a unit vertical assembly is shown in Fig. 7.7. This provides a compact arrangement giving the maximum economy in space requirements. The machine is provided with a three-point resilient mounting inherently isolated from the floor, and the machine cables, protected by flexible conduits are terminated in a junction box mounted on the cable trunking. Fig. 7.8 shows a part sectional view of a motor-generator set in which the armatures of the motor and the generator are both mounted on a one-piece steel shaft held rigidly in line by the outside rolled-steel frame of the set.

Electronic Variable-voltage Control. This system provides a variable-voltage supply to a d.c. motor by means of a grid-controlled rectifier instead of a Ward-Leonard motor-generator set. A mercury-arc rectifier is used, the phase relationship of the grids to the anodes and hence the rectifier output voltage

FIG. 7.7. VERTICAL VARIABLE-VOLTAGE MOTOR GENERATOR SET
(*Express Lift Co.*)

being controlled by a rotary phase-shifting regulator which is driven directly from the lift. The regulator is driven from a camshaft which can be coupled by magnetic clutches, ACCEL, or DECEL, to a chain drive from the lift gear either to run

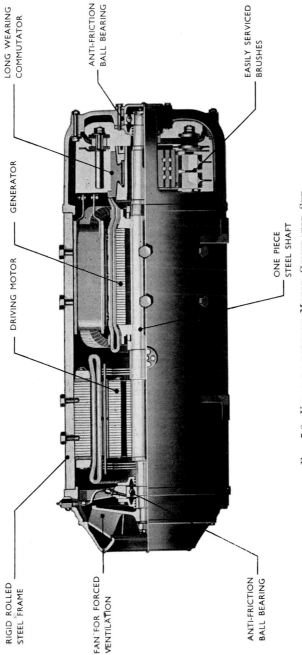

LONG WEARING COMMUTATOR

ANTI-FRICTION BALL BEARING

EASILY SERVICED BRUSHES

GENERATOR

DRIVING MOTOR

ONE PIECE STEEL SHAFT

RIGID ROLLED STEEL FRAME

FAN FOR FORCED VENTILATION

ANTI-FRICTION BALL BEARING

FIG. 7.8. VARIABLE-VOLTAGE MOTOR GENERATOR SET

(Otis Elevator Co.)

forward to accelerate or backwards to decelerate. Thus the movement of the regulator and so the rate of acceleration or deceleration are controlled directly and automatically by the speed and position of the car in the well.

The rectifier equipment consists of a main oil-immersed transformer feeding either a single bulb rectifier for the small equipments or two bulbs for large sizes of over 40 h.p. The

FIG. 7.9. ELECTRONIC CONTROL VARIABLE-VOLTAGE LIFT
(*Wm. Wadsworth & Sons*)

controlling grids are inserted between the cathode and the anode and a potential is applied to these grids through the phase-shifting regulator and the transformer so that the phase relationship between the control grids and the anodes can be varied through 120 electrical degrees. When a negative potential is applied to the grid, the anode, which the grid controls, cannot fire. As soon as the grid potential becomes positive to the cathode, electron flow takes place from cathode to grid and from cathode to anode, which is a positive flow of current from anode to cathode. Thus, by adjusting the moment at which the grid becomes positive to the cathode, it is possible to control the firing of any anode throughout its normal firing period, and this leads to complete control of the output d.c. voltage. A rotary inductor type regulator may be used or a static type electro-magnetic phase shifter. Fig. 7.9 is a typical gear of this type under test, showing the chain drive to the controller camshaft and the rectifier cubicle which contains the transformer at the rear.

Electronic Control of Ward-Leonard Motor-generator Set. In a similar manner the variable-voltage output of the grid-controlled mercury-arc rectifier can be employed for energizing the shunt field of the generator of a normal Ward-Leonard set. The control of the grid, and hence the rectifier output voltage, is by a tachometer generator driven directly by the lift gear. Fig. 7.10 shows the circuits required for this form of control.

A rheostat is connected across a suitable constant voltage source and has contacts connected to as many tapping points as may be required to give smooth control, the contacts being actuated by a camshaft positively driven in unison with the lift car. The voltage developed across the rheostat, which depends on the speed and/or position of the lift car in the well, is applied, when a controller contact is closed, directly to the control grid of a thermionic valve $V1$. Other contacts Up Dn are connected across this contact, as also is a variable resistance connected in one of the supply lines to the armature of the lift motor. The output of a tachometer-generator driven by the lift motor is fed into a conductor from the rheostat to the cathode of $V1$. The anode of $V1$ is supplied with high-tension voltage from a convenient source, and the anode circuit includes a winding of a saturable reactor, the other winding of which is connected to the grid and cathode of a gas-filled rectifier $V2$. A transformer $T1$, the primary of which is supplied with alternating current, has a resistor and the reactor winding connected in series across its secondary. A further transformer $T2$, connected to the same a.c. source, supplies the anode of $V2$, the rectified anode current passing also through the field winding of a generator supplying the lift motor, and driven by a three-phase a.c. motor. The three-phase supply is connected also to a rectifier or rotary converter supplying the lift motor field.

When the lift car is at rest, the rheostat contacts and controller contact are open, and contacts Up Dn are closed. When

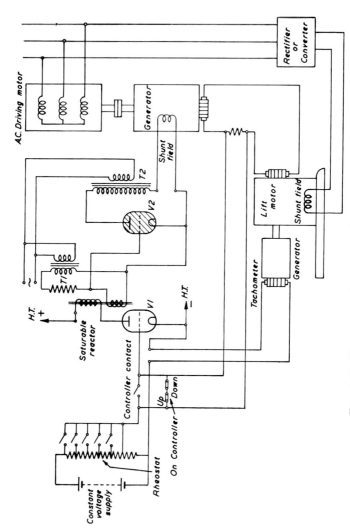

FIG. 7.10. ELECTRONIC CONTROL OF WARD-LEONARD SET

(*Wm. Wadsworth & Sons*)

the Up contactor is energized, contact Up opens, and the lift starts to move, closing the controller contact and the rheostat contacts in succession by means of a camshaft.   A voltage equal to the difference between the voltage across the rheostat and that of the tachometer-generator is then applied to the grid of V1, the rheostat voltage increasing gradually as the contacts are closed. The anode current of V1, flowing through the reactor winding, controls the permeability of the core of the reactor, and hence the inductance of the second reactor winding. This winding and resistor form a circuit in which, by variations in the inductance of this winding, the phase of the alternating voltage (derived from transformer T1) applied to the grid of V2 is varied in relation to the a.c. input and hence in relation to the voltage at the anode of V2. In this way the portion of the a.c. cycle over which V2 is conducting is varied in accordance with the grid voltage of V1, and consequently the mean excitation of the field of the generator, and hence the lift motor input depends on this grid voltage.

As the grid voltage is the difference between the rheostat voltage and the tachometer voltage, any depression of the motor speed due to a load thrown on to the motor, results in an increase of the grid voltage and a corresponding increase in the motor input. On light load, an increase of the motor speed results in a decrease in the motor input. There is thus a stabilizing action, the motor tending to maintain a speed which makes the tachometer voltage nearly equal to the rheostat voltage. As the successive rheostat contacts are closed, the voltage increases and the motor speed also increases, until all the contacts are closed, when the speed is kept constant at its maximum value.

When the lift car approaches a floor, a switch in the shaft, or other equivalent means, causes the camshaft to open the rheostat contacts successively, thereby slowing down the lift, and finally to open the controller contact. The voltage developed across the resistor, depending on the lift motor torque, is then added to the rheostat voltage, so that the speed of the motor depends partly on the torque. The speed now reached depends on the load in the lift car in such a manner as to counteract the effect of the load on the slide of the lift when the brake is applied and the motor de-energized, and consequently to reduce the variation in levelling caused by varying loads in the car.

When the Up contactor is de-energized, the contact Up closes, and when the down contactor is energized the contact Dn opens. The contacts Up and Dn could be replaced by a single contact, opened whenever the lift moves, or further contacts may be added across or in series with the controller contact as required.

With this arrangement the speed of the lift motor is closely proportioned to the voltage developed across the camshaft-operated rheostat, and the input to the motor at starting and stopping is related to the motor torque.

In a development of the above system, the cam-gear which controls the operation of the rheostat tapping contacts is replaced by electrical acceleration and deceleration signals. Speed control is obtained by comparing a reference voltage representing the desired speed at any particular instant during travel, with a voltage proportional to the lift motor speed. This latter voltage is obtained from a tacho-generator mounted on the end of the lift motor shaft. The difference between these two voltages is applied to the grid of a pentode valve which has a known mutual conductance. The grid voltage variations cause corresponding changes in the anode current which are fed to a power amplifier, which in turn supplies the generator field. Hence any deviation in motor speed from the desired speed causes a correcting signal to be applied to the amplifier which results in a change in generator voltage to restore the motor to the desired speed. The reference voltage corresponds to the ideal desired speed at any point during the acceleration, maximum-speed and deceleration periods as shown in Fig. 7.11. This voltage is obtained from a resistance-capacitor circuit or by charging a capacitor through a constant-current pentode valve making use of the characteristic that the pentode anode current is independent of the anode applied potential but is dependent on the screen potential. In this manner the potential slope across the capacitor is proportional to the value of the charging current and produces an ideal velocity/ time curve as a reference for acceleration and deceleration.

This closed-loop type of Ward-Leonard speed control in

which the actual lift motor speed is compared with the ideal speed required, and a correction applied to the generator field, has been developed in many forms. The main differences are in the electronic networks and amplifiers employed to obtain the ideal reference voltage and to integrate it with the motor speed.

A recent high-gain feedback control system, designed by Westinghouse in America* and now developed here by the Express Lift Co., gives high-quality performance by using

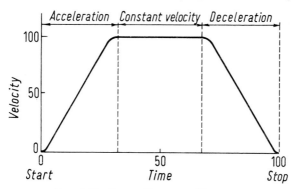

FIG. 7.11. IDEAL VELOCITY PATTERN

transducers to integrate the ideal pattern shown in Fig. 7.11 and feedback signals and to vary special resistors in the generator field.

The acceleration pattern is obtained by regulating the charging rate of a fixed capacitor by feedback circuits which shift the phase of gate signals controlling the firing to two silicon-controlled rectifiers in series with the pattern supply voltage. The required rate of current build-up or car acceleration can be set by a potentiometer. An additional feature which can decrease the acceleration rate as the car approaches its maximum velocity is used to shape the pattern. The acceleration reduction is accomplished by a Zener diode which becomes conducting at a point in the acceleration cycle so as to add a capacitor to the charging circuit.

To follow the pattern velocity/time input curve in Fig. 7.11

* *Electrical Engineering,* March 1962.

a high-gain system is required with adequate damping, and a wide-range velocity transducer of good sensitivity and accuracy has been employed for this purpose.

The transducer is mechanically coupled to the lift motor by two timing belts. Small idlers on the belts control switches which prevent use of the lift if the belts are not in place. The added functions of error measuring, summing input signals

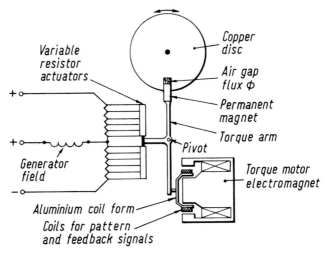

FIG. 7.12. VELOCITY TRANSDUCER

and power amplification are included with the basic transducer design to simplify the integration of the device into the over-all system. A gravity-balanced torque arm in the transducer provides a simple means for combining these functions. Un-balanced torques supply the forces required to actuate variable resistors of sufficient capacity to control directly the maximum field current of most gearless machines, thereby eliminating the need for additional power amplification.

The torque arm consists of a vertical beam member supported on a horizontal pivot near the centre of the velocity transducer. The upper end of the torque arm carries a C-shaped permanent magnet. The moving coils of a torque motor are attached to the lower-end as shown in Fig. 7.12. The entire assembly is gravity-balanced around the horizontal axis and stops are provided to limit the angular motion to approximately 3 degrees. The electromagnet supplies a constant field for the moving-coil windings. Angular motion of the arm actuates two variable resistors which change the excitation of the generator field. One end of the generator field is connected to the centre of the power supply to allow full excitation of either polarity depending upon the angular position of the torque arm. Torques proportional to the lift speed are produced on the arm of the transducer by driving the circumferential edge of a copper disc through the air gap of the permanent magnet. The disc moves in synchronism with the lift drive motor through the magnetic field and generates in the copper, eddy currents which react with the field to give "drag" forces proportional to the angular velocity of the disc. These forces are similar to the familiar damping forces of a watt-hour meter. The eddy-current circuit is completed within the copper disc and slip-rings are not required.

The main function of the velocity transducer is to compare eddy-current-drag torques with the input torques coming from the pattern signals applied to the moving coil of the torque motor. If these two torques are not equal and opposing, the arm deflects to change the excitation of the drive system and bring the speed of the car to a value where the drag torque balances the pattern torque. When the input torque is constant, as during running at maximum speed, the system regulates automatically for constant car velocity. A variation in car velocity results in unbalanced torques on the transducer. The arm deflects and changes the excitation to return the car to the original velocity. The torques are again balanced in the transducer with the arm in a new angular position corresponding to the change in excitation.

During deceleration, a car position pattern signal is supplied by two transducers which require a.c. excitation and generate output signals which are rectified to supply pattern signals to a coil on the torque motor of the velocity transducer. During slow-down to a point 10 in. from the floor a continuous reference pattern is supplied by a transducer mounted on the mechanical car-position selector. This selector is in effect, a miniature lift driven in synchronism with the lift car and includes relays required to open and close the lift doors and to perform other auxiliary functions. The selector has two carriage members, one of which is always in synchronism with the car. The second carriage is initially driven in an advanced position with respect to the first carriage and is stopped when the car begins to decelerate. As the car travels to the floor, the distance between these two carriages is proportional to the

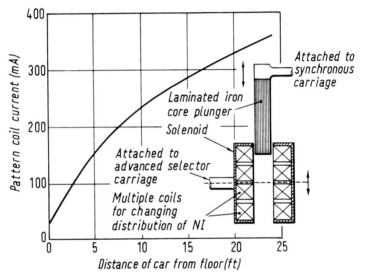

FIG. 7.13. SELECTOR TRANSDUCER AND PATTERN SHAPE

distance of the car from the floor. The relative motion between the carriages is used to change the position of an iron plunger in the solenoid of the selector transducer shown in Fig. 7.13.

During slowdown, the iron plunger moves into the solenoid, changing the coil impedance from a minimum to a maximum. The impedance with respect to the core position is shaped by changing the distribution of turns throughout the length of the solenoid to give the current-position curve shown in Fig. 7.13. This gives the car the desired deceleration characteristic and is such that a $\frac{1}{8}$-in. movement of the plunger corresponds to 1 foot travel of the lift. For a contract speed of 800 ft. per min. the distance travelled during deceleration is about 25 ft.

A second transducer having one member on the car and other members in the well at each floor position, is used to obtain an accurate voltage-position pattern during the last 10 inches of travel. The car member of this transducer, which is shown in Fig. 7.14, consists of two transformers spaced 22 in.

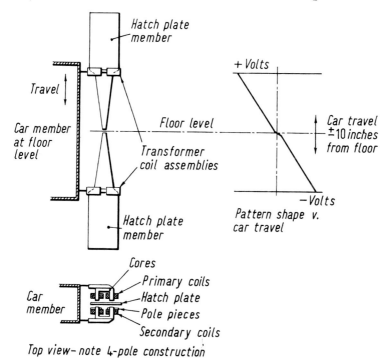

FIG. 7.14. HATCH TRANSDUCER AND PATTERN SHAPE

apart in the direction of car travel. Primary and secondary windings are on core members spaced with a 1-inch air gap. Iron pole pieces on the secondary cores are used to increase the coupling between the primary and secondary windings and the output of the transducer. The hatch (well) member consists of two thin steel plates mounted on laminated plastic. These plates, passing through the air gap of the core members on the car, reduce the coupling flux to a minimum when the plate width overlaps the poles. The relatively large air gap provides sufficient mechanical clearance for car movement normal to the direction of travel. Changes in coupling from sideways motions are minimized by the 4-pole construction. Fig. 7.14 shows the pole arrangement and the general shape of the hatch plates to give the desired pattern signal.

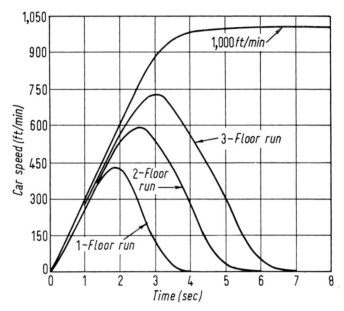

FIG. 7.15. RECORDED VELOCITY/TIME RESPONSE OF 1 000 FEET PER MINUTE LIFT

The secondary voltages from each core assembly are individually rectified and the two d.c. outputs connected to a mixing resistor. When the car is at the floor level the voltages from the top and bottom core assemblies are equal and opposing to give zero pattern voltage across the mixing resistor. When the car leaves the floor the contour of the hatch plates produces unequal voltages in the mixing resistor to give the pattern-displacement voltage shown in Fig. 7.14. Polarity of the pattern voltage is determined by whether or not the car is above or below the floor. When this pattern signal is applied to the torque motor of the velocity transducer, the lift car is position-regulated with reference to the floor level.

The general slope of the hatch transducer pattern is a continuation of the pattern curve of Fig. 7.13 from the 10-inch point to the floor. One inch from the floor the slope of the pattern signal is changed approximately 2 to 1 by variations in the contour of the hatch plates as shown in Fig. 7.14. The greater slope increases the positional stiffness at the floor and as the change occurs at almost zero velocity it is not noticed by passengers. The system stiffness at the floor is made large so that variations in passenger loading have little effect on car position at the floor level.

With this system the car ride is so smooth that it is difficult to tell when the car accelerates and slows down. Actual velocity–time curves of a 1 000 ft. per min. installation for 1-, 2- and 3-floor runs are shown in Fig. 7.15. The 1-floor run takes 4 seconds from start to stop and reaches a maximum velocity of 440 ft. per min. Higher speeds are reached with longer runs until the car reaches its contract speed of 1 000 ft. per min. The top curve in Fig. 7.15 shows the start of a 12-floor run.

CHAPTER VIII

BRAKES

It is important that a lift brake shall be of good design and that the number of wearing parts shall be kept to a minimum in order that it may maintain its adjustment for long periods without attention. Accurate and reasonably constant brake adjustments are essential, as a badly adjusted brake can be responsible for inaccurate floor levelling and consequent "inching," accompanied by greater wear on the controller contacts and an increase in the energy consumption.

Braking Torque. A lift brake must be sufficiently powerful to stop the fully loaded car when travelling at levelling speed, within the distance between the stopping switch and the landing and therefore must be capable of absorbing the kinetic energy of the lift moving parts. In an accurately designed installation, this indicates that the braking torque should be approximately equal to the motor torque at levelling speed.

$$\text{Motor h.p.} = \frac{\text{torque} \times 2\,\pi\,N}{33\,000}$$

$$\therefore \text{torque} = \frac{5\,250 \times \text{h.p.}}{N} \text{ lb. ft.}$$

where N is the revs. per minute of the brake drum when the motor is rotating at levelling speed.

In practice, however, the above braking torque is larger than is necessary, because part of the lift kinetic energy is absorbed by friction in the gear box, bearings and guide shoes, and as the motor is usually the nearest commercial size, its h.p. is more than adequate for the purpose. Further, when the brake operates, it converts the kinetic energy of the lift into heat which is dissipated in the brake and drum. It is therefore necessary to consider also the frequency of operation of the brake, i.e. the number of stops per hour of the lift, to ensure that the brake drum can dissipate the heat without causing the brake lining to be heated beyond its working temperature. It is thus advisable to ascertain the braking conditions in more detail than is given by the above formula and this may be done as follows—

If d ft. is the distance of the car from the landing when the power is cut off, then the time t for the car to reach the landing is $d \Big/ \dfrac{v}{2 \times 60} = \dfrac{120d}{v}$ sec. (which is also the time during which the brake is in operation), where v is the levelling speed in ft. per minute. This is not strictly correct because some small part of t is expended in contactor operation and in applying the brake shoes to the drum.

The brake retarding torque $T = I\alpha$, where I is the moment of inertia of all moving parts referred to the brake drum and α the angular retardation of the drum.

But $I = \dfrac{Wk^2}{g}$ and $\alpha = \dfrac{\omega}{t} = \dfrac{2\pi N}{60t}$ radians per sec. per sec. where W is the weight of all moving parts treated as if they were all at a distance k (the radius of gyration) from the drum shaft, and ω is the drum angular velocity in radians per second.

Hence $T = \dfrac{Wk^2}{g} \times \dfrac{2\pi N}{60t} = \dfrac{0 \cdot 00325\ Wk^2N}{t}$ lb. ft.

The moment of inertia of the equipment referred to the brake drum is calculated as follows—

Let $I_1 =$ the moment of inertia of the motor rotor, brake drum and worm in lb. ft.^{2}

$I_2 =$ the moment of inertia of the worm wheel and sheave in lb. ft.,^{2} and this will be small compared with I_1.

$N_1 =$ the revs. per minute of the worm wheel and sheave.

$L =$ the contract load in lb., and assume that the counterweight is balanced for half load.

$r =$ the radius of the sheave in ft.

The moment of inertia of the motor, brake drum and worm

$$= I_1 \text{ lb. ft.}^2$$

The moment of inertia of the worm wheel and sheave

$$= I_2 \left(\frac{N_1}{N}\right)^2 \text{ lb. ft.}^2$$

The moment of inertia of the car and load

$$= \frac{L}{2} r^2 \left(\frac{v}{2\pi r N}\right)^2 = \frac{Lv^2}{8\pi^2 N^2} \text{ lb. ft.}^2$$

The sum of these three is the total moment of inertia at the brake drum (I). The values of I_1 and I_2 can be calculated from the dimensions and weights of the parts.

If the lift makes S journeys per hour and t_1 seconds is the average time of each trip, then $\frac{S(t_1 - t)}{360}$ is the fraction of each hour that the brake coil is in circuit. For such duties as lift service the brakes usually have intermittent ratings of 33 per cent or 50 per cent which indicate that the coils should be in circuit for not more than 5 minutes in 15 minutes, or 5 minutes in 10 minutes respectively.

The energy which has to be dissipated by the brake drum depends largely on the lift duty, the amount dissipated per hour being $S \times$ the total energy of the moving parts at levelling speed.

As the speed of a lift brake drum does not usually exceed 1 000 r.p.m. the maximum safe speed of the drum will not be exceeded even on the larger size brakes.

After calculating the various factors governing the brake size as outlined above, a suitable brake can be selected from a manufacturer's catalogue which usually specifies such particulars for the different sizes of brakes listed.

Types of Brake. The most common form of lift brake is constructed on the electromagnetic principle, and in the majority of cases is situated between the motor and the gearbox, i.e. on the high speed side, where it has the greatest mechanical advantage. In a gearless lift the brake is fitted between the motor and the driving sheave. So that the brake may be capable of sustaining the car when the motor armature is withdrawn for repairs, the brake drum is fitted on the worm shaft side of the coupling. The shoes, usually in two parts, are lined with Ferodo or similar material and sometimes function as independent units in order that, in the event of a failure of one half, the other will be available for braking. Operation of the shoes is effected by spiral springs, and release by means of an

electromagnet, the solenoid of which acts upon the brake shoes
either directly or through a system of links. The disadvantage of
links and pins is that small wear on the pins may result in a
comparatively large amount of lost brake motion. In some
forms, the electromagnet is placed below the shoes and the
springs above, as in Fig. 8.1, whilst in others this arrangement is
reversed, an example of which is seen in Fig. 8.2. In yet another
type both solenoid and springs are placed above the shoes, as
in the brake shown in Fig. 8.3. A brief description of the brake
seen in Fig. 8.2 will serve to illustrate the principle of operation
and the method of adjustment of a well-designed brake. In the
type shown, wear is possible only at the brake linings and the
pins A and B, and as the radial movement of the pins is less than
$1°$ and the bearing surface $5\frac{1}{2}$ times the pin diameter, the wear
on these pins is very small. Wear on the linings results in an
increase of the air gap at C which is corrected by withdrawing
the set screw D as required. The shoes are applied by a compression spring situated in the brake base-block and released by
a solenoid acting directly upon the shoes. The strength of
the spring and hence the force exerted by the shoes may be
adjusted by the screw E. The adjustment for maximum torque
is made at the works, after which a steel collar F is fitted to
prevent further compression. If the maximum torque is not
required, the adjusting stud can be screwed back. In the 12 in.
size, which develops a torque of 300 lb. ft., the spring is 9 in.
long, $1\frac{1}{2}$ in. diameter and the spring movement during operation
is only $\frac{1}{64}$ in. The solenoid windings are totally enclosed and protected from dirt, damp, and mechanical damage.

The method of adjusting the brake shown in Fig. 8.2 is
typical of the adjustment of any solenoid-operated lift brake.
The brake is first set to the "Off" position by unscrewing the
adjusting screw E after making sure that the electricity supply
is cut off. Adjust the gap between the lining of each brake shoe
and the drum to a minimum by means of the heel screw and
its lock-nut at the bottom of each shoe. The clearance should
be about paper thickness, i.e. between $\frac{1}{64}$th and $\frac{1}{100}$th inch.
Now lock each heel screw and check to see that the brake drum
is still just clear of the lining. Next adjust the air gap C
between the armature and the solenoid by means of the adjusting screw D. The air gap should be reduced to a minimum by

FIG. 8.1. ELECTROMAGNETIC BRAKE
Electromagnet below and springs above
(*Wm. Wadsworth & Sons*)

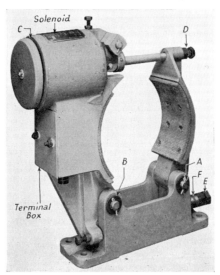

FIG. 8.2. D.C. ELECTROMAGNETIC BRAKE
Electromagnet above and spring below
(*Dewhurst & Partner, Ltd.*)

adjusting the screw D so that while applying hand pressure to the armature, so that the brake shoes are hard against their stops, the armature is just clear of the solenoid. The clearance should be between $\frac{1}{100}$th and $\frac{1}{64}$th inch. This ensures the minimum possible air gap between the armature and yoke and hence the maximum effective pull from the solenoid. Spring

Fig. 8.3. D.C. "Perigrip" Brake

A, rolled-steel shoe pivot; B, armature plate and air gap; C, solenoid pot; D, hand release lever; E, brake-operating spring; F, locking plunger; G, knurled nut; H, operating rod; J, free arm; K, spring-loaded centring screw; L, locknut; M, brake-arm pivot; N, self-aligning brake shoes.

(*AEI*)

pressure should now be applied by the adjusting screw E until it is impossible to turn the lift motor with the winding handle in either direction, with the brake on. After removing the winding handle and switching on, the lift should be run and tested for smooth stopping and accuracy of levelling. The final position of the screw E depends upon the particular stopping characteristics required. For a permanent adjustment, the brake lining must be well bedded in and the drum have a smooth, highly-polished surface. If the coefficient of friction of the linings is too high, the final stopping may be jerky; this is sometimes overcome by using a special zinc-bonded lining in which the zinc acts as a permanent dry lubricant and reduces brake drag to a minimum.

The accuracy of the final levelling depends very largely on the brake adjustment but is also influenced by the speed of the lift at the instant of brake application. The slower this final speed the more accurate the levelling. In practice it is possible to level approximately within $\pm \frac{1}{2}$ in. from a levelling speed of 50 ft. per min., within $\pm \frac{1}{4}$ in. from 20 ft. per min., and within $\pm \frac{3}{16}$th in. from 10 ft. per min.

A.C. and D.C. Brakes. The principles of design and operation of an a.c. brake are somewhat different from those of a brake intended for use on d.c. In the cases of two-speed a.c. motors, and of d.c. motors, the mechanical brake is assisted by the dynamic braking action, but single-speed a.c. motors do not lend themselves to dynamic braking, and consequently the whole of the braking force must be supplied by the mechanical brake which must, therefore, be greater in size. Further, since the rotor of an a.c. machine is larger than that of a comparable d.c. machine, the brake in the former case must be still larger. The operation of a d.c. brake can be made very smooth by connecting a resistance in parallel with the coil, the resistance providing a path for the induced currents and allowing the flux variations to take place gradually. This resistance also prevents the puncturing of the insulation of the brake coil by any rise in pressure due to a supply interruption. It is not practicable, however, to use a parallel resistance with an a.c. brake, and the tendency is for the shoes to be applied more fiercely. A dashpot is often fitted to minimize the brake shock on application of the shoes. The magnet is, in addition, often immersed in oil so as to dissipate the heat developed in the coil and magnet. An oil-immersed a.c. brake is illustrated in Fig. 8.4, and an electro-hydraulic thrustor-operated a.c. brake in Fig. 8.5. The brake in Fig. 8.5 is operated by a compression spring in the base of the brake and released by energizing the electro-hydraulic thrustor which imparts a thrust to a lever from which it is transmitted to a rod and so releases the shoes from contact with the brake drum. Solid magnetic cores are employed in a d.c. brake, and laminated cores

Fig. 8.4. Oil-immersed A.C. Brake
(*R. J. Shaw & Co.*)

Fig. 8.5. Thrustor-operated Electro-hydraulic A.C. Brake
(*Dewhurst & Partner, Ltd.*)

for a.c. brakes in order that the eddy-current losses in the latter type shall be kept to a minimum. Figs. 8.6 and 8.7 show a contactor type a.c. brake with laminated core. Silence is another point in favour of the d.c. brake, although hum from a.c. brakes can be considerably reduced by employing copper loops in the faces of the magnet cores. The current induced in

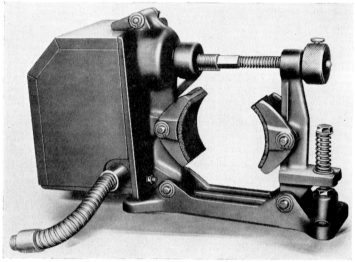

FIG. 8.6. SINGLE-PHASE A.C. "PERIGRIP" BRAKE
(*AEI*)

the loop causes the shaded pole flux to be reduced and to lag behind the main flux, and this results in a more even brake pull. Other differences between the two types of brake are, that whilst the d.c. brake takes the same current regardless of position of the shoes, the a.c. brake takes a much larger current when the air gap is large than when it is small; and, although the pull in the case of the a.c. magnet is nearly constant throughout the travel of the solenoid plunger, in a d.c. magnet the pull is approximately inversely proportional to the square of the air gap. On account of the difficulties associated with a.c. brakes most makers prefer to install d.c. brakes regardless of the nature of the supply, and when this is a.c. the brake supply is obtained from a rectifier.

For gearless d.c. machines the brake must be of larger size because of the absence of gearing. So far as rotor kinetic energy

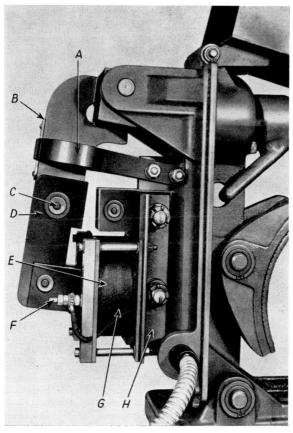

Fig. 8.7. Magnet Assembly of 8-inch Enclosed "Perigrip" a.c. Brake, with Enclosing Cover Swung Clear

A, retaining spring; *B*, copper cushion block; *C*, core rivet; *D*, clapper; *E*, magnet air-gap (inside coil); *F*, terminal for supply cable; *G*, operating coil; *H*, coil supporting angle brackets.

(*AEI*)

is concerned, however, although the rotor is much larger, it runs at a considerably reduced speed. An important factor in the design of gearless motor brakes is the time constant, which should be of such a value that a squeezing action results rather than a direct impact. Its torque should be sufficient to hold $1\frac{1}{4}$ times the contract load in the car. The usual principle

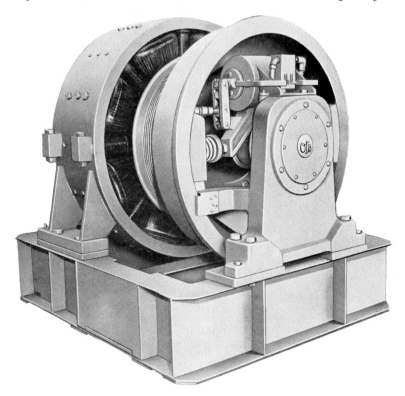

FIG. 8.8. LARGE GEARLESS MACHINE WITH INTERNAL BRAKE
(*Otis Elevator Co.*)

adopted in levelling a gearless lift is to utilize the dynamic braking effect on the armature to bring the lift to rest, after which the machine brake is applied to hold the lift in position. Some firms, however, provide partial braking operation during levelling or during relevelling if required. Gearless machine brakes have large shoe areas and small shoe clearances, and shoe drag can sometimes be caused by voltage fluctuations.

It is wise, therefore, to check brake clearances at intervals with the lift running, particularly during full speed up with full load in the car.

Fig. 8.8 shows a large gearless traction machine for double-wrap drive with a rather unusual brake layout. Instead of the usual method of applying the shoes to the outside of the drum, the complete brake is housed inside the drum, the shoes operating on the inside surface of the drum. The one-piece sheave and brake drum is bolted to the main shaft which is supported by two self-aligning tapered roller bearings. This internal brake provides accurate control for large high-speed lifts.

CHAPTER IX

GEARING

General. If we assume that a typical lift has a lifting rope diameter of ⅝ in., a sheave diameter of 50 rope diameters, i.e. 32 in., and a car speed of 200 ft. per min., then the speed of a suitable motor for a direct drive installation will be about 24 r.p.m. The slow-speed gearless motor, however, is only practicable for speeds above about 350 ft. per min., and for speeds slower than this a general purpose motor running at about 1 000 r.p.m. is used. Hence, with a motor of the latter type, it is necessary to use some form of gearing, the ratio of which, in the above case, should be 42 to 1. In practice, the gear ratios employed vary between about 20 to 1 and 60 to 1. Whilst it is not possible to use the gearless drive for slow car speeds it is also impracticable to employ gearing for speeds much in excess of 400 ft. per min. because of difficulties in manufacture and in maintaining the thrust adjustment sufficiently accurate to ensure smooth operation.

Types. The type of gearing invariably adopted is the worm and worm-wheel, both members being totally enclosed in an oil-tight gearcase. The worm may be either above or below the wheel, and there are probably as many of one type in use as the other. The former is termed the *over-type* gear, and the latter the *under-type*, each having its advantages. Advantages claimed for the over-type are that the low-speed load carrying shaft is lower down on the bedplate, the worm which suffers the most wear is capable of easy inspection, and it is not so difficult to maintain oil-tight joints in the gear casing. With the under-type gear, any wear in the main bearings allows the wheel to drop into closer mesh with the worm instead of drawing the gears apart, and, in addition, the lubrication is probably more complete, since the worm is always immersed in oil. Although, in the over-type gear, the worm is not immersed in oil, sufficient oil to provide ample lubrication is carried upwards by the worm-wheel teeth. The oil is thrown from the wheel and cooled during its passage down the gear case and back to the reservoir. The design of both types should be such that the temperature of the oil in the sump never exceeds 200°F. The worm-wheel is forced on to its shaft by hydraulic pressure and fixed by two keys at right angles to each other or bolted to a cast-iron spider as described below. The hobbing of the worm-wheel after assembly, ensures perfect concentricity of pitch line with journals. The

Fig. 9.1. Over-type Geared Machine with Flange-mounted Motor

(*Marryat and Scott*)

worm shaft and the motor armature shaft are each secured to the brake drum by a key.

An over-type geared machine with a flange-mounted motor for a single-speed lift of contract load 2 000 lb and contract speed 100 ft. per min. is shown in Fig. 9.1. This machine is fitted with rubber anti-vibration mountings. Fig. 9.2 shows an under-type geared machine in which the d.c. motor of a variable-voltage set is also secured to the gearbox by a flange mounting, thus ensuring perfect alignment. The worm-wheel and sheave rims are bolted to a cast-iron driving spider which is mounted on tapered roller bearings. The slow-speed shaft is non-rotating and is securely clamped in its housings, thus retaining alignment and eliminating the necessity for keys on the worm-wheel shaft. In the Otis worm gear the main drive shaft is pressed into the bore of the sheave and worm-wheel spider, and the shaft itself rotates on self-aligning tapered roller bearings.

FIG. 9.2. UNDER-TYPE GEARED MACHINE WITH FLANGE-MOUNTED MOTOR
(*Express Lift Co.*)

Materials. The materials used for worm gears must be capable of being machined and finished satisfactorily, have adequate strength and resistance to wear and a low coefficient of friction when used with a suitable lubricant. These requirements are best fulfilled by steel worms and phosphor bronze wheels. The types of steel and phosphor bronze recommended in B.S. 721 are described below and from these, lift manufacturers invariably select their gear materials.

1. WORM

(*a*) 0·40 *per cent Carbon Steel Normalized.* This has a bending-stress factor of 20 000 lb/in.², good machinability and is a popular material for lift worms.

(b) *0·55 per cent Carbon Steel Normalized.* The bending-stress factor is 25 000 lb/in.^{2}, and although it is stronger than (a) its ductility is lower and it is more difficult to machine.

(c) *Carbon Case-hardening Steel.* The bending-stress factor is 40 000 lb/in.^{2}

(d) *Nickel and Nickel-molybdenum Case-hardening Steels.* The bending-stress factor is 47 000 lb/in.^{2}

(e) *Nickel-chromium and Nickel-chromium-molybdenum Case-hardening Steels.* The bending-stress factor is 50 000 lb/in.^{2}

It will be noted that the case-hardening steels are much stronger than the normalized carbon steels but they are also more costly. In addition, they have a high resistance to surface wear and because of their strength, the worms are smaller in size than normalized-steel worms. Case-hardened worms are therefore used when the highest possible load capacity is required. Case-hardening, however, must be carried out efficiently to obtain a satisfactory worm; the surface must be carburized uniformly to the correct depth and there is always some difficulty in preventing distortion during quenching. These worms are usually finished by surface grinding and polishing and in doing this care must be taken not to remove too much of the hardened surface which would reduce the bending strength.

2. WHEEL

(a) *Phosphor-bronze, Sand-cast.* This has a bending-stress factor of 7 200 lb/in.^{2} and is used for large wheels and for light duties.

(b) *Phosphor-bronze, Sand-cast, Chilled.* The bending-stress factor is 9 100 lb/in.^{2} and it gives good results for most duties.

(c) *Phosphor-bronze, Centrifugally Cast.* This has a bending-stress factor of 10 000 lb/in.^{2} and gives the strongest wheel. It is frequently used for lift worm-wheels and is the best material for high speeds and heavy loads.

For most lift gears the worm-wheel consists of a rim of one of the above phosphor bronzes which is shrunk and fixed by grub-screws or bolted to a steel or cast-iron centre. An example of the latter method is shown in Fig. 9.3. To ensure silent running the gear teeth must be accurately hobbed.

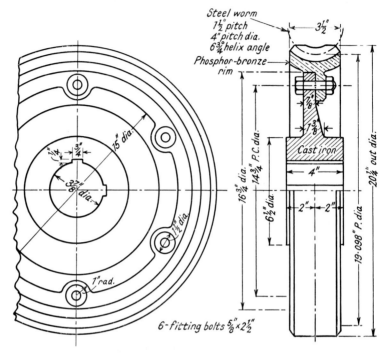

FIG. 9.3. PHOSPHOR-BRONZE WORM-WHEEL
40 teeth, 1½ in. pitch
(*R. J. Shaw & Co.*)

Efficiency. The efficiency of worm gearing, excluding bearing and oil-churning losses is given by the following formula—

$$\text{Efficiency (worm driving)} = \frac{\tan \lambda}{\tan (\lambda + \phi)}$$

$$\text{Efficiency (wormwheel driving)} = \frac{\tan (\lambda - \phi)}{\tan \lambda}$$

where λ is the angle of lead of the worm and $\tan \phi$ the coefficient of friction between the worm and wheel corresponding to the rubbing speed. B.S. 721 gives values of the coefficient of friction for various rubbing speeds.

The relation between efficiency and the worm-lead angle

for various values of the coefficient of friction μ is shown in Fig. 9.4. From this it will be seen that for high-efficiency gears, μ must be small and λ between 20° and 40°. With small values of μ obtained by using high rubbing speeds and

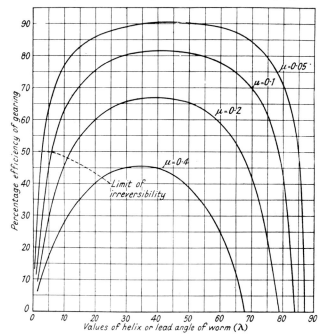

FIG. 9.4. GRAPHS SHOWING RELATION BETWEEN EFFICIENCY AND ANGLES OF LEAD OF WORM FOR VARIOUS VALUES OF μ

efficient lubrication and by employing lead angles of between 30° and 40°, very high efficiencies are obtained. In practice these may be as high as 95 per cent with case-hardened and polished multi-start worms and accurately hobbed and well-finished wheels.

Irreversibility. A worm gear is irreversible, i.e. the worm cannot be rotated by the wheel, if the reversed efficiency (wheel driving) is zero or negative. From the above formula it will be seen that this occurs when λ is equal to or less than the angle of friction ϕ. Fig. 9.4 shows that for irreversibility the efficiency cannot exceed 50 per cent and may be much less. Further, λ must be small, particularly with gearing having a low coefficient of friction, when λ is less than about 5°. The angle of friction, however, changes rapidly with the rubbing speed and the static angle of friction may be reduced by external vibration. Consequently, it is impracticable to guarantee that any gear will be irreversible under working conditions. Even with lead angles as low as 3°, gears have been known to fail to produce irreversibility in service. For example, a lift may be held as rest by a statically irreversible worm gear, but if it were loaded and already descending with a speed giving a value of ϕ less than λ it would accelerate under the out-of-balance load. In the past, many lift gears were manufactured with small lead angles with the object of securing irreversibility but it is now realized that this practice is unsafe. Modern lift gears are highly efficient and the possibility of the lift "running away" is safeguarded by fitting an overspeed governor which automatically cuts off the supply, applies the brake and if necessary, also operates the car safety gear.

Horse-power Rating. B.S. 721 describes how this may be calculated if the gear materials and dimensions are known. The permissible torque is limited either by the surface stress (wear) or the bending stress (strength) in both the worm and the wormwheel. The determination of the load capacity therefore requires four calculations of the wear and strength of the worm and wheel. The permissible torque is the least of these four values. If M_w is the smallest of these four values and N is the revolutions per minute of the wormwheel, the horse-power rating of the worm is $\dfrac{M_w N}{63\,000}$. This is the horse-power to which the gears may be safely subjected for a total running time of 26 000 hours—about 3 years. The rating calculated in this manner is deemed to apply continuously throughout the expected total running time. Lift gears, however, are not subjected to continuous running nor to uniform loading at a constant speed, and for such irregular cycles B.S. 721 describes how the equivalent running times may be calculated.

Lubrication. Efficient lubrication is necessary to maintain an oil film between the working surfaces which will minimize wear, and also to prevent excessive rise in temperature of the worm and wheel. In lift gears, splash lubrication is used in which either the worm or the wheel dips into an oil reservoir in the base of the gearbox and the rotation of the gears flings oil on to the gear teeth. With the under-type gear the oil should submerge at least the bottom half of the worm and with over-type gears the wheel rim should be submerged. Visible external means should be provided on the gearbox to indicate the level of the oil. The working oil temperature should not exceed 200°F.

In the past, castor oil was the favourite lubricant for worm gears but it tends to oxidize rapidly with high temperature and churning, with consequent formation of sludge, and loss of lubricating properties. It is now seldom used as a gear lubricant as modern straight mineral oils are more efficient and less expensive. The values of the coefficient of friction given in B.S. 721 are for gears lubricated with a mineral oil having a viscosity of between 250 and 500 seconds Redwood at 140°F.

Inspection. The lift customer usually leaves the gear design and final testing to the lift manufacturer or to the gear manufacturer and is content to specify that the gear shall conform to the requirements of B.S. 721. Nevertheless, it is frequently the practice to make an examination of the completed gear at works before despatch to site. This is done mainly to ascertain the level of noise and vibration, the teeth-contact pattern, the amount of backlash and the general finish of the gear.

The assessment of noise and vibration when running under load and at the designed speed is a matter of practical experience but a quiet gear is an indication that the teeth have been accurately cut and that they are meshing without binding.

To check the teeth contact, a colouring matter such as red lead is lightly smeared on the worm and after rotating, the pattern on the wheel teeth is examined. This provides information on the width, length and location of the contact, the amount and variation of backlash and the setting of the worm with respect to the wheel. It is desirable that the contact should not start heavily where the worm enters mesh with the wheel teeth as a heavy contact here tends to scrape off the lubricant. In the middle of the face width the contact should spread out uniformly to about two-thirds of the tooth height.

The contact pattern should also be reasonably constant on all teeth.

Some backlash is necessary in worm gearing to ensure that manufacturing variations in pitch and profile do not cause jamming of the teeth. The backlash in lift worm gears must be small, however, because of the reversal of torque as the lift changes direction of motion. The allowable minimum and maximum backlash for various grades of service, as functions of the gear size and number of teeth are quoted in B.S. 721. For lift gears, suitable minimum and maximum values are $0 \cdot 00025m$ $(T + 15) + 0 \cdot 001$ in. and $0 \cdot 00025m$ $(T + 50) + 0 \cdot 0025$ in. respectively, where m is the axial module

$$\left(\frac{\text{Pitch diameter in inches}}{\text{No. of teeth}} \right)$$

and T the number of teeth in the wormwheel. The amount of backlash is obtained by fixing the worm and allowing the wheel to rotate within the backlash limits. This is measured by placing a dial indicator on a wheel tooth.

The lift buyer may confidently leave the manufacture and testing of the worm gear to any lift maker of repute as in the writer's experience, the few cases of lift gears proving unsatisfactory in service have been replaced quickly by the lift manufacturer at no cost to the user.

Thrust Races. The movement of the car up and down the well transmits thrusts in both directions of the worm axis and means must be provided to cater for these thrusts. Single-ball thrust races are sometimes fitted at each end of the worm shaft, but more usually a double-ball thrust as depicted in Fig. 9.5 is provided at the outer end of the shaft, the thrust race being capable of adjustment as wear takes place. In the illustration two angular contact ball bearings are shown, each bearing being capable of carrying radial load and thrust load in one direction.

Tandem Gearing. For heavy duty, tandem gearing, consisting of two worms and two wheels, is sometimes employed, the worms being respectively left- and right-handed. The wormwheels are meshed together as well as the worms and wheels. With this type of gearing, the worm is subjected to two equal

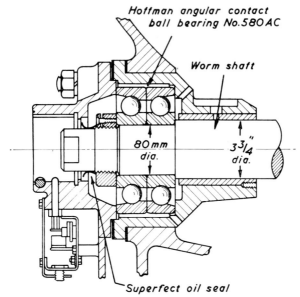

Hoffman angular contact
ball bearing No.580AC

Worm shaft

80mm
dia.

$3\frac{3}{4}$"
dia.

Superfect oil seal

FIG. 9.5. DOUBLE BALL THRUST

FIG. 9.6. DOUBLE-GEARED HEAVY DUTY LIFT MACHINE
(*R. J. Shaw & Co.*)

and opposite thrusts, and separate thrust bearings are therefore unnecessary.

Spur Gearing. This is sometimes used, in addition to a worm reduction gear, for handling heavy loads at slow speeds. A double-geared machine of this type is shown in Fig. 9.6 and consists of the usual worm-geared machine with the addition of an internal spur reduction gearing. The machine shown is capable of lifting 3 tons at a speed of 100 ft. per min., and is fitted with twin oil-immersed brakes for operation from alternating current.

CHAPTER X

CARS, COUNTERWEIGHTS, AND GUIDES

Passenger Cars. The number of passenger car designs is un-limited; the car can be made to almost any specification, the actual finish being usually left to the purchaser. There are a few items of design, however, which should be incorporated in all modern lift cars. The clear height should be at least 6 ft. 6 in., and usually is about 7 ft. 0 in. or even more. A load plate showing the contract load should be fitted in each car. An alarm bell or a telephone should be provided, to enable a call for assistance to be made in case of breakdown or failure between floors. Where only one lift operates in a well, a hinged panel opening outwards should be fitted in the roof, as a means of exit in cases of emergency. This exit should be not less than 18 in. by 24 in. If more than one car operates in a well, side exits are sometimes provided, and these are either removable panels or doors opening inwards. These side exits enable the transference of passengers from one car to another. All emergency exits should be fitted with an electric interlock which will prevent the lift from being operated if an exit is left open. An emergency stop switch should be fitted on the top of the car. Means should also be provided to ventilate the car, particularly if solid car doors are used, whilst artificial lighting is also necessary, and for this almost any form of fitting may be used. Very effective lighting can be obtained by means of tubular fittings at each of the four corners, or by laylights. A car roof unit with combined 9-in. fan and light fitting is shown in Fig. 10.1. The air from the fan is directed into the car through louvres fitted around the light. A 3-pin socket-outlet for a hand lamp should be fitted on top of the car. Cars on some high-speed installations are fitted with load-weighing devices using springs and levers and depend on rope tension for their operation. Others measure the compression of rubber sound isolation pads under the platform and use this to operate calibrated micro-switches.

Lift cars consist of two separate units, namely the *sling* and the *car*, the former being constructed of bolted or welded rolled steel angle or channel sections and must be sufficiently rigid to withstand the operation of the safety gear without permanent distortion. A typical car sling is shown in Fig. 10.2. The side frames are built up of steel angles with gusset plates riveted to each corner, and vertical or diagonal stays to give extra strength and stiffness. The main suspension crosshead is fixed to the top of the frame, and to this is bolted the housing for the two spring-loaded top guide shoes, whilst the bottom

FIG. 10.1. COMBINED FAN AND LIGHT FITTING
(*Express Lift Co.*)

sections carry the car safety gear and the remaining two bottom guide shoes. The crosshead and the bottom platform sections should not deflect more than $\frac{1}{1000}$ of their span with the contract load in the car. The lifting ropes are attached to the top crosshead.

The usual type of guide shoe is spring-loaded and is of phosphor-bronze or cast iron about 9 in. long and shaped to fit the guide, typical examples being shown in Figs. 10.3 and 10.22.

Roller shoes, which have been used extensively in America, have been employed in recent years in this country. These comprise three rubber-tyred spring-loaded rollers, one operating on the head of the guide and the other two on the sides. These transform the lift from a sliding motion to that of a rolling

Fig. 10.2. Car Frame and Safety Gear
(*Marryat & Scott*)

vehicle. The car thus tends to float between the guides and friction is thereby considerably reduced. These rollers, mounted on ball bearings, operate on dry unlubricated guides and there is thus no accumulation of grease on the guides or in the lift well. Each roller is supported by a pivoted rocker-arm which automatically adjusts itself to the guide. A roller guide shoe of this type is shown in Fig. 10.4.

The car is rigidly fixed to the sling in such a manner that unequal loading on the floor cannot throw undue weight on the

FIG. 10.3. GUIDE SHOE
(*Marryat & Scott*)

cabinet work. In modern high-grade cars it is of the practice to insulate the car from the frame and so prevent vibration from the ropes and the guide shoes being transmitted to the passengers. In the Otis passenger car the entire lift car, including the platform, rests on eight "live" rubber blocks mounted on a steel supporting structure. Lateral and upwards motion of the car is restrained by "locking" blocks, but there is no metal to metal connexion between the car and the frame. The top of the car is held firmly to the upright members of the car frame by rubber-faced clamps that are welded to each side of the car's canopy. In addition, the side panels are coated on the outside with a layer of bituminous sound-deadening compound which absorbs the noise of passing air and prevents amplification of sounds originating in the car. All car wiring conduit is similarly isolated from the car platform and

FIG. 10.4. ROLLER GUIDE SHOE
(*Otis Elevator Co.*)

enclosure, and where this is impracticable the wiring is run in flexible conduit.

The floor is made of well-seasoned wood, usually maple, and frequently iron tongues are let in between the planks to afford extra strength. Linoleum, rubber, or parquet, finished in any design to blend with the remainder of the car, may be used as the floor covering. The car back and side framings are often of $1\frac{1}{2}$ in. teak, oak, mahogany or walnut, with panelling of about $\frac{3}{4}$ in. to match, whilst ornamental pilasters or corner posts may also be incorporated, but no glass should be used in the roof or in the car enclosure except for lighting fittings or a door vision panel of the shatter-proof type. Alternatively, a painted or enamelled finish or padded leather sides and back are sometimes adopted; also, laminated boards lined with rubber, the joints being covered with metal strips. Cars lined with Formica are now popular. This material is made in many finishes, is

Fig. 10.5. Passenger Car
(*Wm. Wadsworth & Sons*)

durable and not easily defaced. Occasionally, seats are required, whilst a kicking plate about 9 in. deep and of bronze or stainless steel is frequently fitted. Fig. 10.5 shows a two-entrance car with multi-leaf sliding doors. Fig. 10.6 is the inside of a timber car veneered with a decorative laminated plastic in three colours. The ceiling is fitted with a recessed glass fibre dome housing the lighting fitting, and the ceiling surround trim and the ventilating grilles in the walls and ceiling are all of a natural metal satin finish. The dado moulding and skirting trim moulding are of aluminium finished natural and the floor covering of one-colour linoleum. Fig. 10.7 shows the inside of

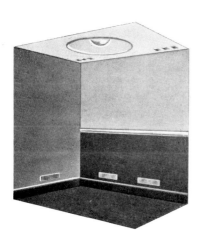

FIG. 10.6. THREE-COLOUR TIMBER
PASSENGER CAR
(*Express Lift Co.*)

FIG. 10.7. PANELLED-STEEL
PASSENGER CAR
(*Express Lift Co.*)

a modern sheet steel passenger car with cellulosed finish to any required colour. The handrail is plastic-covered and the handrail fixings and the lower close-louvred ventilation grilles cellulosed to match the walls. The entrance trims and top ventilating grilles are bronze finished natural satin. The horizontal protection strips are of black rubber carried in aluminium sections finished natural, the skirting being also of black fluted rubber trimmed with aluminium. The floor covering is of one-colour rubber and the outside of the car is treated with anti-drumming material. In these cars, the car control stations, car position indicators and the entrance trims are all finished to match the rest of the car.

Fig. 10.8 is another attractive modern all-metal car finished in cellulose.

FIG. 10.8. PASSENGER CAR
(*Wm. Wadsworth & Sons*)

Car Floor Area. In calculating the floor area necessary, the figure of 2 sq. ft. per person gives a close approximation for contract loads up to 1 500 lb., i.e. 10 persons. As the load increases beyond 1 500 lb., however, the area per person decreases. For example with 20 persons, 33 sq. ft. is required and with 30 persons only 46 sq. ft. of floor is necessary. The contract load for a passenger lift may be calculated from the following formula—

$$L = 60a + 70a \left(1 - \frac{1}{e^{0 \cdot 0166a}}\right)$$

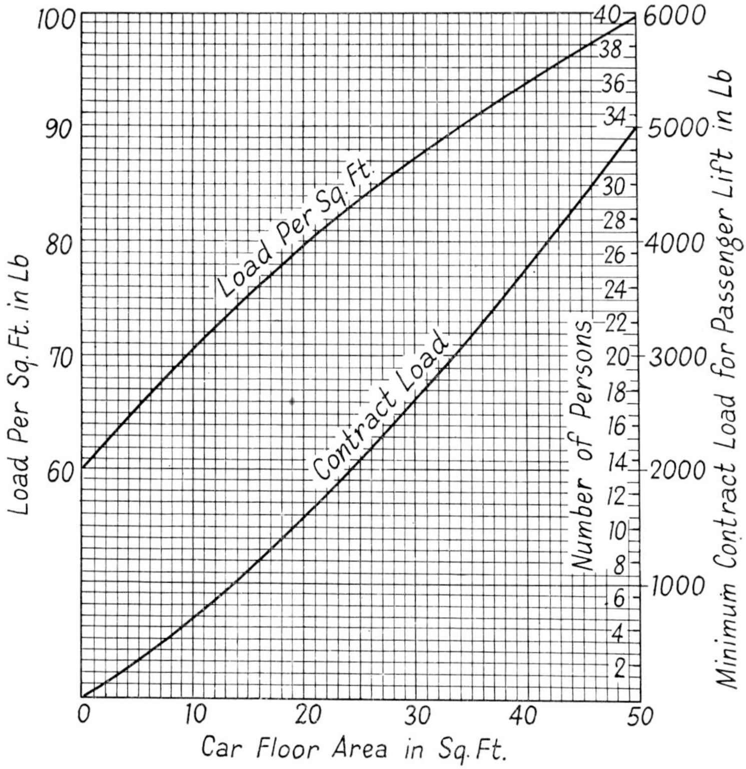

FIG. 10.9. CURVES SHOWING THE RELATIONS BETWEEN LOAD, LOAD PER SQ. FT. AND CAR FLOOR AREA

where

L = minimum contract load in lbs. $e = 2 \cdot 718$.

a = effective platform area in sq. ft. This relation between L and a is shown in the curve in Fig. 10.9. For cars larger than 50 sq. ft. the contract load curve straightens out and above

50 sq. ft. the load is approximately proportional to the floor area up to 25 000 lb at 200 sq. ft. area.

Goods Cars. The sling is of similar construction to that adopted for passenger cars but the car itself is of much rougher construction. The floor is made of double planks at right angles, the top layer being made of well-seasoned maple planks about 5 in. wide and 2 in. thick, and the bottom layer of 2 in. deal, whilst the sides, back, and top are of 1 in. tongued and grooved pitch pine. Wire mesh is sometimes used for the car top instead of wood. A typical goods car is shown in Fig. 10.10, the sides being of tongued and grooved varnished pine and the floor double-boarded at right angles, the top layer of maple. The sides and back are often lined to a height of about 3 ft. with sheet steel $\frac{1}{8}$ in. thick. All-steel cars are frequently used for the transport of heavy or rough materials, the sides being constructed of sheet steel or expanded metal, and the floor of steel chequer plate. An all-steel goods car with chequer plate floor and fire-resisting steel shutter door is shown in Fig. 10.11 and a panelled steel goods car with collapsible gates in Fig. 10.12. In both of these cars it will be noted that the entrance is the full car width. With passenger or goods cars it is good practice to have a plate fitted to the crosshead and engraved to show the weight of the complete car with fittings, the contract speed, and the strength, size and construction of the hoisting ropes. The minimum contract load for goods lifts shall be based on a load of not less than 70 lb. per sq. ft. of car floor area.

When the load in a goods lift consists of pallets, heavy bags, boxes or similar heavy single-piece loads, loaded by a power truck, it is necessary, during loading and unloading, to take into account the total load on the car platform, the brake capacity, and the resistance to slipping of the ropes in the sheave grooves. Similarly, the guides and their fixings, the frame and platform should be designed to withstand the horizontal thrust imposed by trucks and motor vehicles.

Car Travelling Cable. All electrical connexions to the car are made by means of a multi-core hanging flexible cable, one end of which is connected to a terminal box fitted under the car floor, the other to a terminal box fitted in the well at approximately the mid position. Each conductor in the cable consists of 40 tinned copper wires and the number of cores is 2, 3, 4, 6

Fig 10.11. All-steel Goods Car
Marryat & Scott)

Fig. 10.10. Goods Car
(Wm. Wadsworth & Sons)

or 10 of overall diameters varying from 4 in. to 6 in. The 10-core construction is generally used for the higher speeds as the heavier cable gives a better running performance than lighter

Fig. 10.12. Panelled Steel Goods Car
(*Express Lift Co.*)

cables. To remove any twist in the cable it is recommended that it should be freely suspended in the well for at least 24 hours before being fitted to the lift. These travelling cables should fulfill the requirements of B.S. 977. All car connexions from the controller are run to the well terminal box. The length of the flexible cable should be approximately equal to half the lift total travel plus 15 ft. so that the car may travel from end to end of the well without subjecting the cable to any strain.

Car Emergency Telephone. A telephone for emergency use is now generally provided in the lift car either as a direct exchange line or as a P.B.X. extension. A private internal telephone circuit between the lift car and the machine room for lift maintenance purposes is sometimes installed, which cannot be connected to the public network and which the Post Office neither provides nor maintains. Some existing lift telephone installations are mounted in a recess in the wall of the car having an opening 12 in. square and 6 in. deep. The telephone and terminal block are mounted on a hinged wallboard constructed and fitted by the lift contractor in accordance with Post Office requirements. This is not now normally used as it is difficult to provide a recess of 6 in. depth in modern lift cars. To accommodate the present standard Post Office equipment comprising the telephone, dial, buzzer and induction coil, the lift contractor provides in the wall of the lift car a recess $9\frac{1}{2}$ in. high and at least 3 in. deep. The contractor also provides, installs and maintains the travelling cable containing the two telephone wires and all other telephone wiring between the terminal block associated with the telephone in the car and the terminal block situated in an agreed position outside the lift well. The Post Office maintains the rest of the telephone circuit. The words "EMERGENCY TELEPHONE" should be inscribed on the outside of the recess door, and on the inside there should be a notice giving instructions on what to do in an emergency and a suitable telephone number to call.

COUNTERWEIGHTS

The object of the counterweight is to provide traction and to balance the weight of the car plus a predetermined proportion, usually 40 to 50 per cent, of the maximum car load, and thereby to reduce the size of the motor. Incidentally, the counterweight provides a certain measure of safety when landing on its buffer and removing traction from the car. The most usual construction consists of cast-iron sections firmly secured against movement by at least two steel tie rods having lock nuts or split pins at each end, and passing through each section,

FIG. 10.13
COUNTERWEIGHT
(*Marryat & Scott*)

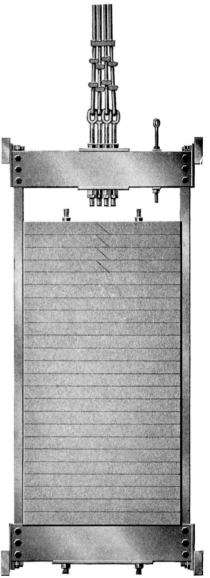

FIG. 10.14. COUNTERWEIGHT
(*Express Lift Co.*)

178

as shown in Fig. 10.13. In addition to tie rods, the sections are sometimes mounted in a steel framework as in Fig. 10.14. For heavy loads the sections are sometimes weighted with lead so as to reduce the size of the counterweight. The hoisting ropes are secured to the counterweight top frame by screwed eyebolts which allow of rope adjustment, and splices or clips. Four spring-loaded shoes, similar to those for the car, are fitted to ensure that the counterweight will travel smoothly in its guides. Safety gear is sometimes fitted and may be any of the types described in Chapter XIII.

GUIDES

Some form of guiding is necessary for both the car and counterweight so that they will travel in a uniformly vertical direction. The guides must be of such lengths that it will be impossible for any of the car or counterweight shoes to run off the guides. In the most common arrangement, two guides are required for the car and two for the counterweight, but the actual number and the relative positions of the guides depend upon the landing openings required. Openings all on one side of the well permit of the simplest guide arrangement; the counterweight guides are fixed to the wall opposite the landings, and the car guides, one on each of the two remaining walls. In some instances it is necessary to cater for landing openings on opposite sides of the well, and in these cases a car guide is supported on each of the two remaining walls and both counterweight guides on one of these walls. Openings on adjacent sides or even on three sides are sometimes required, and this involves guiding the car diagonally at two of its corners, or alternatively at each of the four corners. In the case of this corner guiding, the car frame is cut away at the corners so as to provide fixings for the guide shoes. A three-point guiding is sometimes adopted for cars with adjacent openings, and in this arrangement two guides are placed, one at each corner of one side, with the third guide at the middle of the opposite side.

Material. Well-seasoned teak and rolled steel channel have been extensively used in the past for car and counter-weight guides, but modern practice favours either round steel or tee-section steel guides for speeds exceeding 100 ft. per min. except where the nature of the work carried on in the building renders steel unsuitable, due to acid fumes, when specially prepared wooden guides are installed. Tee-section guides are recommended for both car and counterweight on all lifts travelling at speeds in excess of 200 ft. per min. The advantages of tee-section guides are, that there is no tendency for the shoes to get out of place, as is possible with round steel guides; they afford a larger shoe bearing surface, and result in more even shoe wear. Further, this type of guide is sufficiently rigid to make the use of guide backings unnecessary.

Sizes. Three standard sizes of Tee guides are manufactured by Messrs. British Guide Rails, and these together with data on each type are listed in Fig. 10.15. For comparison purposes corresponding information for round-section steel guides used for lifts is also quoted. Tee guides are being used in increasing numbers and are now rapidly replacing round guides. These Tees are straightened before being machined and when finished are guaranteed to be straight to within limits of 0·002 in. The stock length is 16 ft. As the deflexion of a guide due to a load is inversely proportional to the moment of inertia of its section it will be noted from the table that T160, T161 and T163 guides are approximately equivalent to 3 in., $2\frac{3}{8}$ in. and 2 in. diameter round steel guides. Hence there is a considerable saving in steel by using Tee guides. The relative strengths of Tee and round guides is illustrated in Fig. 10.16 which shows deflexion curves for guide fixings at various spacings, about the two principal axes of a T161 guide, about a diameter of a round guide of approximately the same weight ($1\frac{1}{4}$ in. dia.) and about a diameter of a round guide of approx. the same strength ($2\frac{3}{8}$ in. dia.). A guide length between two clips has been considered and for this purpose it has been assumed that there is a shoe pressure on the guide of 0·25 ton when the shoe is midway between two clips. As these clips fix the directions of the ends of each section, the length between the clips has been treated as a beam with fixed ends, and the formula $\dfrac{WL^3}{192EI}$ has been used.

B.S. 2655 states that the guides shall not deflect by more than $\frac{1}{8}$ in. under normal operation, and it is necessary therefore, to check that the guide size and fixing spacings are such that this deflexion will not be exceeded. An estimate can be made

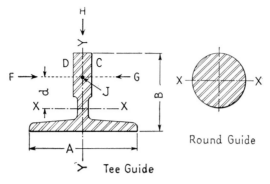

Tee Guide

Round Guide

Steel Tees Nº	Size A (In)	B (In)	Weight Lb per Ft	Area Sq In	I XX (In)^4	I YY (In)^4
T 163	$2\frac{3}{4}$ x	$1\frac{15}{16}$	6·25	1·79	0·75	0·62
T 161	$3\frac{1}{2}$ x	$2\frac{7}{16}$	8·0	2·41	1·4	1·28
T 160	5 x	$3\frac{1}{2}$	14·87	4·417	4·75	5·93
Steel Rounds	$1\frac{1}{2}$ Dia		6·008	1·767	0·248	
,,	$1\frac{3}{4}$,,		8·170	2·405	0·460	
,,	2 ,,		10·68	3·141	0·785	
,,	$2\frac{1}{4}$,,		13·52	3·976	1·258	
,,	$2\frac{3}{8}$,,		15·06	4·430	1·562	
,,	$2\frac{1}{2}$,,		16·69	4·908	1·917	
,,	$2\frac{3}{4}$,,		20·19	5·939	2·807	
,,	3 ,,		24·03	7·068	3·976	
,,	$3\frac{1}{4}$,,		28·21	8·295	5·476	

FIG. 10.15. DATA RELATING TO TEE GUIDES AND ROUND GUIDES

of the maximum normal thrust on the guides and the deflexion calculated by the above formula. E may be taken as 30×10^6 lb per sq. in. and the moments of inertia as 4·75 (about XX) and 5·93 (about YY) for T160 guides, 1·4 (about XX) and 1·28 (about YY) for T161 guides and 0·66 (about XX) and 0·62 (about YY) for T163 guides.

Guide Stresses. During the normal travel of an empty car the thrusts of the shoes on the guides are negligible if the car is of a balanced construction and is centrally slung. Also, if the car is loaded in such a manner that the load is uniformly distributed or centrally placed on the car floor no appreciable thrusts will be imposed on the guides. A concentrated load in the car located at some point other than the floor centre, however, will cause thrusts on the guides in the direction

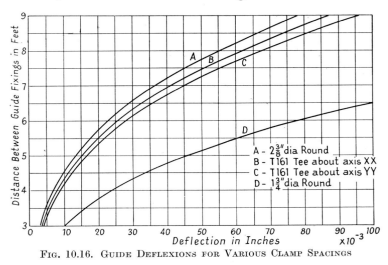

Fig. 10.16. Guide Deflexions for Various Clamp Spacings

F, G or H (Fig. 10.15), depending upon the position of the load. When the car is at a landing and is being loaded the weight of the load on the front edge of the car platform results in a guide thrust in the direction F or G. These thrusts may be of a high value if the load is concentrated and particularly if the car is very deep. Heavy loads are thus imposed on the guides if automobiles or laden trucks are run into the car. To withstand such heavy thrusts it is advisable that guide clamps be fitted opposite the guide shoes when the car is level with the landing. The thrusts are thus transferred to these fastenings but if they are not securely made the clips may be loosened. When the car is at a landing and the guide fixings are not opposite the car shoes the loading thrusts on the guides cause a guide deflexion which will be greatest if the shoe is midway between two clips. The amount of this deflexion will depend upon the guide section, the distance between fixings, the size and weight of the car and on the car load. If the car is large and heavy concentrated loads are handled it is necessary to consider backing or bracketing the guides and adopting close-spaced fixings to prevent the deflexion being excessive. Such deflexion caused during loading or by the load being off-centre in the car, should not exceed $\frac{1}{8}$ in. The opening of a heavy car gate or side-opening car door will cause a side load and result in guide forces in the direction H but these are not usually of appreciable magnitude. During normal operation the thrusts and consequent deflexions on the counterweight guides are small and guide fixings at intervals not exceeding 10 ft. will usually ensure the guide deflexion being within the B.S. requirement of $\frac{1}{8}$ in.

In addition to the normal stresses outlined above, considerable loads are imposed on the guides when the safety gear is brought into operation, the magnitude of the resultant stresses depending upon the contract load and speed and the type of safety gear fitted. Referring to Fig. 10.15, the safety jaws grip the guides on the working faces C and D and this may be considered equivalent to a downward load imposed on the guide section at the point J. This is an eccentric load acting at a distance d from the centre of the section and results not only in a direct compressive stress but also in a bending moment of value equal to the retarding force multiplied by the distance d. The compressive force is transmitted to the guide end in the pit and hence the desirability of resting the guide end on a soleplate in order to distribute this load. The bending moment results in guide deflection about the axis XX, i.e. guide spreading, the deflection being proportional to the resultant safety gear force, the modulus of the guide section and the distance between the guide clips. This bending moment also causes a tensile stress in the guide clamp fixing bolts. In addition, the forces imposed on these fixings due to the normal operation of the lift described above also result in a tensile stress in the clamp bolts. It is, therefore, important to ensure that the clamp fixings are strong enough to withstand these stresses. In practice the diameter of the steel clamp bolts should be not less than $\frac{1}{2}$ in. for T163 guides and not less than $\frac{5}{8}$ in. for T160 and T161 guides. The resultant safety gear force at J is considerable, particularly if the safety gear is of the instantaneous type. For example, if a car with its load weighs one ton and is fitted with instantaneous gear which operates at a car speed of 200 ft. per minute a sudden downward load of several tons will be imposed on each guide. The guide section must be adequate to limit both the compressive stress and the deflection to reasonable values. Gradual wedge-clamp or wedge-clamp safety gear slows the car more gradually and thus the retarding force is much smaller. For this reason instantaneous safety gears are not used for speeds above 200 ft. per minute, and some authorities limit their use to 100 ft. per minute and even then to small lifts only.

It will be appreciated from the above that considerable care is necessary in the selection of suitable guides and clip fixings for any particular type of lift. Except in special circumstances it is good practice to use T163, T161 and T160 guides with cars of maximum contract loads of 1 000 lb., 2 500 lb. and 10 000 lb., respectively. If counterweight safety gear is fitted the size of the counterweight guides should be the same as those for the car. Where counterweight safety gear is not provided, the counterweight guides have to withstand much smaller stresses than the car guides and, therefore, may be of smaller size. In such cases T163 is suitable for counterweights of lifts of contract loads up to 2 500 lb. and T161 where the contract load is as high as 10 000 lb. A guide clamp spacing of 10 ft. is common practice, but if concentrated loads are handled or the contract speed is high it is advisable to use a spacing of 8 ft. or even 6 ft.

Fixing and Jointing. The method of fixing the guides depends upon the construction of the well and the relative positions of the car and counterweight guides, but the number of fixings should be such that the guides will not deflect more than $\frac{1}{8}$ in. under normal operation. In a staircase well or a self-contained steel structure the guides are fixed by plates and bolts with the additional provision of steel stringers if necessary. In a bricked well, the guides are secured either by Lewis bolts, expanding bolts, or by bolts passing through the brickwork and fastened on the outer sides by steel plates. The former method is frequently adopted for outside walls and the latter for inside walls. In either case, a sound fixing in which all the shims are of metal is necessary, and when Lewis bolts are employed they should be sunk in the wall at least 6 in. and firmly bedded. At their bottom ends, the guides should rest solidly on the buffer channel-iron brackets.

Fig. 10.17 shows the method of fixing and jointing channel iron guides for counterweights. This form of guide is sometimes used for counterweights of slow-speed goods lifts. Each guide consists of two channels bolted together, the counterweight shoe travelling in the space formed by the web and two flanges of one of the channels. The various lengths, in each half of the guide, are arranged so that the joints are spaced at intervals of 4 ft., the jointing being carried out by four countersunk bolts passing through the web of each channel. The channels are, in addition, screwed to each other by single countersunk bolts at 4 ft. spacing. The iron strap is bolted to each guide at intervals of 4 ft. throughout the entire length of the guides. These straps are secured to each guide by a tapped bolt and a through bolt, the latter serving as the fixing to the well and therefore may be either a Lewis bolt, expanding bolt, or a through bolt fixed at the outer side by a steel plate.

A method of fixing round guides is shown in Fig. 10.18 in which the guide is secured at intervals of 4 ft. to cast-iron seatings, these being screwed to iron straps. The latter are fixed to the well by Lewis bolts or through bolts.

Fig. 10.19 shows a method of fixing round steel guides in which rolled steel backings are employed. These backings avoid the necessity of using large section guides in order to secure rigidity, and also enable the safety gear to operate on the backings, as shown in Fig. 10.20, instead of on the polished guides. The guide is fixed to cast-iron seatings by countersunk screws, these seatings in turn being bolted to the backings. The various lengths of guide are jointed by screwed socket joints, whilst the lengths of backings are jointed by steel fish-plates. A spacing of about 4 ft. is adopted for the bolts which secure the backings to the well.

Many different methods are employed for fixing tee-section guides, some of these being illustrated in Figs. 10.21 and 10.24. Fig. 10.21 shows the method of jointing tee-guides and of

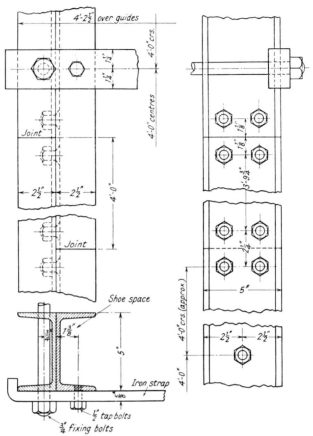

FIG. 10.17. METHOD OF FIXING AND JOINTING
CHANNEL IRON GUIDES
(R. J. Shaw & Co.)

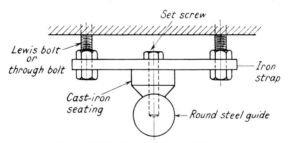

FIG. 10.18. ROUND GUIDE FIXINGS

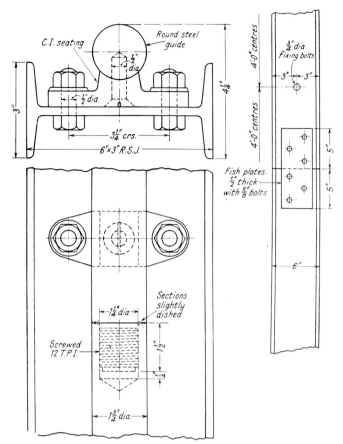

FIG. 10.19. METHODS OF FIXING AND JOINTING ROUND STEEL GUIDE
AND BACKING

(R. J. Shaw & Co.)

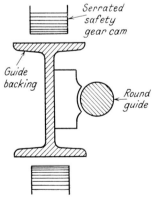

FIG 10.20 OPERATION OF SAFETY GEAR ON GUIDE BACKING

fixing the car guides to the well. The lengths of guide are jointed by machined spigot and socket joints along the web,

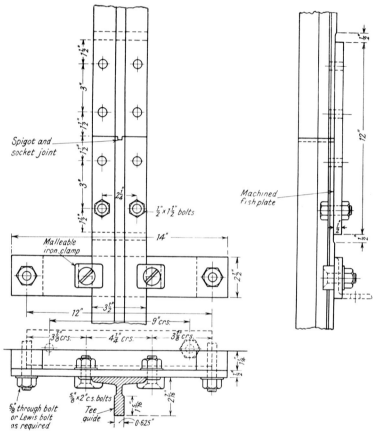

FIG. 10.21. ARRANGEMENT OF CAR GUIDE FIXED TO WALL OR STEEL STRUCTURE
Showing method of jointing
(*R. J. Shaw & Co.*)

and in some cases along the flange as well; the flanges being machined and jointed by a machined steel fishplate, as illustrated in Fig. 10.22. It is good practice to use a short length

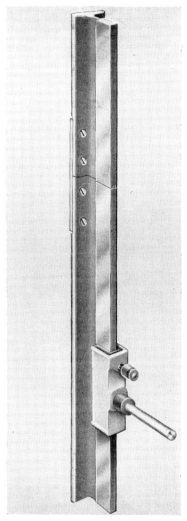

FIG. 10.22. TEE GUIDE, SHOWING METHOD OF JOINTING AND
TYPICAL SHOE
(*Express Lift Co.*)

of guide about 8 ft. long at the bottom of one of the car guides and a similar length at the bottom of one of the counterweight guides to ensure staggering of the car and counterweight joints. For heavy loads or high speeds the guide sections are sometimes jointed by short lengths of tee guide instead of a

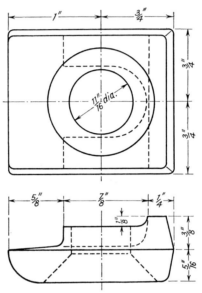

FIG. 10.23. MALLEABLE IRON GUIDE CLAMP FOR TEE GUIDES
(*R. J. Shaw & Co.*)

flat plate to provide extra joint stiffness. The guide is fastened to steel brackets at intervals of about 6 ft. by malleable iron clamps, a typical example of which is shown in Fig. 10.23. The back plates are secured to the well by through bolts or Lewis bolts, spacing pieces being interposed between the plates and the well. Alternatively, the back plates, either flat or angle iron, may be bolted to the steel well structure.

Fig. 10.24 shows methods of fixing car and counterweight guides when landing openings on opposite sides are required, and it becomes necessary to fix a car guide on the same wall as the counterweight guides. In (*a*) the counterweight guides are fixed to channel irons by the usual clamps, these channels

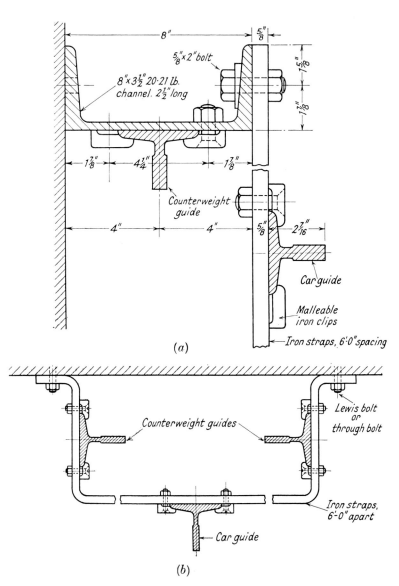

FIG. 10.24. METHODS OF FIXING CAR AND COUNTERWEIGHT GUIDES

(a) Car guide fixed to counterweight guide brackets
(b) Car guide fixed to counterweight guide straps

191

being fixed to the well every 6 ft. Similar spacing is adopted for the car guide fixings, the guides being secured to iron straps

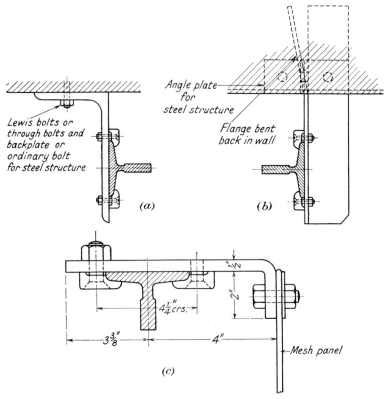

Fig. 10.25. Counterweight Guides and Guards
(a) and (b) Other methods of fixing counterweight tee guides
(c) Bracket to fix wire mesh guards to counterweight guides
(*R. J. Shaw & Co.*)

which brace the two channel irons. In (b) a similar result is obtained by using iron straps instead of angles.

Other methods of fixing counterweight guides and counterweight guards are shown in Fig. 10.25. In (a) the guides are secured to the well by angle irons, and in (b) by short lengths of angle, either embedded in the well or bolted to the steel structure. A bracket for attaching the counterweight guard to the counterweight guides is shown at (*c*).

Metal Inserts. One of the best methods of providing supports in the well for fixing the guide bracket bolts is by the insertion of metal guide inserts in the well face. An insert of this type

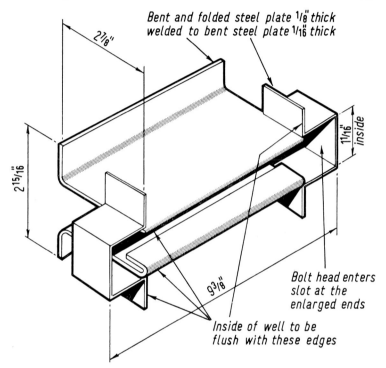

FIG. 10.26. GUIDE BRACKET FIXING INSERT
(*Express Lift Co.*)

is shown in Fig. 10.26 and consists of steel plates bent and welded together to form a secure pocket for the fixing bolt heads. These inserts are built in as the well erection proceeds so that the bolt slots are flush with the well face. Bolts with special heads to prevent turning are slid into the slots, to secure the guide brackets. A short insert may be used for each bracket bolt as in Fig. 10.26 or alternatively a long insert about 24 in. in length may be used to house two bracket bolts. In a brick well these inserts are fixed in concrete blocks built securely in the brickwork whilst in a concrete well they are fitted to, and level with, the inside face of the shuttering before pouring and consolidating the concrete. In the latter case, care must be taken to ensure that the inserts are rigidly fixed to the shuttering and do not move during concrete consolidation. To prevent this, holes are sometimes cast or drilled in the face of the insert so that it can be nailed to the

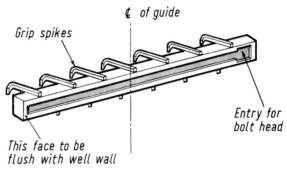

FIG. 10.27. LONG METAL INSERT FOR HOLDING BOTH BRACKET BOLTS

shuttering. These inserts provide a good fixing and also permit lateral adjustment of the bracket bolts during lining up of the guides. Fig. 10.27 shows a long-section metal insert for securing both bracket bolts. Bent spikes are welded to the top and bottom faces to ensure a sound fixing in the well wall. The bracket is secured by two special square-head bolts which enter the insert at the enlarged end of the slot.

Sliding Clips. Specially shaped clips are sometimes used which fit snugly along the guide flanges and permit the guide to slide in the clips whilst still holding the guide to its brackets. These sliding clips give longer bearing surfaces on the guide flanges than are provided with the normal malleable iron clamps and are used in high buildings to avoid distortion of the guides during movements in the building structure. A simple type of sliding clip is shown in Fig. 10.28 and consists of a piece of flat steel plate bent to fit over the guide flange and held to the bracket by a bolt. Another form of sliding clip is shown in Fig. 10.29; this comprises a U-shaped clip bolted to the bracket, the clip holding a U-shaped friction plate around the edge of the guide flanges. This is a flexible

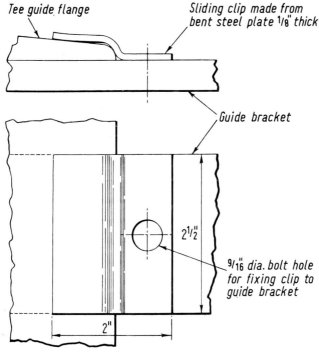

Fig. 10.28. Guide Sliding Clip
(*Otis Elevator Co.*)

arrangement which readily permits the guide, when necessary, to slide through the friction plates.

Guide Lubrication. The guides are exposed to dust and dirt from the well and landings, and to keep them lubricated satisfactorily is not an easy matter. It is important that they should be adequately lubricated to minimize wear of guides and shoes and to ensure comfortable riding conditions. With many of the older slow-speed lifts the guides are painted with heavy oil or grease at regular intervals, but most lifts of recent construction have their guides lubricated by automatic means and this is standard practice to-day. A number of different types of automatic lubricators are in use, some employing oil and others grease. Thin oils tend to run down the guides and wash off the dirt but if the contract speed is high the oil may

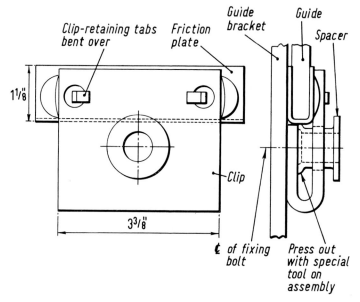

FIG. 10.20. GUIDE SLIDING CLIP
(Express Lift Co.)

be thrown off the guides by the shoes and might find its way to the inside of the car. The design of an oil lubricator must therefore incorporate some means which will ensure that too much oil is not used. Although grease adheres to the guides better than oil it has the disadvantage of holding dust and dirt.

Some guide lubricators are of the stationary type consisting of an oil reservoir mounted at the top of each guide, the oil being fed to the guide by wick feeds. It then runs down the guide to the bearing surfaces. The most popular lubricators, however, are of the travelling kind and these consist of two lubricators mounted on the car for lubricating the car guides and two on the counterweight for lubricating the counterweight guides. A lubricator fitted to a shoe is shown in Fig. 10.3 Another type for attachment to the car and counterweight guide shoes and which uses heavy oil is shown in Fig. 10.30.

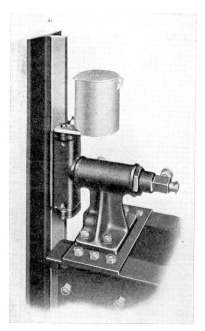

FIG. 10.30. GUIDE LUBRICATOR
(*Otis Elevator Co.*)

This latter lubricator holds about $1\frac{1}{2}$ quarts of oil, the rate of flow of which may be regulated by a screw adjustment.

A recent development is the use of dry guides with which carbon lined shoes are employed. Messrs. Morgan Crucible Co. have experimented with this type of shoe on some recent lifts and the results are promising. The lining is composed of three flat carbon plates held against the three sides of the shoe by metal keep plates screwed to the top and bottom of the shoe. The linings are backed by thin sheets of rubber bonded to the carbon plates. This minimizes the effect of shock on the carbons and also retains them in position if they crack. The main advantage of this type of shoe is that lubrication is not required and thus the guides and the outside of the car can be kept clean and maintenance reduced. Renewable nylon linings are also in use.

CHAPTER XI

GATES, DOORS, AND LOCKS

Gates. Until recent years the most common form of protection for lift car and landing entrances was the collapsible steel gate of the overhung type. The gate is supported by ball-bearing rollers running on an overhead track, the pickets being guided by a self-cleaning channel-shaped bottom track. The gates are of the close picket type in which the openings between pickets do not exceed $2\frac{1}{2}$ in. in width when the gate is fully extended. When used on landings the leading picket is widened to accommodate the electro-mechanical lock. Where a particularly silent gate is required, as in hospitals, rubber buffers are fitted between the pickets. A hollow tube gate construction, which eliminates the shearing hazard, is sometimes adopted, the tubes frequently being arranged to run around the side of the car when open, thus giving an opening equal to the full car width. Gates can be made of bronze, aluminium alloy or stainless steel and may be chromium-plated if required. Sometimes they are finished in cellulose, copper- or brass-plated, or bronzed. Most gates are of the single side opening type, although for large cars double centre-opening gates are sometimes provided. To ensure sufficient rigidity the gate should withstand a pressure of 75 lb. applied at points on two adjacent pickets so as to divide the load equally. A typical midbar gate is shown in Fig. 11.1. Gates have now largely been superseded by doors, particularly as protection for landings, and on the higher-grade passenger lifts.

Doors. The advantages of doors over gates are that they are quieter, they eliminate draughts emanating from the well, prevent noise being transmitted to various floors via the well openings and also improve the appearance of a lift. The rigidity of doors enables them to be operated by power in a more satisfactory manner than gates. To prevent the possibility of a person being trapped between the landing and car doors, the distance between the well side of the landing door and the lift-well edge of the landing threshold should not exceed 4 in. for hinged doors or $2\frac{3}{4}$ in. for sliding doors or gates. The distance between the well side of the landing door and the well side of the car door should not exceed $5\frac{1}{2}$ in. The distance between the car and landing sills should not exceed $1\frac{1}{4}$ in.

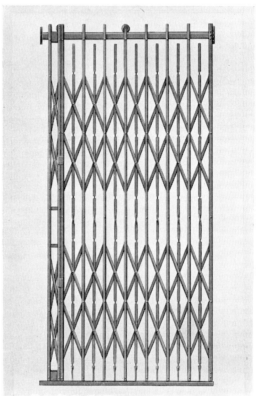

Fig. 11.1. Collapsible Midbar Gate
(*Express Lift Co.*)

Hinged Doors. These are used on landings and may be of the single pattern or of the double centre-opening type and can give a clear opening equal to the full car width. These doors, however, require lobby space in which to swing and are now seldom used, having been replaced by sliding doors.

Single-panel Sliding Door. A typical door of this type is

shown in Fig. 11.2. The vision panel should be of safety or wired glass. This entrance is used where moderate openings of

FIG. 11.2. SINGLE-LEAF (FLUSH) SLIDING DOOR
(*Marryat & Scott*)

about 3 ft. are required, this usually being a little more than half the car width.

TWO-PANEL SLIDE. One panel slides behind the other, the former at twice the speed of the latter, and both panels therefore arrive at the fully-opened position simultaneously. The clear opening is about two-thirds of the car width. A greater depth of sill is required with this type than with a single-panel or two-panel centre-opening sliding door. A two-panel sliding door is shown in Fig. 11.3.

TWO-PANEL CENTRE-OPENING SLIDE. One-panel slides to the left and the other to the right. Rubber bumpers are fitted

FIG. 11.3. TWO-PANEL SLIDING DOOR
(*Marryat & Scott*)

on the edges of each leaf to minimize the shock when closing. The clear opening is about half the car width. An important advantage of these doors is that the operating time is half that of other types as the distance to be moved is only half the opening width. Since these doors have two mechanical connexions to the operator, there are two self-cancelling reactions, no twisting action is imparted to the car, and high door speeds can be used. For these reasons this type of door is very popular for the wider passenger cars and also where high car speeds are employed and it is desirable to take full advantage of such speeds by cutting down the door operating time to a minimum. A door of this type is shown in Fig. 11.4.

FIG. 11.4. CENTRE-OPENING DOOR
(Wm. Wadsworth & Sons)

TWO-PANEL, SLIDING AND STATIONARY. The sliding panel moves behind the stationary panel and the opening is therefore equal to the width of the sliding panel which is about half the car width.

THREE-PANEL TWO-SPEED, SLIDING AND STATIONARY. The two sliding panels move behind the stationary panel, one at twice the speed of the other so that both panels arrive at the fully open position simultaneously. This gives the entrance a characteristic pleasing stepped appearance. The door is sometimes used where it is desirable to produce architectural symmetry in the corridor treatment for landing entrances which

FIG. 11.5. THREE-PANEL, TWO-SPEED DOORS
(*Wm. Wadsworth & Sons*)

would be off-centre or improperly located with any other type of entrance. The opening is approximately two-thirds of the car width. Fig. 11.5 illustrates this entrance.

FOUR-PANEL SLIDING, TWO-SPEED CENTRE-OPENING. Two panels slide to the left, one at twice the speed of the other, the remaining two leaves slide similarly to the right. This is sometimes used on large passenger cars.

MULTI-PANEL SLIDING. This consists of any number, between about three and twelve, of narrow panels. The panels telescope behind one another as the door opens and give an

FIG. 11.6. PASSENGER LIFT WITH MULTILEAF DOOR
(*Marryat & Scott*)

opening of about half of the fully-extended width for the small sizes and three-quarters for the larger sizes. The door, which is attractive in appearance, is sometimes used on wide passenger cars and is shown in Fig. 11.6.

COLLAPSIBLE STEEL-SHUTTER DOOR. This is a sliding door

comprised of a large number of narrow panels which collapse into a small space. It is very popular for the entrance of goods lifts because it gives a particularly wide opening. They

Fig. 11.7. Collapsible Steel-shutter Door
(*Marryat & Scott*)

exclude draughts and are fire-resisting. As an alternative to steel, they are available in aluminium alloy. The individual panels may be 4 in., 6 in., or 9 in. wide. An inspection window of shatter-proof glass is usually fitted to the leading panel. This door is illustrated in Fig. 11.7.

Vertical Bi-parting Door. This consists of two panels

which slide vertically, one upwards and the other downwards. The panels are connected so that they move simultaneously. These doors are robust and strongly constructed and will stand more rough treatment than any other form of door or gate, consequently they are usually fitted on large goods lifts and, because of their weight, are generally power-operated. The doors, however, are counter-balanced so that they can be moved with the minimum effort. They are expensive, but as they cannot be damaged by trucks during loading and unloading, their maintenance costs are small. The rising car gate used with these power-operated doors is generally formed of wire-mesh panels in a steel section frame. Another advantage is that they provide a full car-width entrance. The standard type of bi-parting door requires a minimum floor to floor height of about 12 ft., but for restricted floor heights, a special pass type, in which the rising top portion of one door passes the lower panel of the door above, is available.

The risk of injury between the two panels is lessened by providing cushioned edges and having a slipping clutch in the door-operating mechanism.

The push-buttons for operating power-operated vertical bi-parting landing doors and power-operated vertical car doors or gates should be fitted in a conspicuous position, such that the doors or gates are in direct sight of the user during closing. The door-closing button should be of the continuous pressure type, the release of which should cause the landing doors and car gate or door to stop and re-open.

Where power-operated vertical bi-parting landing doors are used with power-operated vertical sliding car doors or gates and the lift serves more than one entrance at any level, the lift car should be provided with separate door-operating push-buttons for each car door or gate and its adjacent landing door. Any door-operating push-buttons at a landing should control only the landing door and its adjacent car door or gate for the entrance where the push-buttons are fitted.

The lift car of lifts fitted with power-operated vertical sliding car-doors or gates, or power-operated vertical bi-parting landing doors should be provided with a momentary-pressure push-button, the operation of which will cause the car door or gate and landing door to open, if the lift car is in a landing zone.

Where power-operated vertical-sliding car doors or gates are used with power operated vertical bi-parting landing doors,

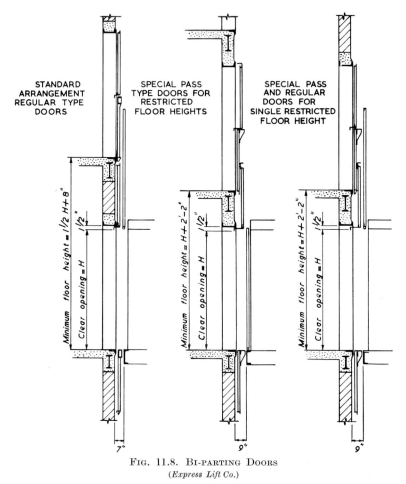

FIG. 11.8. BI-PARTING DOORS
(*Express Lift Co.*)

the car doors should close before the landing doors start to close, and open after the landing doors have opened.

Fig. 11.8 shows the space required for the various arrangements of bi-parting doors in which both the landing doors and

FIG. 11.9. POWER-OPERATED BI-PARTING DOORS

Marryat & Scott

the car gate are power-operated by the same operator. Fig. 11.9 is an entrance with power-operated bi-parting landing doors, and manually-operated collapsible car gate, whilst Fig. 11.10 shows manually-operated bi-parting steel doors for the landing and car entrances.

Sliding doors are invariably overhung, the bottom edges

FIG. 11.10. VERTICAL BI-PARTING DOORS
(*Wm. Wadsworth & Sons*)

being provided with rectangular guides that travel in the machined grooves of the sills. Rubber bumpers are fitted on the top and bottom door stops to minimize shocks when opening and closing. Doors may be constructed of wood and finished to harmonize with the surrounding landing architecture. Probably the best practice, however, is to use sheet steel No. 14 or No. 16 gauge, suitably framed and filled and reinforced if necessary to receive the operating mechanism. The advantages of sheet steel doors are that there is no tendency to lose shape, as is the case with wooden doors, and the fire risk is lessened. Door hangers consist of ball bearing hardened steel rollers mounted on steel brackets fixed on the top of each panel. The rollers are about 3 in. in diameter, two being provided for each panel. Specially shaped top steel tracks on which the door rollers run are fixed to the door frame. A small ball bearing check roller is mounted immediately below each main roller and on the same bracket and engages the under side of the track. These check rollers prevent the doors from being lifted upwards. A large number of door designs is available, but every landing door should be provided with a " fire-resisting " vision panel. The decorative designs may be applied by stippling, stencilling, or a transfer process. Plain colours, grained, or bronze enamel finishes are very popular, whilst further decoration is sometimes obtained by means of strips of bronze, monel metal or non-oxide steel applied to the doors.

B.S. 2655 states that each car entrance shall be provided with a car gate or door. It is interesting, however, to note that many lifts in other European countries have cars without gates or doors. The wells are smooth faced on the landing entrance side and the insides of the landing doors are flush with the well face.

Methods of Operation. Collapsible gates or solid doors may be operated by one of the several methods detailed below.

MANUALLY OPERATED. The landing and car doors are opened and closed by hand after the car has arrived level with the landing and released the locks.

One method of reducing the labour involved in opening and closing the car and landing gates makes use of an electro-magnet fitted to the car gate. This magnet, when energized, magnetically connects the landing and car gates together through the medium of a hinged iron plate fitted on the inside of the landing gate. Thus both gates are opened in one movement instead of two.

SELF-CLOSING. In this form, the door or gate is opened manually but closes automatically when released. The closing is usually effected by means of springs, the action of which is cushioned by a dashpot, or alternatively by a falling weight connected to the door by a light chain. Sometimes light single

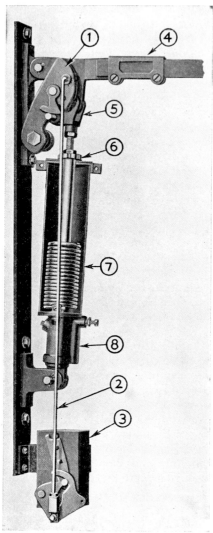

FIG. 11.11. DETAILS OF CLOSER
(*Otis Elevator Co.*)

doors are linked together when the lift stops at a landing so that movement of one door operates the two, and by giving a mechanical closing bias to the doors they automatically close when released.

Fig. 11.11 shows details of a closer in which (1) is the mechanical locking device which prevents the re-opening of the door when closed to within $2\frac{1}{2}$ in. of the door jamb. The connecting rod (2) operates the electric interlock (3), this occurring when the door is within $2\frac{1}{2}$ in. of the door jamb, at which point the door is mechanically locked and the interlock contact closed. The adjustable closer arm is at (4) and the connecting piston yoke at (5). Adjustment for the tension spring (7) is provided at (6), whilst the oil dashpot (8) cushions the door without noise or shock.

POWER OPERATION. In this form the door is both opened and closed by power, other than by hand, gravity, springs or the movement of the car. The power employed is either an electric motor or compressed air, the latter acting on a plunger in a cylinder, the plunger being connected to the door through a system of levers. If a failure of the electricity supply or air pressure occurs, the operating mechanism is so arranged that it may be readily disconnected and the doors operated by hand. High door-operating speeds are employed only if means are adopted to prevent passengers from being struck by doors. In one method of achieving this, photo-electric cells are used. The exciting lamp is fitted on one side of the entrance and illuminates a photo-electric cell fitted on the other side. When a person passes and interrupts the beam of light, the cell operates a control relay and prevents the closing of the door. There is now a tendency to use higher speeds for opening than for closing.

In one form of power operator, OPEN and CLOSE buttons are fitted in the car and at each landing. Momentary pressure of a button will cause the doors to open or close. In this case arrangements must be made to ensure that the landing buttons are only operative when the car is level with that particular landing.

The Express Lift Co.'s door operator is shown in Fig. 11.12 (doors closed) and in Fig. 11.13 (doors open). The motive power is a small squirrel cage motor coupled by a vee-rope drive to a worm reduction gear. This turns an operating lever through a half-revolution, the power being transmitted through an adjustable friction clutch. The operating lever is connected to the doors so that this half revolution causes the doors to complete an opening or closing movement. The motion approximates to simple harmonic, and thus no additional means, mechanical or electrical, are required for accelerating or retarding the motion of the doors. The operating lever assumes a dead-centre position with respect to the connecting rods, both when the doors are fully opened and fully closed. This is necessary to obtain the harmonic motion, but it also constitutes a lock on the doors. On centre-opening doors the levers are attached directly to the door hangers without the interposition of connecting links, thus involving a minimum of levers and bearings. The mechanical connexion between the car and landing doors consists of a retractable driving vane on the door which engages with a fixed driving block on the landing door, this vane being withdrawn by a solenoid

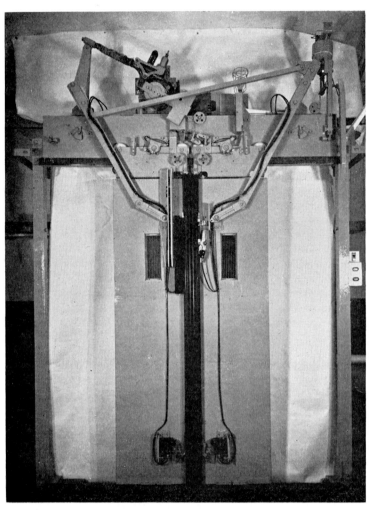

FIG. 11.12. POWER DOOR OPERATOR WITH LIGHT-RAY
DETECTING EQUIPMENT

(*Express Lift Co.*)

FIG. 11.13. POWER DOOR OPERATOR WITH SENSITIVE-EDGE
EQUIPMENT

(*Express Lift Co.*)

215

before the car can move away. The vane also operates the landing-door prelocks.

To enable passengers to open the car and landing doors if the car stops near the landing and the doors do not open, a spring-loaded ball catch is incorporated in the connecting rods between the operating lever and the door. When the doors are closed, it is possible to release the ball catch and telescope the connecting rods by pulling the doors open. The spring load on the ball catch is adjustable and is set so as to transmit a reasonable margin of power above that necessary to operate the doors. In the event of power failure resulting in the de-energizing of the magnet, the retractable vane is ejected under spring pressure so that the landing door opposite the car would be engaged by the vane and unlocked and so be free to move with the car door.

The mechanism will not cause injury to a passenger caught by the closing doors as these are driven through a friction clutch which transmits only sufficient torque to operate the doors, and the drive would slip if the doors met an obstruction.

The closing of the doors is initiated by pressure on any car floor button and their motion can be reversed by using a button marked DOOR OPEN. A limit switch stops the motor automatically when the doors are closed. As the landing doors are closed, the pre-lock locks the doors mechanically and then completes the electric lock circuit which causes the vane to retract and the lift is then able to start.

When the lift enters the landing zone the driving vane solenoid is de-energized and the vane is spring-ejected into the driving block on the landing door. Simultaneously the lock is released. The door motor then starts and the doors commence to open whilst the lift is levelling to the floor. Fig. 11.13 also shows the car apron below the car entrance which prevents a person's foot being trapped between the landing sill and the car floor whilst levelling in the down direction when the doors are opening.

The doors remain open for a short period before re-closing to start the car in response to other calls. This is called "non-interference" time and it must be long enough to permit passengers to leave the car or enter but not cause unnecessary delay to the detriment of the lift service. Two non-interference times are used, a short one when the car stops to unload passengers where landing calls have not been registered and a longer one when passengers are expected to enter the car. Passenger movement across the car threshold is detected by an invisible beam which falls on to a sensitive photo-electric receiver. When the light beam is interrupted as passengers enter or leave the car, the door delay time is curtailed to less than 1 second, i.e. long enough to permit a succession of passengers to cross the threshold without interference but short enough to close the doors immediately behind the last. The photo-electric receiver is shown at the bottom of the door in Fig. 11.12 and incorporates a germanium photo-transistor which drives another transistor, this operating a control relay without the necessity of employing an amplifier. As the photo-transistor is sensitive to light in the infra-red range, an invisible beam can be employed. This equipment supplements the doors safety edge in preventing the doors closing on passengers entering or leaving the car. The retractable safety edge mechanism shown adjacent to the photo-electric equipment in Fig. 11.12 and also at the bottom of the doors in Fig. 11.13, operates to cause the doors to re-open if they close on a passenger. However, the interruption of the invisible light beam also causes closing doors to reverse immediately, thus minimizing the possibility of passengers being struck by the safety edge. Operation by an attendant is similar but in this case higher door-operating speeds are employed.

The Otis high-speed door operator shown in Fig. 11.14 uses a $\frac{1}{2}$-h.p. d.c. motor. The doors are connected to the motor through a gear-reduction unit and suitable linkage. The supply to the motor is taken through a saturable reactor and rectified. Door speed is regulated by a controlled voltage applied to the reactor control winding and throughout the door motion the motor is adjusted to develop only the torque required to obtain the desired door speed. The horizontal thrust is about 30 lb and reversal takes place within the detection zone. A hydraulic cushioning device built into the gear unit brings the doors smoothly to rest, the motor torque being reduced during this stage. The landing doors are operated by a vane attached to the car doors, which is also shown in Fig. 11.14, and door operation begins in the levelling zone. As the car door starts to open the vane is engaged by rollers on the landing door interlock. The vane remains engaged by the rollers throughout the opening operation and until the doors approach the fully closed position, when the vane is freed from the landing-door rollers.

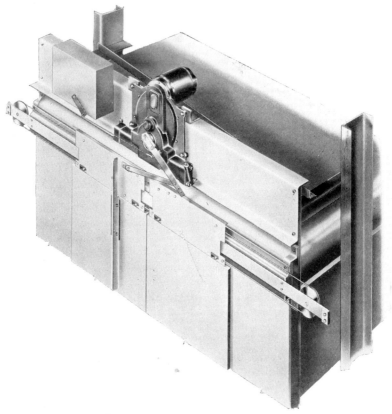

Fig. 11.14. Power Door Operator
(*Otis Elevator Co.*)

This operator also permits the car and landing doors to be opened from the car by hand in the event of a power failure. With attendant-operation the doors open to give a clearance of 3 ft. 6 in. in 1·1 seconds, and closing occupies the same time. A slower closing speed is used without an attendant, the corresponding time being 2·2 seconds.

The doors are equipped with an electronic detector which provides a three-dimensional zone of detection extending along, across and a short distance in front of the leading edges of the car and landing-doors. This detector senses the presence of a passenger between the doors and immediately stops and reverses them. Cold-cathode tubes are used in which firing takes place between the cathode and grid and the capacitance effect is established between the entering passenger and antenna plates fitted on the door edges. Door reversals do not continue indefinitely but are over-ridden after a preset time by a condenser-controlled timing relay. This causes a buzzer to sound and the doors to resume closing at slow speed so that a passenger still between them is "nudged" out of the way.

Car Gate or Door Locks. Every lift car gate or door is fitted with an electric interlock which prevents the movement of the car by cutting off the supply to the control circuit, unless the gate or door is properly closed. It is permissible however, for the car or landing gate or door to be open when the lift is levelling at slow speed within 10 in. above or below landing level. A car gate interlock is shown in Fig. 11.15. This lock is fitted to the top gate track in such a

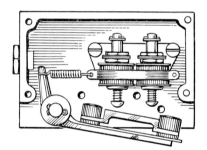

FIG. 11.15 CAR GATE INTERLOCK
(*Marryat & Scott*)

position that the leading picket raises the lock contact arm and short-circuits the two contacts when the gate is in the closed position. Immediately the gate commences to open, the arm drops due to the combined effects of gravity and a spring, and the control feed is cut off. Another contact suitable for a car gate or car door is shown in Fig. 11.16. The roller striker fixed to the door presses home the rocking lever when the door is closed and thus completes the interlocking circuit. Opening the door breaks the circuit positively as the striker bears on the tail of the rocking lever.

Cars having an entrance which opens into a space in any

portion of the lift travel which is in excess of 5 in. should have the gate or door to that side of the car protected by a mechanical locking device with an electrical interlock.

Car Gate Delayed Contact. It is frequently the practice, when a lift is worked on the automatic control system, to arrange that after a passenger has entered the car and closed both car

FIG. 11.16. CAR GATE CONTACT
(*Dewhurst & Partner*)

and landing gates, a period of 5 to 10 sec. is allowed during which the landing buttons are cut out of circuit. During this period the car is under the sole control of the passenger in the car who is thus given time in which to select and operate the desired car button. If a car button is not pressed within the period allowed, the car may be called to any landing in response to pressure on a landing button. This facility is generally provided by incorporating an auxiliary delayed contact in the car gate lock box, this contact disconnecting the landing buttons from the control circuit for the desired period, immediately the gate is closed.

Landing Gate or Door Locks. Several different types of lock are employed for landing gates or doors, the particular lock used depending upon the method of operation of the gate or door. The main objects of all types, however, are to ensure that the gate or door cannot be opened until the car is either level with the landing or within that particular landing zone, and that the car cannot be moved unless all gates are closed. Arrangements are made for the opening of the gate or door, in case of emergency, by means of a special key, irrespective of the position of the car.

The most common form of locking system for landing gates and doors comprises a lock box which is fitted to the gate or door framing. The lock consists of a mechanical lock combined with an electric interlock, the control circuit being completed (except for pressure on a car or landing button or the operation of the car switch) when the electrical interlock is operated. The door is mechanically locked before the interlock makes contact. Hence, before the mechanical lock can be released to open the door, the electrical interlock must be broken, thus disconnecting the control circuit.

An example of a landing door lock is shown in Figs. 11.17 and 11.18. This incorporates a pre-locking feature in which the lock must be proved to be mechanically locked before the lift can start. The lower contact is the electric interlock and the upper contact the counterlocking or prelocking contact. When the beak is moved into the lock to close the door the lower contact makes after the beak is in the locked position and ready to be counterlocked. This lower contact closes certain preparatory circuits, including the retiring cam coil. The cam on the car can now be electrically retired by the operation of the push button or car switch. This causes the roller arm spindle to rotate under the action of a spring and so lowers the counterlocking plate which first performs the mechanical locking and a little later closes the counterlocking contact. This contact completes the circuit for the up and down contactor coil and so starts the lift. As the lift cannot start until this contact has been made this proves that locking has actually been performed. The lock beak is counterlocked in the latched position until the counterlock plate is withdrawn by the release of the cam on the car when the lift comes to floor level. As it nears the stopping

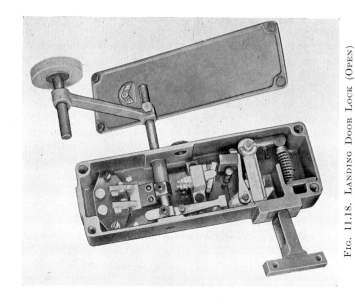

Fig. 11.18. Landing Door Lock (Open)

Etchells, Congdon & Muir

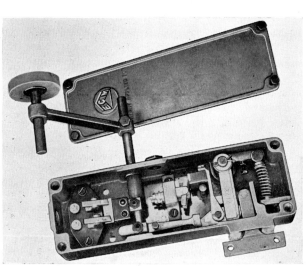

Fig. 11.17. Landing Door Lock (Closed)

Etchells, Congdon & Muir

floor the cam on the car is spring-ejected, and this turns the roller arm, so raising the counterlock plate and breaking the upper switch and thus preventing the lift being restarted. The door handle is now free to unlatch the lock and open the lower contact. When the beak is withdrawn it is impossible for the counterlocking action to take place, and hence the upper switch cannot be made by tampering and so start the lift. Another landing gate or door lock of the "pre-locking" type is shown in Figs. 11.19, 11.20, and 11.21. Fig. 11.19 shows the lock in the unoperated position. In Fig. 11.20 the door beak has entered, and the first or upper contacts are made. In Fig. 11.21 the cam has retired, and both pairs of contacts are made.

Another locking system employed is that in which the lift is operated only from the car, and is provided with solid sliding landing doors which are equipped with a door closer and are locked with a mechanical contrivance actuated by means of levers operated from the car side only. The door is considered closed and the car may be moved away from the landings when the door is within $2\frac{1}{2}$ in. of the jamb, or in the case of centre opening doors, when these are within $2\frac{1}{2}$ in. of each other, provided an approved attachment is fitted which will effectively prevent the doors from being re-opened after they have reached a limit of $2\frac{1}{2}$ in. and provided also that the door closer is of such a type as will eventually carry the door to, and lock it in the closed position.

When an automatic type of control is employed and the landing doors are of the solid sliding type, equipped with a door closer, the doors may be locked mechanically by means of levers.

For service lifts, the mechanical locking unit may be separate from the electrical interlock.

Lock Requirements. All parts of the lock should be of substantial design and construction, and the locks must be so arranged that the electrical interlock is not made until the gate or door is closed. The contacts of the interlock should be of the solid type, and the insulation should be capable of withstanding for 1 min. a pressure of 2 000 volts a.c. The levers operating the mechanical part of the lock should be protected from interference from the landing side of the enclosure. Any spring used in the lock construction should preferably be in compression and the contacts should be opened

Fig. 11.20. Landing Door Lock
(*Wm. Wadsworth & Sons*)

Fig. 11.19. Landing Door Lock
(*Wm. Wadsworth & Sons*)

positively by a lever or other device operated by the door or gate. When the emergency release push is in temporary use, or when the car is being moved under the control of the levelling gear, the lock should not prevent the operation of the lift. A lock key must be provided to enable the lock at any landing to

FIG. 11.21. LANDING DOOR LOCK
Wm. Wadsworth & Sons)

be released to provide an emergency entrance to the well. When the door is open it should be impossible to close the contact by a screwdriver, pencil or anything other than the shaped beak. As the lock is fixed in the well, maintenance is facilitated if the cover is held in position by captive screws or nuts.

In addition to the above, provision should be made on lifts operated from the car and landings to prevent the opening of any landing gate or door when the car is passing that zone in response to a call from another landing. The gate lock is operated by a cam fixed to the car, the cam hitting the gate lock arm when the car reaches a landing. On a number of old lifts the cam strikes the gate lock arm as each landing is reached, irrespective of whether a stop is required. With a lock of this type, it will be appreciated that it is possible for a person at a landing to snatch open the door or gate as the car passes. This bad practice, apart from the possibility of temporarily putting the lift out of service, is liable to cause damage, since the gate contacts are not designed to break current. The above requirement is therefore highly desirable and on modern lifts having a speed exceeding 120 ft. per min. and a travel of more than two floors a device known as a *retiring lock release cam* (Fig. 11.22) is fitted, thus making it impossible for the cam to operate the gate lock, except at the landing where a stop is to be made. The cam is withdrawn beyond reach of the gate lock arm by means of an electromagnet which is energized at all times when the car is outside the particular landing zone where a stop is required. When stopping, the electromagnet is de-energized and this releases the cam and enables it to operate the gate lock. When power operated doors are employed, the retiring cam is frequently operated by the door motor and a separate electromagnet is therefore unnecessary.

Wiring of Locks. The electric cables to the lock contacts should be run in conduit, which should be securely fixed to the lock box and be in electrical and mechanical continuity. The contacts must be wired in the control circuit in such a manner that the car cannot be started or kept in motion between landing zones, unless all landing and car gates are closed and their interlocks consequently made. The landing zone is the space between positions not more than 10 in. above and below the landing level.

The usual method of wiring the lock contacts is to arrange for the supply to the control circuit to pass through the contacts, and this is satisfactory under normal conditions. With three-wire systems of supply, however, it is possible, if an earth fault develops, for the lift to be operated with the gates open, which is a highly dangerous condition. Assume that the locks are joined in the positive or line feed to the control circuit which is fed from the outers of a three-wire d.c. or a.c.

supply, the centre or neutral wire being earthed. An earth fault occurring between the locks and the controller contactors will blow the + ve or line fuse but will leave half the control voltage fed to the coils, and this feed will not be controlled

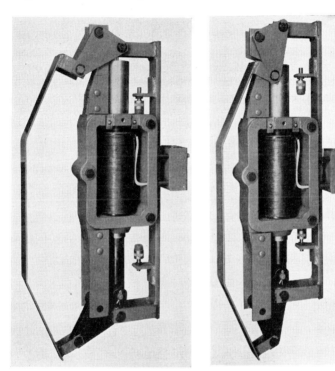

Fig. 11.22. Retiring Lock Release Cam
Left, Normal Position; *Right*, Retracted
(*Dewhurst & Partner*)

by the gate locks. This voltage may be sufficient to operate the contactors, but if d.c. will certainly maintain them if the lift is in operation when the fault occurs. A solution of this difficulty, and one which renders an earth fault harmless, consists in operating the control circuits from the low voltage side of the supply, i.e. between + ve or − ve, and earth or line and neutral. If the low voltage main is not available, fuses of a special type may be used in the control circuit, these fuses being constructed so that the fuse wire is connected in one side of the supply and supports a contact connected in the other side. If one fuse is in each main, the blowing of either fuse will cause the opposite side to be disconnected.

The general practice now is to use a d.c. controller, the supply for which is obtained from rectifiers when the mains are a.c. The gate locks are usually connected in one of the rectifier d.c. feeds to the control circuit. The rectifier may consist of a three-phase network, a single-phase double-wave network from outer to outer, or a single-phase double-wave network from outer to neutral, but in each of these cases an earth fault may cause the contactors to be fed by half-wave impulses not controlled by the locks. Complete safety may be obtained by joining the locks in the a.c. live main before it reaches the rectifier. Alternatively, if the rectifier is fed from the outers of the a.c. supply via a transformer, the locks may be joined in the + ve d.c. main from the rectifier and the − ve rectifier main earthed.

CHAPTER XII

CAR OPERATING AND INDICATING EQUIPMENT

Car-operating Equipment. In attendant-controlled lifts the car is operated by a car-switch or by push-buttons fitted in the car. With car-switch control the car is started by the attendant moving the switch from the OFF position to the UP or DOWN position depending on the required direction of travel. A car

FIG. 12.1 TWO-SPEED CAR SWITCH
(*Dewhurst & Partner*)

switch suitable for operating a two-speed motor, with SLOW and FAST positions for each direction of travel is shown in Fig. 12.1. For stopping, the switch is returned to the OFF position. With high-speed attendant lifts, automatic stopping is provided, in which the attendant merely releases the switch when approaching the desired floor. A switch of this type is shown in Fig. 12.2 in which the panel also incorporates a lighting switch, an emergency stop switch and glass-enclosed emergency

buttons which permit the lift being run with either the landing
or car doors open.

On slow-speed automatic lifts, car control is by a set of simple
push-buttons, one for each landing served, with the addition

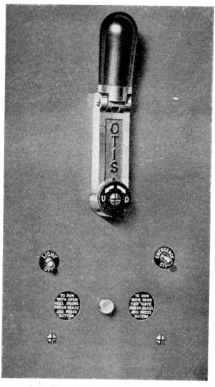

FIG. 12.2. CAR SWITCH
(*Otis Elevator Co.*)

usually of an emergency stop button. On car control stations
of modern lifts, however, each button is arranged to light up
when pressed, thus indicating to passengers in the car, the
floors at which calls have been registered. The panels usually
include additional features such as start, alarm and door-
operating buttons. The Express Lift Co.'s car control station

is shown in Fig. 12.3 in which the face of each button is of transparent plastic, the button being illuminated when pressed. This panel also incorporates a two-way switch for emergency-stop and alarm, door-open, start and pass buttons and also buttons for permitting non-stop travel up or down. The Otis car-operating panel for use on their high-speed automatic lifts is shown in Fig. 12.4 and includes door-operating and alarm buttons and an emergency-stop switch. For attendant operation, a lockable attendant panel is also fitted which provides various other facilities such as start, up or down reversal, motor generator, lighting, fan and non-stop buttons or switches. This car-operating panel is fitted with touch buttons of plastic squares arranged in two columns which are slanted towards the door opening for quick reading. The recessed portion of the button is touched to register a call. The button is a gas-filled cold-cathode triode and is connected through a hair-spring with a plate on the tube envelope. A voltage of 150 V a.c. is applied to the cathode of the tube, while an a.c. component in the anode voltage causes the latter to vary so that it is always 135 V positive to the cathode. This anode/cathode potential is insufficient to strike the tube. On the positive peak of each cycle the anode is 285 V positive to earth. If, therefore, the plate of the envelope is earthed by a person touching the square with a finger, a discharge takes place between anode and plate, which initiates an anode/cathode discharge. This is maintained as in a thyratron, after the earth has been removed and the tube continues conducting until the arrival of the car at the floor trips a switch in the cathode circuit. This system saves a relay and reduces wiring compared with a normal push-button. A car control station for use on the Express Lift Co.'s banks of high-speed lifts is shown in Fig. 12.5. In addition to an illuminated button for each floor served, it includes start, doors and alarm buttons and at the top, three signals, respectively, "THIS LIFT NEXT, STARTING SOON", "USE ANOTHER LIFT", and "CAR OVERLOADED".

Car Indicators. Some form of indicator in the car is necessary with car switch controlled lifts so that the attendant shall be made aware of the floors at which persons are waiting. It is also necessary, with some forms of control, that indications be given of the directions in which the persons wish to travel.

FIG. 12.3. CAR
CONTROL STATION
(*Express Lift Co.*)

FIG. 12.4. CAR CONTROL STATION
WITH TOUCH-BUTTONS
(*Otis Elevator Co.*)

FIG. 12.5. CAR
CONTROL STATION
FOR HIGH-SPEED
LIFT
(*Express Lift Co.*)

Several different forms of car indicators are available, some employing electromagnetically operated targets, and others electric lamps.

In the target indicator, the magnet targets are operated and made visible when the landing call pushes are pressed. The unit comprises one magnet-operated target for each floor served, a controlling relay, and a buzzer. On the operation of a landing call button, a magnet moves the corresponding target into view, thereby registering the signal in the car. After the call has been answered, pressure of the reset button (fitted on the car indicator unit) energizes another magnet and moves the target out of sight. Alternatively, the resetting magnet can be energized by the arrival of the car at the desired floor, in which case the target remains set until the call has definitely been answered. The buzzer gives an audible signal when the call is made and registered. Fig. 12.6 shows an indicator of this type.

Another form of car indicator for use in attendant operated lifts is shown in Fig. 12.7, and comprises a luminous bull's eye and a cancelling button for each floor. Each bull's eye glows when operated by the corresponding landing button, and the calls may be cancelled singly by the attendant as and when answered. A buzzer is provided to draw attention to the fact that calls are being made.

A typical wiring diagram for an indicator similar to that in Fig. 12.7 is shown in Fig. 12.8, the operation of the circuit being as follows. Pressure on, say, the ground floor landing push completes the circuit from +ve line through the ground floor relay, ground floor push, ground floor reset button to −ve line. The operation of the ground floor relay closes its contact and thus provides a holding circuit for the relay via its contact, bottom contact of ground floor push and ground floor reset push, this circuit being independent of the operation of the landing push. A parallel circuit via the relay contact illuminates the ground floor indicator lamp. During the operation of the landing push, a circuit is provided for the buzzer via the ground push and reset button, this circuit being broken when the push is released. After the call has been answered, the reset button is operated and this disconnects the relay circuit, opens its contact and extinguishes the lamp.

Another type of indicator somewhat similar to that above dispenses with reset buttons, the extinguishing of the lamps being performed automatically as the calls are answered. The relay and lamp circuits are disconnected by the opening of contacts on the landing gates, and the calls therefore remain

FIG. 12.6. TARGET INDICATOR
(*Otis Elevator Co.*)

FIG. 12.7. CAR
INDICATOR
(*J. & E. Hall*)

registered until the corresponding landing gates are opened. The circuit is the same as that shown in Fig. 12.8, except that the reset buttons are replaced by the landing gate contacts. As in the case of the target indicator, a double row of lamps may be installed in the car when it is desired that the attendant shall know the direction in which a person wishes to travel.

An intercommunication system between each car and the landings is sometimes used when a bank of attendant-controlled cars is operating in a building, and this ensures that only the nearest car travelling in the desired direction will answer a landing call. With this system, pressure on any up or down landing button illuminates the corresponding white up or down lamps in all cars. The attendant of the nearest car, which has sufficient room and is travelling in the desired direction, then

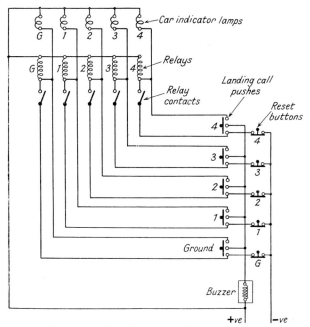

FIG. 12.8. CAR INDICATOR WIRING DIAGRAM

presses a button corresponding to the floor. The operation of this button cancels the calls in all other cars and, in addition, signals to the landing which car will answer the call and illuminates a red lamp in the car, thus telling the operator where to stop. The car and landing signal lamps are automatically extinguished when the landing door is opened. For a signalling system of this type each intermediate landing requires an up and down call button, whilst the top and bottom terminal landings require only a down and up button respectively. Each car requires an up and a down signal lamp at every intermediate landing with single lamps at the terminal landings. On each car indicator, every intermediate floor position is equipped with a red and a white light for each direction of travel, i.e. a total of four lights, whilst the terminal landing positions on the car indicator have a red and a white light for only one direction of travel.

A car position indicator is usually fitted in the car of modern lifts to indicate the car's position to passengers. Fig. 12.9 shows an illuminated indicator of this type arranged for horizontal mounting above the door exit.

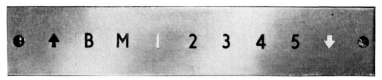

FIG. 12.9. CAR POSITION INDICATOR
(*Marryat & Scott*)

Landing Call and Indicating Equipment. Calls to summon a lift to a landing are made by pressure on a call button fitted adjacent to the landing entrance. With slow-speed lifts serving a small number of floors, one call button at each landing is often adequate. Sometimes this button is associated with an illuminated indication that the car is moving to answer the call as shown in Fig. 12.10. On long-travel high-speed lifts, two buttons are fitted at each intermediate landing, one for each direction of travel, and one button only at each terminal landing. An example of the Otis landing call buttons for terminal and intermediate floors is shown in Fig. 12.11. These have associated illuminated arrows to indicate the desired direction of travel. On their high-speed lifts, however, this firm uses their square-shaped slanted touch-buttons as described on page 231, each of which is marked with an arrow to indicate the direction of travel. An example of the Express Co.'s intermediate landing call pushes is shown in Fig. 12.12. This push illuminates as confirmation that the call has been registered and the illumination is cancelled as soon as the lift answers the call.

FIG. 12.10. CALL
PUSH AND SIGNAL
(*Express Lift Co.*)

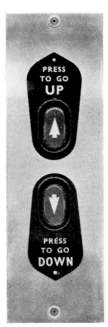

FIG. 12.11. "LANDING" PUSH BUTTONS
(*Otis Elevator Co.*)

FIG. 12.12.
INTERMEDIATE
LANDING CALL
PUSH-BUTTONS
(*Express Lift Co.*)

To avoid impatience on the part of waiting passengers it is desirable to give some indication of the car's motion at the various landings. Indicators are available which show when the car is in motion, the direction of travel, and the actual position of the car in the well.

An illuminated bull's eye installed either adjacent to the call button or above the landing entrance, is sometimes fitted to indicate when the car is in motion, and is useful particularly

Fig. 12.13.
Direction
Indicator
(*Express Lift Co.*

Fig. 12.14. Illuminated Hall
Lanterns with Directional
Arrows
(*Express Lift Co*).

when solid type landing doors are installed, and the well is, therefore, not visible to waiting passengers. A "LIFT COMING" indicator is shown in Fig.12 .10.

DIRECTION INDICATORS. An additional feature, in which the light is arranged to glow either red or white, indicates when the car is travelling in the down or up directions respectively. Alternatively, the direction of travel may be shown by illuminated arrows or by illuminated signs, an example of the former type for mounting at the side of the entrance being shown in Fig. 12.13. After the lift has passed each floor, the corresponding lamp is extinguished until the lift is again approaching.

Illuminated hall lanterns as shown in Fig. 12.14, with

directional arrows for mounting above the lift entrance, are popular on modern lifts. The arrows illuminate to show the direction in which the lift will leave the landing, and with a bank of lifts, which lift will be the next to leave. A single-stroke gong is usually provided which sounds immediately an arrow is illuminated as the car approaches the floor.

POSITION INDICATORS. The tendency on most modern installations is to indicate the position of the car in the well, either with, or without, a separate direction indicator. A landing position indicator with lift direction arrows and call pushes is shown in Fig. 12.15, the indicator being marked with embossed transparent letters, each illuminated singly as the car passes or stops at the various floors. A position indicator mounted horizontally and incorporating illuminated direction arrows, is sometimes mounted above the landing entrances, as shown in Fig. 12.16.

The wiring diagram for a landing indicator system is shown in Fig. 12.17. Switches $U1$–4, DG and $D1$–3 are switches of a floor selector machine installed in the main machine room. The selector machine is similar in construction and operation to the floor selector described in Chapter XIV. The selector shaft together with its operating arms is fixed to a pulley which is driven from the car by a steel flyrope, and each switch is arranged to close and open when the car approaches and recedes respectively from the corresponding landing. Switches $U1$–4 close during the upward motion of the car, whilst DG and $D1$–3 close during the downward journey, the switches opening and closing when the car is at half the distance between floors from the respective floors. As the car approaches the first floor in the downward direction, switch $D1$ closes and the

FIG. 12.15.
COMBINED
LANDING CALL
AND POSITION
INDICATOR
(*Express Lift Co.*)

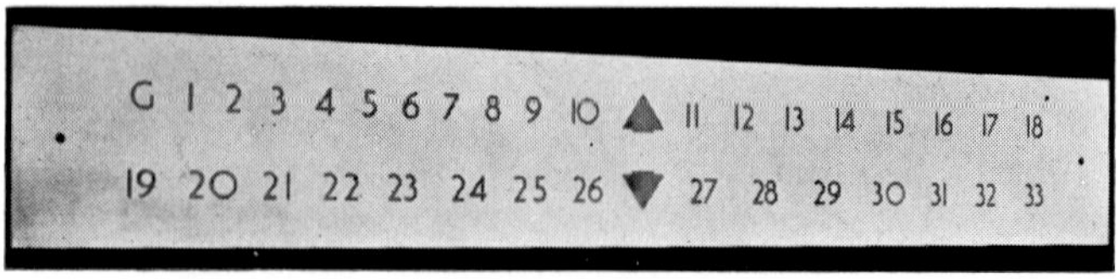

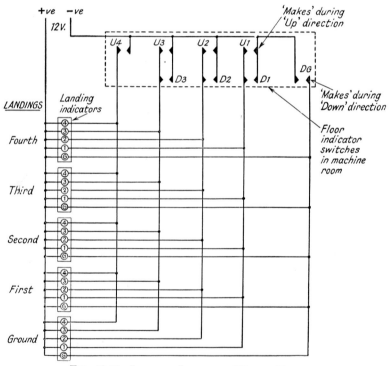

Fig. 12.16. Double-row Landing Position and Direction
Indicator
(*Express Lift Co.*)

circuits for the first floor lamps on each landing are completed via *D*1, this switch opening after the car has left the
landing. Similarly, when the car is approaching any landing,
the corresponding landing lamps are illuminated, the circuits

Fig. 12.17. Landing Indicator Wiring Diagram

having been completed through the *U* or *D* floor switches. If additional lamps showing the direction of travel are required on the indicators, then it is necessary to fit direction switches which may be operated by the main reversing contactors, one or other of these switches closing as the lift changes its direction of travel. The UP lamps are connected to one switch and the DOWN lamps to the other.

With a bank of high-speed gearless lifts operated on one of the modern group-control systems such as the Otis Auto-tronic, the Express Traffic E.T.6 or Mark IV, a combined starter's station and supervisory indicator is fitted in the ground-floor entrance lobby. A supervisory panel for an Express Traffic Control Mark IV installation for a group of four lifts (Nos. 5, 6, 7 and 8) is shown in Fig. 12.18. The four centre vertical rows of indicators are the position and direction indicators for each of the four cars. The left outer row shows the DOWN landing calls registered and the right outer row, the UP landing calls registered. Other signals give possible fault conditions. Below the indicators are the generator switches and car cut-out switches which enable any car to be taken out of group service for special duty.

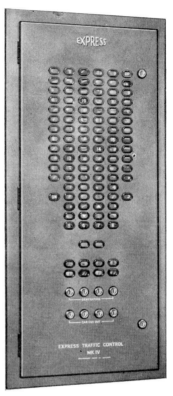

FIG. 12.18. STARTING AND SUPERVISORY PANEL FOR BANK OF FOUR HIGH-SPEED GEARLESS LIFTS
(*Express Lift Co.*)

CHAPTER XIII

SAFETY FEATURES

Car and Counterweight Safety Gears. Safety gear is invariably fitted to lift cars, with the object of preventing any uncontrolled movement of the car in a downward direction, by clamping the car to its guides should the lifting ropes break or the speed of the car, in descent, exceed a certain predetermined value. The gear is sometimes arranged to operate and clamp the car to its guides if the lifting ropes stretch unequally, as an additional safety to those mentioned above. Car safety gear is always fitted under the car framing, as cases have been recorded where the car has broken away from the safety gear when the latter has been fitted to the top of the car. Counterweights are sometimes provided with safety equipment in addition to that installed under the car, in order to stop an overspeeding ascending car and this is invariably the practice when the lift does not extend to the basement, and occupied spaces are under the well. Every lift which has a travel exceeding 35 ft. should have its safety gear operated by an overspeed governor and when any safety gear is applied, the motor and brake circuits should be opened before or at the same time. All safety gears should be applied positively and none should depend upon an electric circuit for its operation. Ropes for operating safety gears should be supported by their own pulleys and these ropes should not be less than $\frac{1}{4}$ in. in diameter. Safety gears must operate on both guides simultaneously and if a rope release carrier is used, it should be carried on the car frame and not on the enclosure. After the tripping of the governor jaws the car or counterweight should not travel more than 2 ft. 6 in. before the application of the safety gear jaws. It is essential for this time interval to be small, otherwise if the ropes break a dangerous speed may be attained before the safety gear operates. Three main types of safety gear for car and counterweights are in use, and the type adopted for any particular installation depends chiefly upon the maximum car speed employed.

Instantaneous Type. This gear, which is almost instantaneous in operation, is limited to speeds not exceeding 160 ft. per min.

when fitted to car frames and 250 ft. per min. on counter-weights as, on account of the small stopping distance, the car and guides are subjected to heavy strains, whilst at speeds much in excess of 160 ft. per min. passengers would experience severe shock if the gear came into operation. No delayed action of the jaws is purposely introduced when this gear operates.

For light duty, the gear sometimes consists of two serrated steel cams fitted one on each side of the car and keyed to a connecting camshaft. The cams are eccentric in shape and normally travel just clear of the guides, being held away from them by a light return spring. When the connecting shaft is slightly rotated, the leading edges of the cams are brought into contact with the guides, the remaining motion of the cams necessary to clamp the car to the guides being automatically provided by the falling car. Either the camshaft, or one of the cams, is connected to the counterweight, or to the counterweight safety gear, if fitted, by means of a flexible steel flyrope which passes over an idle pulley at the top of the well. The safety flyrope is normally slack, but if the lifting ropes break or stretch unduly, the resultant tension placed upon the safety rope causes the cams to be drawn into contact with the guides and thus clamp the car. The safety rope passes from the car safety gear over a top idler, and is then secured to one of the counterweight cams. A direct pull on the safety rope, such as would be caused by a breakage or slackening of the lifting ropes, would cause both cams to operate. Another single cam type gear fitted on a counterweight travelling in tee-section guides is shown in Fig. 13.1. In this case, one cam is keyed to each end of a connecting camshaft which passes through the counter-weight bottom framing. The safety rope passes from the car safety gear over a top pulley, and is then fastened to an opera-ting arm (not shown) keyed to the centre of the counterweight camshaft. Tension on the safety rope causes a pull on the arm, which in turn rotates the shaft and brings each cam into contact with its guide, the latter being clamped between the cam and a buffer plate on the counterweight.

A better type of cam safety gear makes use of two cams on each side of the car, the cams operating on opposite sides of the same guide as in Fig. 13.2, or on the guide backing, and thus equal pressure is applied in opposite directions to each guide and there is no tendency for the guide to be forced out of the vertical plane. In this type, two connecting camshafts are used, each having a cam keyed at either end. The two shafts

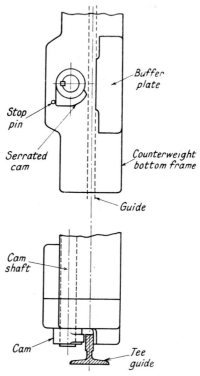

Fig. 13.1. Instantaneous Cam Safety Gear, Fitted to Counterweight

are linked together and rotate in opposite directions when the safety rope is pulled. The rope may be fastened to one of the cams, as in Fig. 13.2, or to one of the camshafts, but in either case it is arranged that tension on the rope will cause all four cams to operate simultaneously. The principle of operation of another system of double cams is shown in Fig. 13.3, although the method of operation in this case is somewhat different from those already described. The gear shown will be brought into operation if either or all of the lifting ropes break or if one of the ropes stretches unduly owing to weakness or to a faulty rope fastening. The sketch illustrates the gear fitted to the top of a counterweight, although it may equally well be employed under the car. Two lifting ropes are attached to a double lever *F* which is free to rotate about its supporting shaft *A*.

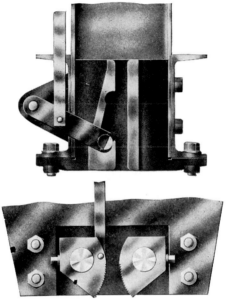

FIG. 13.2   INSTANTANEOUS SAFETY GEARS
(*Express Lift Co.*)

The safety rope passes from the car safety gear to the counterweight, being attached to the latter at the end of the rope lever *F*. Two operating levers are keyed to each camshaft, the lever arms normally being clear of the rope equalizing lever, which, if the ropes are of equal length, should be in the horizontal position. If both lifting ropes break or stretch unduly, a tension is placed upon the safety rope which causes the equalizing lever to rotate in a clockwise direction, and push upwards the arms of the operating levers *B* and *C*, and these in turn rotate both camshafts and cause the four cams to clamp the guides. A similar result would follow the breaking of rope Q alone, whilst if rope P broke the cams would be operated by the upward movement of levers D and E. If either rope stretches unduly, the extra tension placed upon the other rope would cause the equalizing lever to rotate, and again the cams would be operated.

Another type of instantaneous safety gear sometimes used

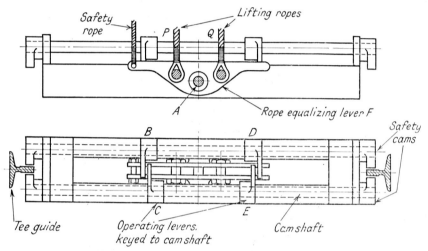

FIG. 13.3. INSTANTANEOUS CAM SAFETY GEAR

on large lifts consists of a serrated roller operating in a tapered steel block as shown in the upper illustration of Fig. 13.2.

The methods of operation of the cams which have been described all depend upon the lifting ropes breaking or slackening. The cams, however, may be rotated by connecting the safety rope to an overspeed governor which brings the cams into contact with the guides when a predetermined excess speed in the down direction is reached. At the same time a switch in the safety circuit is opened. By this method the operation of the cams is thus independent of any possible failure of the lifting machine. The action of the governor is described later in this chapter.

When set, the cams are released from the guides by raising the car, this being done by rotating the motor by hand with the handle usually provided for this purpose.

A repeat-action safety gear is shown in Fig. 13.4 in which, owing to the design of the double toggle link spring assembly, the cams once operated, are kept in contact with the guides independently of any possible failure of the safety rope. The

FIG. 13.4. REPEAT-ACTION SAFETY GEAR
(*Wm. Wadsworth & Sons*)

safety gear will then repeatedly act to sustain the car no matter how many times it may attempt to fall. The cams cannot be rotated out of active position by raising the car until the manual resetting procedure is carried out.

Gradual Wedge-clamp Safety Gear. On most modern lifts running at speeds in excess of 200 ft. per min., the gradual wedge-clamp type safety gear is employed. The clamps exert a gradually increasing retarding force upon the guides, and the car is thus brought to rest smoothly, without fear of damage to the guides or shock to passengers. The gear may be operated on the broken rope principle as described above, but the more usual method is by means of a speed governor. A view of safety gear of this type is shown in Fig. 13.5. The governor rope passes round the governor sheave to a special carrier fitted to the car frame, thence round a weighted pulley in the bottom of the well, and finally back to the governor sheave, as shown in Fig. 13.6. Instead of employing a weighted pulley to ensure a definite drive on the governor sheave, it is sometimes the practice to dispense with this pulley and insert a screw adjustment to give

Fig. 13.5. Gradual Wedge-clamp Safety Gear
(*Etchells, Congdon & Muir*)

the necessary rope tension. The governor rope is, therefore, driven by the car rope carrier at a speed equal to that of the car, and the speed of rotation of the governor balls or weights, which are driven from the governor sheave, either direct or through gearing, is proportional to that of the car. The safety rope is attached to the governor rope and passes round a drum mounted under the car frame. The diameter of this safety gear drum should not be less than 5 in., and after the safety clamps have operated there should be at least three complete turns of the rope left on the drum. Under normal conditions the carrier grips the governor rope, and when the car speed reaches, say, 10 per cent above the maximum speed, the governor balls have risen sufficiently to operate a switch which cuts off power from the lift and applies the brake. If, however, the speed continues to rise to, say, 20 per cent above the maximum value, the governor balls will rise still higher, and this causes a pair of governor jaws to grip the governor rope. The rope is then jerked free of its carrier and the resulting tension on the safety rope causes the safety drum to revolve. This operates right- and left-handed screws connected to the drum and results in two powerful smooth jaws gripping each car guide and gradually bringing the car to rest.

Another type of wedge-clamp safety gear is illustrated in Fig. 13.7 in which the jaws are split and the clamping pressure on the guide is controlled by a spiral spring which is preset to give the requisite pressure on the guides according to the mass to be de-celerated. Since the load on the guides can be controlled, the possi-bility of the jaws seizing on the guides or the car being distorted is eliminated. With this type of gear the retarding force is practically constant as also is the rate of deceleration.

Governor Rope Carriers. Several different forms of rope carriers are in use, but the principle is the same in each case. They all perform the functions of attaching the governor rope to the car frame under normal running conditions, and of detaching the rope from the car when the governor jaws grip the rope.

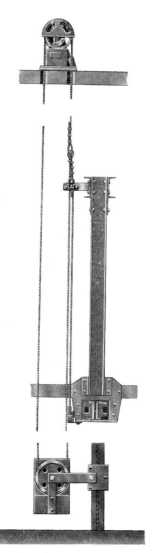

FIG. 13.6
GOVERNOR AND SAFETY ROPES
(*Express Lift Co.*)

One type is shown in Fig. 13.8 in which a stop, secured to the rope, is held between two clamps A and B, these being fixed to the car framework. The top clamp is free to turn

FIG. 13.7. WEDGE-CLAMP SAFETY GEAR
(*Express Lift Co.*)

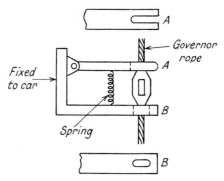

FIG. 13.8. GOVERNOR ROPE CARRIER

upwards and is normally held against the rope stop by the tension of a spring. When the governor jaws set, the pull on the rope rotates the top clamp against the spring tension and the rope stop is jerked free of the carrier.

Another safety rope release, known as a torpedo rope release, is shown in Fig. 13.9. A $\frac{1}{2}$ in. dia. steel pin passes lengthwise across a recess in one face of the "torpedo," whilst pressure is brought to bear on the opposite face by an adjustable spring in a steel cup. The spring pressure is adjustable from zero to a maximum when a pull of 500 lb. is required on the rope in order to release.

FIG. 13.9. TORPEDO ROPE RELEASE
(Dewhurst & Partner)

Governors. It will be gathered from the foregoing that the object of a governor is to cause the safety gear to operate, if from any cause the speed of the car exceeds its normal maximum speed by more than a certain predetermined value. In addition to operating the safety gear, the governor is generally arranged to cut off the control circuit before or at the same time that the safety gear sets. It is recommended that a governor be fitted to all power-driven lifts which have travels in excess of 35 ft. With the greater use of high-efficiency gearing a governor has become a necessity, since a modern lift will accelerate under conditions which the older self-sustaining gears would not permit. Car speed governors should be set so as to cause the application of the safety gear at a speed not less than 15 per cent above the contract speed and at not more than the tripping speeds in the table below. Intermediate values are shown in Fig. 13.10.

GOVERNOR MAXIMUM TRIPPING SPEEDS

Contract Speed in ft. per min.	Maximum Governor Tripping Speed in ft. per min.
0–125	175
150	210
200	280
250	337
300	395
350	452
400	510
500	625
600	740
700	855
800	970

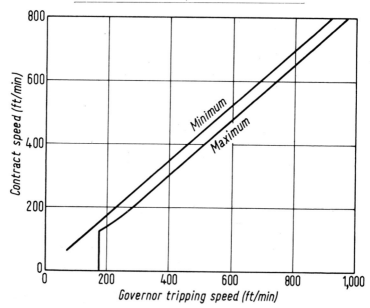

FIG. 13.10. RELATION BETWEEN CONTRACT SPEED AND GOVERNOR TRIPPING SPEED

No governor should be required to trip at a speed less than 140 ft. per min., and with the instantaneous type gear, should trip at a speed not above 200 ft. per min. If counterweight safety gear is provided, this may be operated by the same

Fig. 13.11. Vertical Shaft Governor
(*By courtesy of the Postmaster-General*)

governor and governor rope used to operate the car safety gear. Provision should be made to cause the counterweight gear to be applied at a speed in excess (not more than 10 per cent) of that at which the car safety gear applies. The governor ropes which should be not less than $\frac{5}{16}$ in. in diameter should run free of the governor jaws during normal operation of the lift, and it is recommended that the governor gears have self-lubricating bearings which do not require frequent attention. The machine room is probably the best, and is certainly the most usual, location for the governor. Several different types of overspeed governors are available for lifts, but they are

Fig. 13.12. Automatic Speed Governor and Switch
(*J. & E. Hall*)

almost all of the centrifugal pattern, having either a vertical or horizontal governor shaft.

A geared vertical shaft governor is shown in Fig. 13.11. As the car speed increases, the balls open and raise the operating plate against the spring tension and thus causes the jaw to grip the rope.

Fig. 13.12 illustrates a governor of the horizontal shaft pattern with the weights secured to the governor sheave and revolving in a vertical plane. As the speed rises, the rotating diameter of the weights increases due to centrifugal force operating against the pull of two equalizing springs. When a speed of, say, 10 per cent, in excess of the maximum speed is attained, the weights open the control switch shown at the right of the illustration, and thus cut off the control circuit. If the speed continues to rise to, say, 20 per cent above the maximum speed, the weights open still further, and release the pawl which normally holds the governor jaws clear of the governor rope. The tension of the two springs at the left-hand side of the illustration then pulls the jaws into contact with the rope and causes the safety gear to operate.

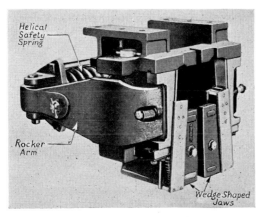

FIG. 13.13. FLEXIBLE GUIDE CLAMP SAFETY GEAR
(*Otis Elevator Co.*)

Flexible Guide Clamp Safety Gear. This is a further type of safety gear sometimes employed on high-speed lifts and, when operated, applies a constant retarding force to the car's motion. The gear is illustrated in Fig. 13.13 and consists of two clamps, one for each guide, bolted to the bottom member of the car frame, the wedge-shaped jaws of each clamp being connected by a system of rods and linkages to the governor rope. Each clamp has two of these wedge-shaped steel jaws to grip the guide and a heavy flexible spring to regulate the pressure exerted by the jaws. When the car's downward speed exceeds a predetermined value, both clamps operate simultaneously by direct pulls on the jaws, which thus grip the guides to bring the car to a swift, sliding stop. When the cause of the overspeed has been corrected the safety gear may be released by running

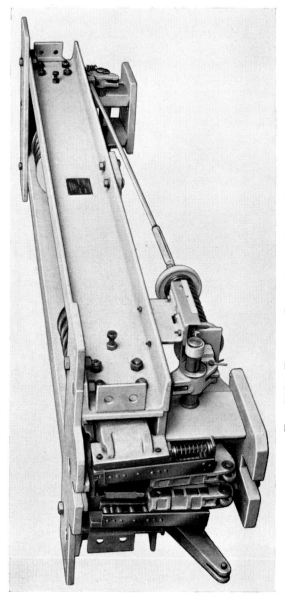

Fig. 13.14. Flexible Guide Clamp Safety Gear

(*Express Lift Co.*)

the car up a few inches. This reverses the wedging action, and the safety jaws slide back in their ready position without the use of a winding wrench. This gear requires a pull on the governor rope of less than 3 in., and the maximum speed which the car can attain before the jaws take hold is thereby reduced —an important feature in the operation of safety gear for high-speed lifts. After the safety jaws touch the guide, the final stopping of the car is independent of the governor and governor rope. This safety gear cannot be released from the car, unless the hoisting machine is in a condition to lift the car and thus the danger of a premature release is minimized. The use of this gear ensures that the car floor is solid and free of wrench holes and cover plate.

A view of the Express Flexible Guide Clamp Safety gear is shown in Fig. 13.14. When the governor trips, the governor rope is clamped and this operates the lever shown on the left which, through the linkage underneath, causes the notched cast-iron gibs to override the safety gear jaws which are lined with chromium-plated and hardened rollers. The equalizing spring is precompressed and on the setting of the safety gear, it is further compressed by the wedging action of the gibs between the jaw rollers and the web of the guide. The gear can be adjusted to suit the load to be arrested and can be used on dry or lubricated guides.

Safety Gear Stopping Distances. The stopping distance with the instantaneous type safety gear is very small but the stopping distances of the wedge clamp types and the flexible guide clamp type are much greater. The maximum and minimum distances for the latter types are as stated in the table on page 258, the stopping distance being defined as the actual slide, as measured by the marks on the guides.

Car and Counterweight Clearances. Clearances must be allowed for the car at the top and bottom of the well so as to permit it to slow down if, for any reason, a terminal floor is overrun and the final terminal switch is operated. Suitable top and bottom clearances must also be provided for the counterweight.

The top car clearance is the distance between the uppermost part of the car or any rigid attachment thereto and the nearest fouling point of the overhead structure, when the car is level with the top landing.

The bottom car clearance or overtravel is the distance, including the buffer compression, between the lowermost part of the car or any rigid attachment and the top of the car buffers, when the car is level with the bottom landing.

The top counterweight clearance is the shortest vertical distance between any part of the counterweight assembly and the nearest part of the overhead structure when the car is level with the bottom landing.

The bottom counterweight clearance or overtravel is the shortest vertical distance including any buffer compression between any part of the counterweight assembly and any fixed obstruction below it when the car is level with the top landing.

SAFETY GEAR STOPPING DISTANCES

Governor Tripping Speed in ft. per min.	Stopping Distance	
	Minimum	Maximum
	ft. in.	ft. in.
175	0 6	1 3
200	0 6	1 4
250	0 8	1 7
300	0 10	2 0
350	0 11	2 4
400	1 1	2 9
450	1 3	3 4
500	1 6	4 0
550	1 8	4 7
600	1 11	5 3
650	2 2	6 0
700	2 6	6 9
750	2 9	7 7
800	3 2	8 9
850	3 7	9 10
900	3 11	10 11
950	4 3	12 1
1000	4 8	13 2

The clearances allowed, in practice, should be as stated opposite.

(*a*) *Clearances for car and counterweight when spring buffers are employed.*

Contract Speed Ft. per. min.	Minimum Bottom Clearance for Car and Counterweight	Minimum Top Clearance for Car and Counterweight
	ft. in.	ft. in.
0–100	1 1	1 6
101–200	1 4	2 0
201–300	1 8	2 6

(*b*) *Clearances for car and counterweight when oil buffers are employed.*

To ensure satisfactory bottom clearances the depth of the pit should be at least the sum of the following—

(1) The overall length of the fully extended car buffer used.

(2) The distance between the upper surface of the car floor and the lower surface of the buffer striking plate.

(3) Six inches.

Where the bottom car clearance is greater than the minimum specified above then the top counterweight clearance must be increased by a similar amount.

The top car clearance should be at least the sum of the following—

(1) The distance between the counterweight buffer and its buffer block, which should be at least 6 in.

(2) The stroke of the counterweight buffer.

(3) Twelve inches.

(4) The counterweight buffer stroke corresponding to the governor tripping speed less one half the stroke of the counterweight buffer used.

Item (4) may be omitted if provision is made to eliminate the jump of the car at counterweight buffer engagement.

The top counterweight clearance should be at least the sum of the following—

(1) The distance between the car buffer and the buffer striking plate which should be at least 6 in.

(2) The stroke of the car buffer used.

(3) Six inches.

(4) The car buffer stroke corresponding to the governor tripping speed less one half the stroke of the car buffer used.

Item (4) may be omitted if provision is made to eliminate the jump of the counterweight at car buffer engagement.

The top overtravel is defined as the distance provided for the car floor to travel above the level of the top terminal landing when the car is stopped by the normal terminal stopping switch.

The bottom overtravel is the distance the car floor can travel below the level of the bottom landing when the car is stopped by the normal terminal stopping switch.

Normal Terminal Stopping Switches. Every electric lift should be provided with an upper and a lower normal terminal stopping switch, these being arranged to stop the car automatically within the top and bottom overtravels, from the contract speed. These switches act independently of the normal operating device, the ultimate or final terminal stopping switches, and the buffers. They may be fitted to the car and be operated by a ramp in the well, or in the well and operated by a car ramp, the former usually for high speeds and the latter for low speeds. Alternatively, the switches may be fitted in the motor room, but in all three cases they must be operated by the movement of the car. In some switches the contacts are of the knife pattern and, in others, of the plate contactor type.

A normal terminal stopping switch of the knife pattern for cutting off the control circuit and suitable for well or car mounting is shown in Fig. 13.15 and one of the contactor type in Fig. 13.16.

Fig. 13.17 depicts a terminal stopping switch arranged for mounting in the machine room.

On high-speed lifts it is usual to install slowing limit switches which automatically reduce the speed of the car before the terminals are reached. The slowing down is accomplished by means of a multiple slowing switch, the contacts of which are opened by ramps fitted in the well. An example of a switch of this type is shown in Fig. 13.18.

Final Terminal Stopping Switches. These are arranged to stop the car automatically within the top and bottom clearances, independently of the normal operating device and the normal terminal stopping switches, but with the buffers operative. A normal terminal stopping switch and a final terminal stopping switch of the type shown in Fig. 13.16, both mounted

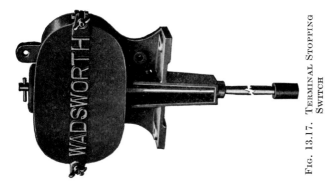

Fig. 13.17. Terminal Stopping
Switch
(Wm. Wadsworth & Sons)

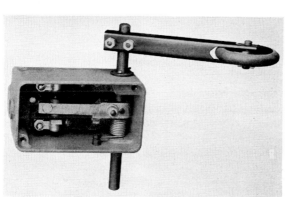

Fig. 13.16.
Terminal Stopping Switch
(Dewhurst & Partner)

Fig. 13.15. Terminal Stopping
Switch
Cover removed
(J. & E. Hall)

FIG. 13.19. TERMINAL AND FINAL
LIMIT SWITCHES
(*Marryat & Scott*)

FIG. 13.18. MULTIPLE SLOWING SWITCH
With cover removed
(*J. & E. Hall*)

in the well, are shown in Fig. 13.19. The switches should be so arranged that the opening of the switch takes place with the car as close to the terminal floor as practicable without interfering with the normal operation of the lift. With spring buffers, however, the switch should open before the buffers are engaged. After operating, these switches remain open until the car has been moved by hand winding to a position within the limits of normal travel. These switches are operated by the movement of the car in the well and the general practice is for them to open simultaneously both the control and motor circuits.

With variable voltage control the operation of either final limit switch should cause the lift motor to exert its full dynamic braking effort with the motor field winding energized and the motor generator running.

One method of operation is shown in Fig. 13.20, this being used for speeds up to 400 ft. per min. A steel flyrope, passing round the switch operating drum, is anchored in the machine room at one end and has a balance weight attached to the other end. Rope stops

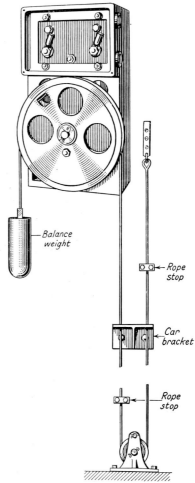

Fig. 13.20. Final Terminal Switch
(*Wm. Wadsworth & Sons*)

are secured to this rope at the top and bottom limits of travel, and a bracket, fixed to the car, slides freely over the rope. When the car bracket comes into contact with either rope stop, the drum is rotated and the switch opened. This type should have an automatic safety switch which will stop the lift if the flyrope breaks. A final stopping switch is shown in Fig. 13.21. This is for a 3-phase supply with rectified control circuit.

Although a rope-driven mechanically-operated final stopping

FIG. 13.21. FINAL STOPPING SWITCH
Cover removed
(*Etchells, Congdon & Muir*)

switch is satisfactory at low speeds, it is not suitable for high-speed variable-voltage gearless drives for which an electrically-operated device is required. With high speeds it is necessary to provide protection at a point some distance in advance of the terminal floors and also to keep the machine energized in order to reach the floor safely at levelling speed. For high speeds the final stopping switch is fitted in the well and is operated directly by the gear in the event of overtravel. The operation of this switch trips both the reversing contactors and the line contactor and so cuts off current to the motor and applies the brake. Hence there must be simultaneous failure of two contactors to create an unsafe condition. To eliminate the possibility of welding of contacts they are usually carbon to copper. Some manufacturers fit this direct-action final switch on slow-speed lifts (instead of a rope-driven limit switch) as well as on high-speed installations.

BUFFERS

In the event of the final terminal stopping switch operating, and the car not coming to rest within the clearance provided, either the car or counterweight will make contact with buffers which lessen the impact. Two buffers are fitted in the well under the car, these being placed symmetrically with respect to the centre of gravity of the car, and, in addition, one or two buffers are fitted under the counterweight. These buffers therefore constitute the final emergency device.

Spring Buffers. For speeds up to 300 ft. per min., preferably not exceeding 200 ft. per min., buffers may consist of helical springs of the type shown in Fig. 13.22, or volute springs, an

Fig. 13.22. Spring Buffers Mounted on Fixing Plate
Note registration piece for tee guides
(*Express Lift Co.*)

example of which is reproduced in Fig. 13.23. The rubber buffers shown are used with lifts of contract speed not exceeding 75 ft. per min. or any service lift. Their size must be such that they are capable of stopping the loaded car from contract speed without permanent distortion, and the maximum rate of retardation with 150 lb. in the car should not exceed 80·5 ft. per sec. per sec. All spring buffers have an increasing rate of deceleration as the spring is compressed.

Oil Buffers. These are fitted for car speeds above 300 ft. per min., the piston stroke depending upon the maximum car speed. The minimum total stroke is calculated upon an average retardation of 32·2 ft. per sec. per sec., at 115 per cent contract speed. The maximum retardation, based on 115 per cent contract speed, should not exceed 80·5 ft. per sec. per sec., i.e. $2\frac{1}{2}$ times the retardation due to gravity. These buffers are designed so that the rate of deceleration is as nearly as possible constant and are self-resetting.

FIG. 13.23. TYPES OF BUFFER
(*Marryat & Scott*)

SPRING RETURN OIL BUFFER. The oil buffer usually employed for lift cars is of the spring return type. The following is a brief description of Messrs. Otis's spring-return oil buffer, a drawing of which is shown in Fig. 13.24 and an illustration in Fig. 13.25. When the car strikes the buffer, a spring in the upper part of the piston rod is compressed, and acts as a cushion for the rod and piston, thus preventing them from receiving too sudden a blow. A rubber cushion, at the top of the buffer, also further deadens the blow. As the piston starts its downward movement, the oil beneath it is forced from the cylinder into the outer casing, and also into the upper chamber,

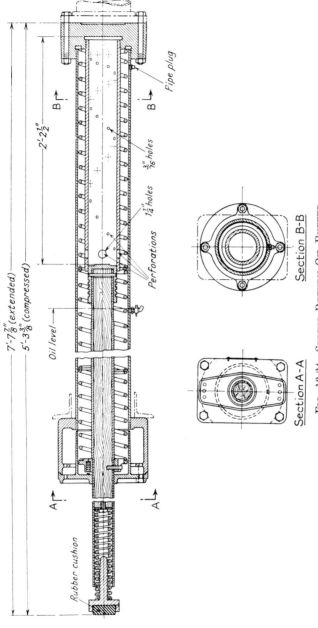

7'-7⅞" (extended)
5'-3⅜" (compressed)

2'-2½"

Pipe plug

B

B

¾" holes

1¼" holes

Perforations

Oil level

Rubber cushion

A

A

Section A-A

Section B-B

Fig. 13.24. Spring Return Oil Buffer
28 in. stroke
Otis Elevator Co.

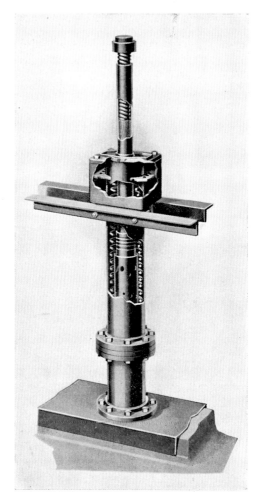

FIG. 13.25. OIL BUFFER
Spring return
(*Otis Elevator Co.*)

through a series of holes in the cylinder. The area of the upper holes is large, and therefore little resistance is offered to the flow of oil during the initial piston movement. Hence, the piston accelerates fairly rapidly and takes up the running speed of the car. Any back pressure due to the inertia of the piston and rod in starting is absorbed by the spring. As the piston descends, the number and size of the holes below the piston become less and the restriction to the oil flow increases. When the piston has closed the last hole the car comes to rest.

As soon as the car is moved away, the piston is forced upwards by the large spring, and the moving parts are restored to their normal position, whilst at the same time the oil flows back into the cylinder.

Fig. 13.26 shows the Express Lift Co.'s oil buffer for speeds above 350 ft/min. It has a fabricated-steel cylinder housing the mild-steel plunger which is forced downwards into the oil chamber by the descending car or counter-weight. The oil escapes through an annular orifice into the plunger, the rate of flow being controlled by a tapered steel pin which automatically reduces the size of the orifice as the plunger is forced downwards, resulting in a constant retarding force. The spring holds the plunger in the operating position and also returns it to its original position after operation. The centre enlarged portion is the oil foaming chamber and the top is capped with a tough rubber buffer.

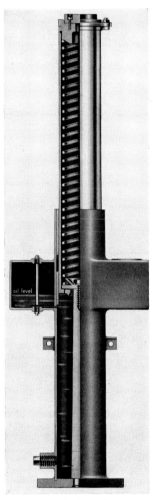

Fig. 13.26. Oil Buffer
(*Express Lift Co.*)

Gravity Return Oil Buffer. The following description relates to Messrs. Otis's oil buffer, shown in Fig. 13.27. This type is generally used as a

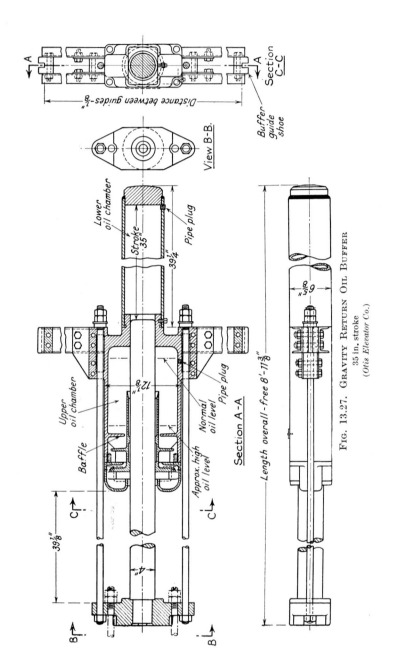

FIG. 13.27. GRAVITY RETURN OIL BUFFER

35 in. stroke

(Otis Elevator Co.)

Section C-C

Distance between guides—1⅜"

Buffer guide shoe

View B-B.

Lower oil chamber

Stroke 35"

Pipe plug

39¼"

Upper oil chamber

12⅛"

Baffle

Normal oil level

Pipe plug

Approx. high oil level

Section A-A

Length overall - free 8'-11⅜"

39⅛"

4"

6⅝"

counterweight buffer, as it can be attached to the bottom of the counterweight and so form a useful part of the counterweight itself. The cylinder consists of upper and lower oil chambers, and when the buffer is stopped by striking a block at the bottom of the well, the oil contained in the lower chamber is forced into the upper one, through an area formed by lnngitudinal slots in the plunger and its guiding sleeve. These slots are of varying lengths and so arranged as to provide a reasonably constant retardation of the moving counterweight. By the time the longest slot is covered in the sleeve, the energy of the counterweight has been entirely absorbed. Any further downward movement of the plunger is prevented since the oil has no additional outlet. The cover of the upper oil chamber is provided with baffles to prevent the oil from being sprayed out when the buffer is in operation.

Guarding. Lift well enclosures should be provided and should extend on all sides from floor to floor or stair to stair. If the enclosure is of the open type any mesh used should not be greater than $1\frac{1}{4}$ in., but if the clearance between the enclosure and any moving part of the lift is less than 2 in. the mesh should be not greater than $\frac{1}{2}$ in. square and of wire not smaller than 20 s.w.g. Counterweight guards of wire mesh should be provided, except where the use of compensating cables makes this impracticable. One of these guards is fitted at the position where the car and counterweight pass each other and another in the pit, extending to a height of about 7 feet above the pit floor. They eliminate the possibility of injury to maintenance engineers when travelling on the car or working in the pit respectively.

Gates. Many accidents have been caused in the past by persons' feet or arms being crushed when protruding through the space between gate pickets, the usual spacing of which, with ordinary type gates, is about 5 in. Consequently, gates for landings and the car are now invariably of the close picket type usually known as mid-bar gates, in which the distance between adjacent pickets is not more than $2\frac{1}{2}$ in.

A practice which is sometimes adopted on older lifts to obviate the need to replace completely wide-spaced picket gates, is to fit short mid-bars about 12 in. long to the bottom of the gates to prevent toes from being trapped, and also a safety plate near the gate handle so that it will be impossible for an arm to reach through the gate to the car push button plate.

Gate or Door Locks. The locks fitted to the landing and to the car entrances comprise one of the most important safety items in a lift installation.

The landing gate or door locks should be such that the gate or door cannot be opened from the landing unless the car is at that particular landing and also that the car cannot be moved unless all landing gates or doors are closed and locked. The lock must therefore comprise a mechanical lock which is capable of being released only when the car is at the landing, and in addition, an electric interlock, the contacts of which are open when the landing gate or door is open. It is thus impossible for a landing opening to be left unprotected when the car is away from that landing.

The car gate or door lock consists of an electric switch which opens its contacts immediately the gate or door is opened, thus ensuring that a lift cannot move unless the car gate is closed.

Landing Gate or Door Clearance. Accidents have occurred due to persons standing in the space between the landing and car gates whilst opening one of these gates and the car then being called to another landing. To prevent this happening the distance between the well side of the landing gate or door and the well edge of the landing threshold should not exceed 4 in. for hinged doors or $2\frac{3}{4}$ in. for sliding doors or gates.

Car Apron. On many modern lifts the doors begin to open when the car enters the levelling zone, which may be 10 in. from the floor. To prevent a person's foot from being trapped between the landing and the descending car during this levelling period, an apron should be fitted to the car platform of such a depth that no gap exists at any time when a landing door is opening.

Car Emergency Exit. This should be fitted with an electric switch, which will cut off the supply to the lift when the exit door is open and thus prevent the lift from moving when a person may be using the exit and is consequently then in an unsafe position.

Emergency Stop Push. All passenger operated lifts, except those fitted with sliding doors on both the car and landing, should be fitted with a stop push in the car, momentary pressure on which will bring the car to rest.

To provide the maintenance engineer with protection when working on the top of the car an emergency stop switch should be fitted on the car top.

FIG. 13.28. SLACK ROPE SAFETY SWITCH
(*Dewhurst & Partner*)

Slack Rope Safety Switch. This switch is fitted to drum-drive lifts and automatically cuts off the supply to the motor if the car meets an obstruction and the ropes become slack. This is shown in Fig. 13.28.

Wiring. From the safety aspect it is essential that all electric cabling and wiring be carried out in a sound and efficient manner and be in accordance with the Institution of Electrical Engineers "Regulations for the Electrical Equipment of Buildings." The following regulations affect lifts and should be noted.

REG. 209 (B). All cables, other than travelling cables installed for any purpose in a lift or hoist shaft, shall normally be armoured, or enclosed in steel conduit, duct or trunking, or copper conduit, or be of the mineral-insulated metal-sheathed type or of the aluminium-sheathed type.

REG. 214 (A). This indicates that if the travelling cable contains any conductor operating at a voltage in excess of 30 a.c. or 50 d.c., no other conductor in that cable shall be used in any system operating at a voltage lower than 30 a.c. or 50 d.c. or in any radio, telephone, bell and call or sound distribution circuit.

REG. 315. This requires that there shall be, preferably in the machine room, efficient means whereby all voltage can be cut off from the motor, controller and automatic circuit breaker. If the means of isolation is not in the machine room, provision shall be made for it to be locked in the OFF position.

The usual method of effecting this is by a separate switch in the machine room controlling the incoming mains to the lift. If the machine room houses several lifts, the incoming mains may be terminated at a distribution board with a separate set of removable fuses for each lift circuit. Alternatively, a separate switch for each lift supply may be fitted in the machine room near each lift.

REG. 402. This specifies that the metal framework of the motors, controllers, switchgear, electrical equipment in the car, and the electric conduits shall all be effectively earthed.

REG. 404. This requires that the car metal framework shall be bonded and connected to earth.

The best method of achieving this is by connecting the car frame to the earthing system of the metalwork of the machine room via a conductor in the travelling cable.

REG. 407 (B). This specifies that the cross-sectional area of the earth conductor in the travelling cable shall be equal to that of the current-carrying conductors.

Earthing. The earthing system for a lift installation is usually taken from the earth on the building main switchboard from where it is extended to the lift isolating switch in the machine room by continuous screwed conduit, the sheath of the lift supply cable or a separate earth wire. The earth from the isolating switch should be extended to the whole lift installation as specified in the above paragraphs. If screwed steel conduit is used for this purpose it should have sound clean threads and the joints made tightly. Where flexible conduit is used this should be bridged over by a separate earth wire clamped to each end of the length of flexible conduit. Neither the armouring of cables nor the lift guides should be relied upon for earthing. Suitable sizes of earthing leads for motors and generators are 7/·029 in. for machines up to 50 A, 7/·064 in. for machines between 50 and 100 A and 19/·064 in. for machines taking more than 100 A. If it is necessary to provide a separate earth lead to the controller, this should be terminated at an earth terminal on the controller and the size of the lead should be the same as that required for the lift motor. For earthing the metal cases of other lift electrical apparatus, the minimum size of the separate earth lead, if this has to be provided, should be 7/·029 in.

Factories Acts. Other regulations affecting the safe working of lifts are in Appendix I.

CHAPTER XIV

FLOOR LEVELLING

ACCURATE levelling of the car at the various landings is an essential to good lift service; with passenger lifts the danger of passengers tripping when entering or leaving the car must be minimized, whilst with goods lifts, particularly if trucks or trolleys are carried, easy entry and exit are desirable. In-accurate levelling also results in "inching" of the car to obtain better final levelling, and a consequent increase in energy consumption and wear on the controller contacts.

With simple car switch control, stopping is performed by the operator returning the car switch to the centre position when the car is at the appropriate distance from the floor. The centring of the car switch cuts off the supply and causes the brake to operate, in the case of a single-speed lift; but if a two-speed motor is employed, an intermediate car switch position enables the final stopping to be performed from slow speed. The accuracy of levelling therefore depends upon the judgment of the operator in returning the switch to the slow and stop positions at the correct moments. In some forms of car switch control, stopping is performed automatically by centring the car switch when the car has passed the floor immediately preceding that at which a stop is desired. The levelling, in these cases, is dependent upon the settings of the automatic slowing and stopping devices, and the adjustment of the mechanical brake.

Several methods are employed to stop an automatic lift at the desired landing, the efficiency of the schemes depending upon the accuracy of timing of the automatic slowing and stopping devices in reducing the car speed and cutting off the motor supply respectively before the operation of the mechanical brake.

Besides good brake adjustment and accuracy of setting of the slowing and stopping equipment, another factor which affects levelling is the variation of load in the car. Some small consideration will show that the distance required for the stopping of the car, after the application of the brake, will be greater if full load is being lowered than with an empty car, on the assumption that the brake retarding force is the same in each case. Similarly, the distance in which the car will come to rest when raising full load is less than when raising an empty car. To ensure good levelling it is essential that the distances travelled by the car after the application of the brake should be the same, irrespective of load, and to effect this it is desirable to arrange that the final cut-off speed of the car is slightly higher when lifting full load than when lifting an empty car. In other words, the speed should slightly increase with the load, and this means a rising motor characteristic. This characteristic, whilst unstable with an ordinary shunt machine, may readily be obtained with variable voltage control. The rising characteristic is only required at the slow levelling speed, which is from about one-sixth to one-twentieth of the maximum speed.

This may be illustrated by considering the following two conditions.

(a) *The fully loaded car travelling upwards.*

Let the contract load and the weight of the car be 1 200 lb. each and the weight of the counterweight 1 800 lb. When the power is cut off by the stopping switch, the car is slowed down by the application of the brake and the decelerating force of the moving system. If the brake retarding force and the frictional resistances are constant, the distance moved after the power is cut off depends upon the retardation due to the moving car and counterweight.

$$\text{The retarding force} = \frac{\text{total mass} \times \text{retardation}}{g}$$

$$\therefore \qquad 600 = \frac{4\ 200 \times f_1}{32}$$

and $f_1 = \dfrac{32}{7}$ ft. per second per second.

(b) *The car $\frac{3}{4}$ loaded travelling upwards.*

Under this condition the motor is exerting approximately half of its full load torque.

$$\text{The retarding force} = \frac{3\ 900 \times f_2}{32}$$

$$\therefore \qquad f_2 = \frac{300 \times 32}{3\ 900} = \frac{32}{13} \text{ ft. per second per second.}$$

For accurate levelling the distance s moved after the power is cut off must be the same in each case.

But $v_1{}^2 = 2f_1 s$ and $v_2{}^2 = 2f_2 s$.

where v_1 and v_2 are the respective levelling speeds in the two cases.

$$\therefore \qquad \frac{v_1}{v_2} = \sqrt{\frac{f_1}{f_2}} = \sqrt{\frac{13}{7}} = 1 \cdot 36.$$

Hence the levelling speed with contract load (motor at full load) should be 36 per cent greater than that with $\frac{3}{4}$ contract load (motor at $\frac{1}{2}$ load), i.e. the levelling speed should increase with the load.

The principal methods used for automatically slowing, stopping, and levelling lift cars are as follows.

Direction Switches. Direction switches, sometimes used on single-speed automatic lifts, are either single- or two-way switches fitted in the well, one at each landing. A two-way switch is necessary for each intermediate landing (one way for each direction of travel) but a single-way switch only is required at each terminal landing. Each switch is operated by a ramp, fitted to the car, when the car reaches the appropriate landing. When passing landings in the upward direction, the switches are in turn operated, and thereby change the floor relay connexions from the UP contactor to the DOWN coil ready for operation during the down direction. Hence each direction switch which is below the car is connected to the DOWN contactor coil. Similarly, during the downward travel the ramp moves the switch operating arms to the up direction, and therefore all direction switches above the car are joined to the UP contactor coil. When the car is stationary at a landing the corresponding direction switch is in the centre or OFF position, and current is thus cut off from the main contactor coil. Each terminal direction switch is opened and closed when the car arrives at and leaves the terminal landing respectively. A typical operating ramp is shown in Fig. 14.1, the top throw-in horn engaging the switch arm roller during the upward journey, and the bottom horn returning the roller to its original position during the downward journey. When

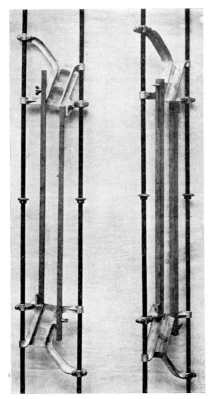

Fig. 14.1. Direction Switch Ramp
(*Marryat & Scott*)

the roller is in contact with the centre vertical track, both contacts of the switch are disconnected. An intermediate floor direction switch is shown in Fig. 14.2.

Floor Selectors. A floor selector switch for installation in the motor room, as shown in Fig. 14.3, is frequently used to perform

FIG. 14.2. DIRECTION SWITCH
Intermediate floor
(*Marryat & Scott*)

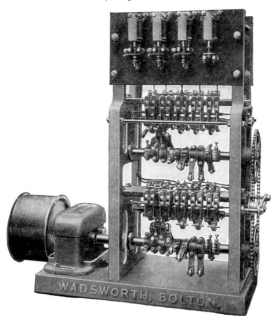

FIG. 14.3. FLOOR SELECTOR FOR AUTOMATIC LIFT
(*Wm. Wadsworth & Sons*)

the functions of slowing and stopping. The slowing and stopping contactor coils are controlled by a system of contacts operated by selector arms fixed to a shaft, on one end of which is keyed a drum, the drum being driven from the car by a steel flyrope. Hence the selector arms rotate in unison with the movement of the car in the well, and open-circuit the slowing and stopping contactor coils when the car is at the appropriate positions in the well. For each intermediate floor served, two switches and two operating arms are required for stopping in the two directions, whilst only one switch and one arm are necessary for each terminal floor. With two-speed motors, additional switches and selector arms are necessary for each floor so as to perform the slowing down prior to the operation of the stopping switches.

Another type of floor selector for installation in the motor room is shown in Fig. 14.4 and consists of two copper drums insulated from one another. These drums are rotated by the motion of the car, being driven through gearing from a pulley which is connected to the car by means of a flat steel tape. In addition to their circumferential movement, the drums rotate on a screw which causes them to travel in the vertical direction. The result is that a high degree of accuracy is obtained, a movement of about $\frac{1}{4}$ in. on the drum being obtained for every 1 in. travel of the car. For tall buildings, the dimensions of the selector increase vertically and not horizontally, and for ten floors and two speeds the height is about 4 ft.

The Otis floor selector used for speeds of 300 ft. per min. and above is shown in Fig. 14.5. This is mechanically driven from the car by a toothed steel tape; ensuring accurate and positive operation. The vertical movement of the selector crosshead therefore corresponds on a smaller scale to the movement of the car in the well. In addition to slowing down and stopping at the required floor this selector controls the car and landing signals and indicators and the opening of the car door. Distributed up and down the selector are "floor bars" corresponding to the floor served, and on each bar is a group of electrical contacts. The sliding contacts are silver to impregnated carbon and the butting contacts silver to silver. As the crosshead is driven up and down it engages these contacts and so operates the relays which perform the various functions referred to

above. For example, if the car of a collective control lift is
travelling downwards at full speed and a down landing call is
registered at the sixth floor, as the car passes the eighth floor
position, the crosshead completes circuits which light the hall
lantern at the sixth floor and extinguish the signal light in the

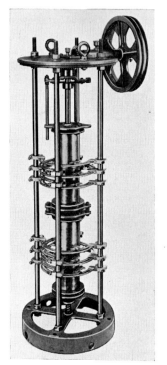

Fig. 14.4.
Two-speed Floor Selector
(*Etchells, Congdon & Muir*)

Fig. 14.5. Floor Selector
(*By courtesy of
H.M. Postmaster-General*)

sixth floor landing button. As the car enters the slowing zone
for the sixth floor, the selector initiates the operation which
slows down the speed of the car, and subsequently a levelling
device on the selector takes over to level the car and open the
car and landing doors. After the call has been answered, the
selector extinguishes the hall lantern, lights the proper signal
lights on the landings ahead and in the car.

The floor selector employed by the Express Lift Company is shown in Fig. 14.6, the overall dimensions of the selector being about 12 in. × 6 in. × 6 in. In addition, this selector performs the functions of establishing direction of travel, initiating stopping, and of operating car position indicators. It is a multi-contact rotary switch which can be fitted with a multiplicity of

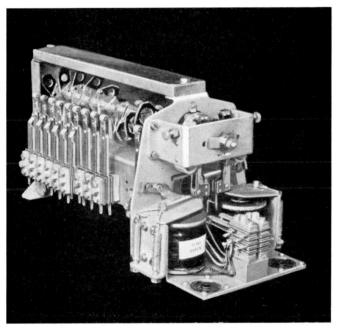

FIG. 14.6. FLOOR SELECTOR
(*Express Lift Co.*)

contacts and operating cams and is operated by two driving magnets which drive the cam shaft in either direction of rotation, depending on the travel of the lift. The principle of operation is similar to that adopted in telephone exchange selectors, although this selector has been specially designed to meet lift requirements. The UP and DOWN direction driving magnets are operated by impulses received from the inductor fitted on the car, a description of which appears later in this chapter. The chart shown in Fig. 14.7 gives the sequence of contact operation as the selector steps one step in the appropriate direction when the car passes the inductor plates fitted in the lift well. The plates are fitted on either side of the floor datum line at such a distance as is required to bring the car to rest at floor level from either direction. Considering a single-speed lift to illustrate the principle of operation of the selector, one plate will be required on each side of each floor

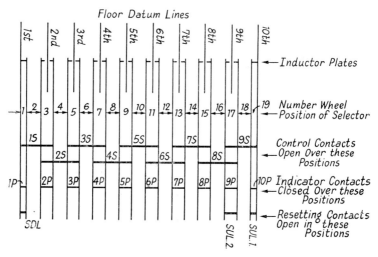

FIG. 14.7. FLOOR SELECTOR CONTACT SEQUENCE

level. Three sets of contacts are provided on the selector as shown, control contacts, indicator contacts, and resetting contacts.

The circuit for driving the selector is shown in Fig. 14.8, the selector being impulsed by the inductor contact *ZS* which closes the circuit to the UP or DOWN driving magnet *DMU* or *DMD* via change-over contact 83. Relay 83 is a change-over directional relay which is operated when the lift travels UP and remains operated until the lift reverses to travel DOWN. The resistance *R3B* in this circuit, which is inserted by contact T1 after each stop, is a protective resistance to prevent overheating of the driving coils should the lift stop for a long period with the inductor operated. Resetting is achieved by the release of relay 23 or 24 at the bottom and top floor respectively, contacts of which relay operate the Up and Down driving magnets via the interrupter contacts *dmu* and *dmd* and the resetting limit contacts *SDL* and *SUL1* or *SUL2* on the selector. The two upper resetting contacts *SUL1* and

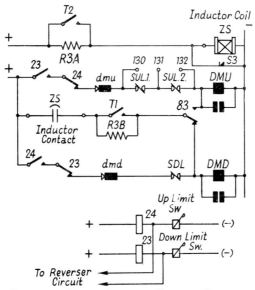

FIG. 14.8. FLOOR SELECTOR DRIVING CIRCUIT

SUL2 are provided in the driving magnet so that a given selector can cover two floor combinations (e.g. 9 or 10 floors). Terminals 131 and 132 are jumpered for an even number of floors and 130 and 131 for an odd number of floors. The resetting limit contacts will normally open if the selector is in sequence before 23 or 24 releases.

The directional and stopping circuit of the selector is shown in Fig. 14.9. At any lift position at least one control contact is open, and at floor level it will be seen from Fig. 14.9 that two contacts are open. Hence any call relay above the open contact is energized via the Up directional relay 81 and any call relay below is energized via the Down directional relay, e.g. with the car at the 4th floor the selector contacts 3S and 4S are open and this isolates the 4th floor call relay 104 and cuts off all call relays 105 to 110 from the DOWN circuit to relay 82. This enables these relays 105 to 110 to operate by pressure on a car or landing button for any floor above the 4th floor only, via the UP circuit to relay 81. Similarly, landing or car buttons below the 4th floor can only operate

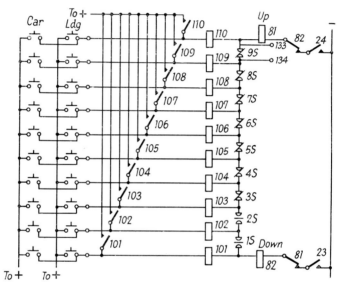

FIG. 14.9. FLOOR SELECTOR DIRECTION AND STOPPING CIRCUIT

their respective relays via the DOWN circuit to relay 82. As the lift travels to answer a call, the relative call relay remains energized until the selector contacts on either side of its negative coil feed are opened, when the car reaches floor level. This opens the circuit to the call relay and the direction relay and stops the car. At a terminal floor, stopping is effected by the opening of the limit switches *UL* or *DL*, which also releases relay 24 or relay 23.

Corrective Levelling Systems. Various means have been adopted for securing accurate final levelling in which the levelling resulting from the operation of the floor selector stopping switches is corrected, if necessary, by independent means. With some of these corrective levelling devices it is possible to secure final levelling to within $\pm \frac{1}{8}$ in., irrespective of load variations or rope stretch during loading.

In one method a three-level ramp is fitted at each landing, this ramp engaging a three-position switch fitted to the car, at those landings where stops are made. If the car stops below the landing, after the floor selector switch has operated in the normal manner, the three-position switch will engage with the bottom level of the ramp. This results in the motor being re-started and run at the slow speed, in such a direction that the car travels upwards. When the switch engages with the centre level the car is correctly levelled, the motor supply is cut off and the brake applied. Similarly, if the levelling is too high, the switch on the car engages with the top ramp level and this causes the motor to run at slow speed in the opposite direction and the car thus travels downwards until the centre cam level engages the switch, when the motor supply is cut off and the brake applied. The levelling switch, which is automatically withdrawn beyond reach of the ramp when passing landings, is of special construction and correctly engages the ramp irrespective of any car side movement due to shoe wear.

A two-position switch operating on this principle is shown in Fig. 14.10. This is fitted to the car, and during normal travel the rollers are withdrawn so that they are clear of the levelling ramps. These ramps are fixed in the well at each floor so that the mid-position between the ramps at each floor corresponds to the rollers on the switch and to the correct levelling position. If the lift under normal operation comes to rest at a distance not exceeding about 11 in. either above or below the floor, one of the two levers will be operated, so causing the lift to move at slow speed in the required direction to floor level where the switch will be central between the two ramps.

THE "TRULEVEL" GEARED MACHINE. This machine has been patented by the Express Lift Co., and is used for car speeds up to a maximum of 200 ft. per min. Two views of the equipment are shown in Figs. 14.11 and 14.12 and it consists of a main driving motor coupled to the worm shaft of a worm reduction gear, by a multi-vee rope drive, an auxiliary levelling motor mounted above the gear box and coupled to a friction clutch

FIG. 14.10. LEVELLING WITHDRAWAL SWITCH
(*Dewhurst & Partner*)

FIG. 14.11. "TRULEVEL" MACHINE
(*Express Lift Co.*)

by a second multi-vee rope drive, an electro-mechanical brake, and the main driving sheave. The main motor is of the single-speed type of voltage to suit the supply characteristics

Fig. 14.12. "Trulevel" Machine
(*Express Lift Co.*)

and is mounted on a platform hinged to the bedplate. Means are provided for adjusting the positions of the two motors to take up initial stretch in the rope drives. The driving sheave and worm-wheel rims are both secured by fitting bolts to a spider thus eliminating keys, whilst the multi rope drives enable the motors to be more completely isolated from the rest of the machine than is possible with fixed couplings. One side of the single plate type friction clutch forms part of the brake drum and the other Ferodo-lined side forms part of the multi-vee rope pulley of the auxiliary levelling motor drive. Both the clutch and the brake are released by solenoid energized magnets and engaged by springs, the magnets being operated by recti-fied direct current. A brake switch ensures prompt application of the brake shoes when the solenoid is de-energized and a clutch switch ensures that the main motor cannot be connected to the line until the clutch has been fully disengaged and also that the clutch cannot be engaged until the main motor has been disconnected from the line.

The main motor is used for accelerating and running the lift at full speed when the clutch is disengaged. The lift is slowed by the application of the brake, the main motor then being disconnected from the line. At the appropriate moment the single speed auxiliary motor is switched on, the clutch engaged, the brake relifted, and the car brought to the landing at the levelling speed, the motor being finally switched off and the brake applied when the landing is reached. The reduction in speed required for levelling is obtained by the rope drive with a maximum reduction of 6 : 1 and also by a difference in speeds of the main and auxiliary motors of up to 2 : 1. Hence the overall speed ratio may be as high as 12 : 1, giving a landing speed of about 16 ft. per min. with a maximum running speed of 200 ft. per min.

INDUCTOR AND INDUCER SYSTEMS. The signals required by the controller to initiate the slowing and stopping, i.e. the change over to auxiliary motor and the final brake application are provided by inductors or inducers mounted on the car. The Express Lift Co. also employ these means for effecting slowing and stopping on types of lifts other than their "Trulevel" lifts. For car speeds up to about 300 ft. per min. permanent magnet inductors have been used in the past, as many of these as necessary being fitted on top of the car. Each inductor is constructed so that a gap exists, on one side of which is a flux generating device such as a permanent magnet and on the other side is a flux collecting pole. The flux so collected is used to actuate a small armature which operates current-carrying contacts wired to the controller switches. The inductors are operated by steel plates fixed in the well as shown in Fig. 14.13 and so situated that they pass through the inductor gaps at the correct moments for initiating deceleration and stopping. Hence when an iron plate passes through the gap the flux is diverted from the collecting pole and the armature becomes free to move under the influence of a spring, but as soon as the plate has left the gap the armature returns to its normal position. Usually three inductors per car and three associated rows of iron plates fixed in the well are provided, one for counting the zone changes, and the others each for controlling up or down slowing or stopping.

The inductor system described above is unsuitable for high speed operation, however, and for speeds in excess of 300 ft. per min. the inducer is used. This consists of a pair of iron cored coils fixed to the car as shown in Fig. 14.14, a gap between the coils permitting of the entry of an iron plate so that alternating flux generated by a.c. in one coil may be temporarily prevented from reaching the other pick-up coil. This latter coil is coupled to a valve amplifier so that only anode current flows to operate controller relays when a plate is in the gap which occurs when deceleration, stopping, or zone counting is required.

Whilst both the above systems have proved mainly satisfactory the former has a speed limitation and the latter, operating on the thermionic principle, is not quite as stable or flexible as would be preferred. Consequently a more reliable and robust inductor known as the solenoid inductor has been produced and this also permits a simpler control circuit. The inductor is solenoid excited as against the previous permanent magnet type. This solenoid inductor, which is shown with case removed in Fig. 14.15, will replace the two earlier types on all installations provided by this company and is suitable for use on lifts of all speeds.

The Pliotron unit, manufactured by the British Thomson-Houston Co., Ltd., is a proximity-type limit switch similar

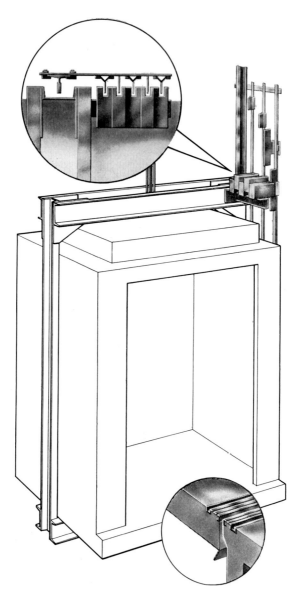

FIG. 14.13. INDUCTOR SWITCHES
(*Express Lift Co.*)

FIG. 14.14. INDUCER BOX
(*Express Lift Co.*)

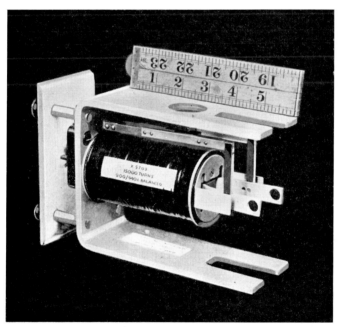

FIG. 14.15. SOLENOID INDUCTOR
(*Express Lift Co.*)

in principle to the inductor described above. The circuit includes two separate components, the unit illustrated in Figs. 14.16 and 14.17 and an electro-magnetic relay shown in Fig. 14.18. The unit consists of a sheet-steel case housing a triode valve and some of its associated circuit components. Two coils separated by an air gap are enclosed in moulded insulation projections on the front of the case. Normally the coils are

FIG. 14.16. THE PLIOTRON UNIT
The operating coils are housed in the moulded projections
(*B.T.-H.*)

magnetically coupled but can be decoupled by the interposition of a metal vane which does not touch any part of the Pliotron unit. This "proximity" effect serves to actuate the relay which is of simple construction, compact and quick-acting, and fitted with silver contacts. The relay is mounted on the controller panel. The overall dimensions of the Pliotron unit are approximately $8\frac{1}{2}$ in. high, $5\frac{1}{2}$ in. wide, and $11\frac{1}{2}$ in. deep. The coils are normally magnetically coupled and the relay is then de-energized. Decoupling is effected by interposing a metal vane between them and this energizes the relay. The operation of the relay coincides for all practical purposes with

FIG. 14.17. REAR VIEW OF THE PLIOTRON UNIT
Cover removed
(*B.T.-H.*)

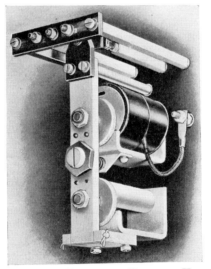

FIG. 14.18. RELAY FOR PLIOTRON UNIT
(*B.T.-H.*)

the passage of the edge of the vane across the centre line of the two coils on the unit. Fig. 14.19 shows two pairs of units mounted on top of a lift car, one of the operating vanes being

Fig. 14.19. PLIOTRON UNITS ON PASSENGER LIFT
(*B.T.-H.*)

seen just above the unit. In some lifts it might be more convenient to mount the vane on the lift car and the unit in the well. Any number of units can be operated in succession by a single vane, or a number of vanes can successfully operate one unit.

CHAPTER XV

CAR CONTROL SYSTEMS

A NUMBER of different methods of controlling the movement of the car from floor to floor are available, and the particular one selected depends upon the service required, the type of building, and the car speed. The various methods may be grouped, broadly, into two systems as follows—

(a) Those requiring an attendant.

(b) Those in which the car is completely under the control of passengers in the car or persons on the landings.

In group (a) we have systems of control known as *car switch*, *departmental stores*, and *signal*, whilst under (b) come *automatic*, *semi-automatic*, and *automatic collective*. *Dual control* may be used with or without an attendant.

Car Switch Control. This requires a single call button at each landing, pressure on which registers a call on an indicator board fitted in the car, thus making the attendant aware of the floor at which service is required. Movement of the car switch by the attendant to the UP or DOWN position starts the car in the required direction, stopping being performed by returning the switch to the OFF position as the car approaches the desired floor. Good floor levelling is dependent upon the judgment of the attendant in returning the switch to the OFF position when the car is at the correct distance from the required floor. The car switch automatically returns to the OFF position under the action of a spring and locks itself there when released. For car speeds up to about 100 ft. per min. a single-speed motor is used, and for speeds up to 200 ft. per min. a two-speed motor is necessary so that levelling can be done at the slower speed. In the latter case the car switch has a fast position and a slow position for each direction of travel. Although this manual stopping of the car was used extensively in past years, it is now mainly employed for passenger lifts serving a small number of floors and where traffic is light, or for goods lifts. A gate is essential as a door would not allow the attendant to judge the position in the well for operating the switch for slowing and stopping. The attendant resets the indicator after calls have been answered.

An improvement on the above control is available if the lift is fitted with a self-levelling device. In this case the attendant merely releases the car switch so that the car stops somewhere within the levelling zone, and the self-levelling equipment then automatically stops the car at the landing.

A further improvement is the provision of automatic stopping in which the attendant releases the switch after passing a floor and the car automatically slows and stops at the next floor, provided that the switch has been released soon enough for the automatic slowing and stopping equipment to function. For passenger lifts provided with car switch control, automatic stopping, car position indicator in the car, and power-operated doors, the doors are closed by the car switch and opened automatically as the car stops.

Departmental Store Control. This is the name given to a system which was developed for use in departmental stores where the lift normally stops at each floor and the attendant is relieved of any effort to make accurate landings. Buttons are not required for this control, but the landings have illuminated car position and direction indicators. When the car switch is moved to the START position, the power operated car and landing doors close automatically, and the car commences to accelerate. The centring of the car switch causes the car to stop level with the next floor, and both car and landing doors open automatically. If it is desired to pass floors, the car switch is held in the start position.

Signal Control. This is used on banks of high-speed lifts, such as are installed in large offices and hotels, employing variable voltage gearless drives, automatic levelling, and power operated doors, and travelling at speeds up to about 1 000 ft. per min. in this country, and 1 600 ft. per min. in America. A person at a landing requiring a car, presses either the UP or DOWN button, depending upon the direction in which he wishes to proceed, and the first car travelling in the desired direction and that can accommodate him stops at the landing. As the car approaches the landing at slow speed (under the control of the automatic levelling gear) the landing light is illuminated, either UP or DOWN, depending upon the direction of travel, thus indicating that the car will stop at that floor. On reaching the landing the car levels automatically, the doors open, and the waiting person enters and calls the number of the floor to which he wishes to proceed. The operator then presses the corresponding car button and moves the car switch to the start position, after which the doors automatically close. The car accelerates and stops automatically at the next floor where a car or landing call has been registered. Car or landing buttons may be pressed in any order, and whether the car is stationary or in motion. A non-stop button in the car enables the attendant to pass any landing if the car is full; the stopping of the first available car resets the control so that unnecessary stops are not made by other cars. A special emergency control lever in the car provides for travel at the slow levelling speed for inspection and maintenance purposes.

Automatic Control. With this type, a single call button is fitted at each landing (Fig. 15.1) and a set of buttons in the car. The car buttons comprise one for each floor and an emergency stop button, as shown in Fig. 15.1. A passenger, after entering the car and closing the gates, presses the car button corresponding to the desired floor, and the car then starts and automatically stops when the floor is reached. Any other calls which may be made from a landing when the car is in motion are ignored. Whilst the car's ultimate destination is controlled by the passenger, the accuracy of levelling is dependent upon the setting of the slowing and stopping switches and adjustment of the brake, unless automatic levelling gear is installed. This control is used when the traffic is occasional and not sufficiently heavy to justify the employment of an attendant. Except in special circumstances this control should not be used for lifts of capacity more than about ten persons as it is not economical to have a large lift answering single calls. A button system consisting of a full set of landing buttons (one for each floor) as well as a full set of car buttons, is sometimes employed. This is useful when the lift is used for both passenger and goods traffic and it is desired to send the lift to any floor without a passenger.

Semi-automatic Control. UP, DOWN, and STOP buttons are fitted in the car and at each intermediate landing, as in Fig. 15.2, whilst UP and STOP buttons are provided at the top terminal landing, and DOWN and STOP buttons at the bottom terminal landing. The momentary pressure of a button, either in the car or on a landing, causes the car to move in the direction indicated, whilst pressure on a stop button stops the car. A modification of this scheme consists of UP and DOWN buttons at each landing (Fig. 15.2) and in the car, the car only

FIG. 15.1. OPERATING BUTTONS
(*Wm. Wadsworth & Sons*)

travelling when the button is actually depressed, and stopping on the release of the button. This results in more accurate floor stopping than with UP, DOWN, and STOP buttons. These systems are sometimes employed in preference to the fully automatic system for slow-speed goods lifts which are not provided with corrective levelling, so as to secure more accurate levelling under varying conditions of load.

The semi-automatic control, however, is now mainly employed on lifts serving two floors only. In this case UP, DOWN, and STOP buttons are fitted only on the landings, when the lift is used solely for goods traffic, but UP, DOWN, and STOP buttons are also required in the car if passengers are to be carried. When two floors only are served, the stop button is

FIG. 15.2. CALL BUTTONS
(*Wm. Wadsworth & Sons*)

for emergency purposes only, as the lift stops automatically on reaching the floor.

Automatic Collective Control. This system has been developed during recent years with the object of providing a fully automatic push button system giving superior service to that obtainable with the ordinary automatic system. The main disadvantage of the automatic control is that it is possible for passengers waiting at a landing to see the car pass, and travelling in the direction in which they wish to proceed, without their being able to gain admittance. This, however, is not possible when collective control is installed, provided the landing button is pressed sufficiently in advance of the arrival of the car.

UP and DOWN buttons are fitted at each intermediate landing, single buttons at the terminal landings and a full set of buttons, together with an emergency stop button, in the car. Every button pressed, whether the car is stationary or in motion, registers a call. Up and down calls are answered during the up and down journeys respectively, in the order in which the floors are reached. If the car is at, say, the ground floor and one or more persons enter the car, close the landing and car doors, and press the appropriate car buttons, the car will start, and stop at the floor corresponding to the lowest numbered car button which has been pressed. The order in which the buttons are pressed has no bearing on the sequence in which the car will stop at the various floors, providing that the button for a given floor is pressed sufficiently in advance of the arrival of the car at that floor to enable the stop to be made. The car, while travelling in the up direction, will also stop in the same manner in response to all up landing button calls. The upward direction is continued until the uppermost call has been answered. If a down call is registered at a floor above the highest up stop, the car will continue in the up direction to the floor at which a down call is registered. It will then reverse automatically, even though not at the top terminal floor, provided car buttons for floors above this point have not been pressed, and will start down, stopping in response to all down calls registered at lower floors in advance of the arrival of the car at these floors. The car will continue in the down direction until the last floor is reached for which a car or down landing button has been pressed. If, after all down calls are answered, the car is not at the bottom landing, it will continue down in response to the pressing of an up button at a floor below the last stop. If, however, no button is pressed below this floor the car will reverse automatically, and start up in response to the pressing of either up or down buttons above this floor. In other words, the car will travel in either direction until the last floor has been reached at which an up or down button has been pressed, or for which a car button has been pressed. It is not necessary that the car make complete round trips.

If traffic is intermittent and only one call is registered at any one time, the lift will operate in the same manner as a single automatic push button lift, but will become collective automatically when more than one call is registered at one time.

Precedence is given to the passenger in the car in the following manner. If, after the car has started to slow down at a given floor in response to the pressing of a button at that floor (this call being the last call registered in the direction in which the car is going) another call is registered at a floor beyond the one at which the car is in process of stopping, the stop will be completed. After the passenger has entered the car he may dispatch the car in accordance with his wishes, provided he presses the proper car button before closing the car gate or within about five seconds after the car gate has closed. He is, therefore, spared the inconvenience of being taken in the direction opposite to that which he expected to go. For example, if the car is at the first floor and down buttons are pressed at the third, fifth, and seventh floors, the car will travel to the seventh floor. If, after the car has commenced to stop at the seventh floor, an up or down button is pressed at the ninth floor, the stop at the seventh floor will be completed and the call at the ninth floor will not be answered until after the car has reversed at the seventh floor and completed its downward trip in response to the car buttons pressed by the passengers who entered the car at the seventh, fifth, and third floors. If the ninth floor button had been pressed before the car commenced to stop at the seventh floor, the car would have proceeded to the ninth floor before answering the down calls at the lower floors. The same operation is effective in both directions. If, after the car has stopped at a given floor in response to the pressing of either a car or landing button, the car or landing door is not opened within about five seconds, the car will respond to any other calls which may be registered, or if no calls are registered, the car will return to the home landing.

Arrangements may be made to enable the car to park automatically at any preselected floor, after all calls have been answered, this facility being particularly desirable when important persons are located on a certain floor or when heavy traffic is expected from a particular floor at certain periods of the day. For instance, the service may be greatly improved in a block of offices if, during heavy morning traffic, the car is made to return automatically to the ground floor or to park at, say, the sixth floor in the evening, to give prompt service in the downward direction instead of remaining at the floor to which it was last called.

The above form of control, which is sometimes referred to

as *directional collective*, may be modified to enable calls to be answered in the same sequence in which the buttons are pressed. This is known as *interceptive collective* control.

The system may be still further modified to work as an automatic push button control in the up direction and as directional collective in the down direction only. This *down collective* control is sometimes employed in flats and is based on the assumption that, whilst tenants and their visitors travelling upwards usually wish to travel directly to their own floor, all persons travelling downwards wish to alight at the ground floor. Hence, an upward travelling passenger is enabled to travel directly to his own floor without interference, whilst in the downward direction there is no reasonable objection to the lift stopping at intermediate floors to pick up other passengers who almost certainly also wish to travel downwards to the ground floor.

The lift normally stands idle with the doors closed, from which position it moves in either direction in response to landing calls or will respond to the landing call for the floor at which the lift stands, thus opening the doors. If the lift car moves upwards in response to a call, all such calls are registered, and the lift proceeds to the highest call, such calls which are passed during this journey to the highest call being retained for service during the next downward journey. On stopping for the highest call, the doors open, and the landing pushes are disconnected to prevent higher calls being registered. Any calls registered during the upward journey are stored and held inoperative until the doors have opened and then reclosed. This gives car preference, and a passenger may enter the car and select a direction of travel without interference from further landing calls or calls already registered. If the passenger operates a car push for an upward direction before the doors close, the landing pushes will be disconnected, and any registered calls will be held inoperative during this journey. Where the car pushes are operated for more than one floor in the upwards direction, the lift will proceed to the highest floor, retaining the others for attention during the next downward journey. When the car starts downwards, both car and landing pushes are operative to give service below the lift car, and the car stops automatically and in sequence in response to such calls made during the downward journey. On the downward journey it is impossible for the direction to be reversed by car or landing pushes until all down calls have been answered.

Duplex Control. This is used when two lifts are installed in adjacent wells. A common landing button serves for both lifts, and a passenger wishing to make a call momentarily presses the button. A CALL-ACCEPTED signal is immediately illuminated, and when a car answers the call a LIFT-COMING signal is illuminated. With power-operated doors, pressure on a car button automatically closes the doors, and then the lift starts. Landing calls are registered and are allocated one at a time as the cars become free from car calls.

Triplex Collective Control. Triplex Collective Control is for three adjacent collective control lifts with a common landing call button system, and the nearest car travelling in the required direction answers the call.

Dual Control. This is a combination of car switch control and one of the automatic forms of control, and requires a car switch and indicator and a full set of call buttons in the car. The lift may be worked either by an attendant operating the car switch or by the passengers operating the buttons, depending upon the position of a transfer switch. For light traffic the lift is operated by passengers, but during periods of peak load, by an attendant, with a consequent improvement in the service.

A full dual system of control may also be provided without the use of a car switch. In this case the operation of a transfer switch in a locked box on the landing converts the lift from automatic to a service equivalent to car switch, but by still using only the car buttons. Pressure on a landing button illuminates the corresponding floor button in the car and sounds a buzzer. The call is cancelled only when it is answered. The attendant can cause the car to stop at intermediate floors in response to landing calls by pressing the appropriate floor button while the car is in motion. Thus, while the lift is travelling to the fifth floor, a call from the third floor can be answered by pressing the third-floor button, provided the car has not already entered the third-floor stopping zone.

Dual Collective Control. This is a collective control with the additional facility of being operated by an attendant during very busy periods or on special occasions. The closing of the doors and the starting of the lift are controlled by the attendant's special Up and Down buttons in the car. For starting, the calls are dealt with automatically and in order as in collective control, but the attendant may by-pass calls if the car is full. This is done by either a by-pass push or an automatic micro-switch operated by full load on the car floor. The operation of a small change-over switch converts the lift to collective control when the attendant leaves the car. Automatic parking of the car, usually at the ground floor, is frequently provided, so that between calls the car is waiting at the floor where service is most likely to be required.

A variation of this control comprises down collective operation when under passenger control and interceptive collective when operated by an attendant.

Signal Collective Control. This combines the features of signal and collective controls, and therefore may be used with or without an attendant. It can also be used with one, two, or three cars and is intended to be used normally as signal control and only occasionally, e.g. at night, by passengers. Passengers wishing to call a car momentarily press an Up or Down button on the landing, and the button is illuminated until the call is answered. When on attendant control, the attendant uses a continuous press start button to close the doors, and after these are locked the car starts automatically. Without attendant, the car starts after the passenger has pressed the appropriate floor button and the doors have closed. Once started, the car answers the nearest call from the car or landing in the direction in which it is travelling. All other calls are answered strictly in rotation. When the car reaches the highest floor, its direction is automatically reversed. For three-car banks, hall lanterns are fitted over each entrance, and direction arrows are also fitted to the car control panel.

With a two- or three-car bank the landing call system is common to all lifts, and each call is automatically allocated to the best-placed car. At the parking floor a START signal and a gong denote the correct time of leaving that floor for each lift. The START signal is received according to the registered calls in the system and the position of the other cars. These signals are controlled so that the cars are effectively spaced and so give the best possible service. When operated by passengers, a car at a parking floor always responds to a car call.

A starter panel is mounted outside the lifts, and this houses the car light switches, transfer from signal to automatic collective control switches, motor generator switches, and door pushes for parking the cars out of service during slack periods.

The car control panel contains in addition to the floor buttons a combined stop and alarm switch and, for attendant use, reversing buttons and a PASS button. A DOOR OPEN button is also fitted.

When operated by passengers the cars are subject to an over-riding control which puts them out of service or restores them to use separately as the traffic falls off or increases. A combined indicator and service control panel for a bank of three lifts using signal collective control is shown in Fig. 15.3. The top panel indicates the movements of all the cars and the location of landing calls. The lower panel can be opened by the supervisor so that he can make the necessary adjustments to the service of the lifts.

Modern Practice. Car switch was one of the earliest forms of control and had the advantage of requiring a simple controller and also of giving an attendant full control over the starting, stopping and levelling. A good attendant could undoubtedly provide an excellent service which, by carefully watching the indicator and observing the position of passengers, was some measure of collective control. The advent of automatic control enabled a service by passengers themselves when the duty

FIG. 15.3. COMBINED INDICATOR AND SERVICE CONTROL PANEL FOR A BANK OF THREE LIFTS USING MULTIPLE CALL CONTROL FOR THREE CARS
(Express Lift Co.)

was light and it was uneconomical to employ an attendant. These two controls, together with the combination of the two (dual control), were in general use for many years. However, as the size of buildings and the intensity of lift traffic increased, the shortcomings of both controls became apparent. Automatic control gave a service to an individual passenger or to one group at a particular landing, and there was no means for service to be given to passengers at intermediate floors even if they too wished to travel to the same floor. The need for an improved automatic service resulted in the introduction of collective control, which has now practically entirely replaced the original automatic control for passenger service, even in the smaller and less busy offices. The difficulties experienced with car switch control were the considerable variations in the service if the attendants were not efficient. In recent years it has become difficult to recruit and retain good attendants, and the wages aspect has further resulted in a considerable falling off in popularity of car switch control, even with such features as automatic levelling and power doors. There has been, too, a corresponding increase in the facilities available, notably the feature of interconnecting a bank of lifts, and in the reliability of automatic forms of control. As a result, experience tends to show to-day that it is unlikely that a team of efficient lift attendants can, throughout a day, give such a good service as would be provided by a corresponding bank of modern interconnected collective control lifts. These problems have already been encountered in America where in many large buildings of up to thirty storeys the lifts are operating successfully without attendants. Even in this country where lifts have been changed over from car switch to automatic collective control, there has been an improvement in the service. There are, therefore, clear indications that control without attendants may, in the near future, replace control with attendants even in the large densely-populated office buildings, particularly as there has been advancement in the technique of enabling lifts to adapt themselves automatically to varying traffic conditions.

Groups of High-speed Lifts. Systems, without car attendants, have been developed in recent years to control automatically the individual lifts in a bank, depending on the traffic requirements during different times of the day. These control systems are usually based on the principle of dispatching the cars from the main terminal floors according to a number of predetermined operating programmes designed to cater for the various patterns of traffic flow and intensity of traffic. The appropriate operating programme is automatically selected by a traffic analyser which continuously measures automatically the rate at which calls are being answered and summates the differences in Up and Down stops of all cars in the group. Examples of this type of control are the Otis "Autotronic" system, so called because it comprises certain electronic features, and the Express Traffic E.T.6 system. Both of these normally cater for either four programmes or six programmes. The former comprises programmes 1–4 below and is normally used for banks of up to three cars. Programmes 5 and 6 are usually added for banks of four or more cars of rises exceeding about 14 floors and speeds above about 500 ft. per min.

1. "Up–Down" or Balanced Programme. This condition of balanced traffic prevails during most of the day when the passenger movement is substantially the same in both directions. Cars are dispatched automatically at controlled intervals from both terminal floors, each car making a through trip. A car which is fully loaded (detected by weighing contacts under the car floor) by-passes landing calls until its load is reduced. If Down calls exist behind a down-travelling car, an up car that is late may be reversed at its highest call in order to answer them. A car is dispatched ahead of time from the lower terminal if the preceding car has completed its round trip and Up calls exist behind it.

2. "Up" Peak Programme. One of the heaviest demands on lifts is the morning Up peak. This calls for quick filling of the building, practically all traffic being upwards. Cars are time-dispatched from the lower terminal only, and reverse automatically when empty. Each car leaves as soon as it is full, and indicators are illuminated automatically to show passengers which car to enter.

3. "Down" Peak Programme. This caters for the home-going evening rush, cars being time-dispatched from the upper terminal and starting back from the lower terminal as soon as they are empty. A zone return system operates if waiting times at lower floors reach a predetermined limit. The building

is divided into a high zone and a low zone, and all cars reverse at the highest call in their respective zones. To assist low-zone cars to relieve congestion, unfilled high-zone cars answer low-zone calls until filled. When the down traffic lessens, regular down-dispatching of cars is resumed.

4. INTERMITTENT OR LIGHT TRAFFIC. This caters for nights, week-ends and holidays. As the traffic diminishes, the motor generator set of each lift automatically switches off in a pre-controlled manner. As the demand increases other cars are brought into service as required. All cars are arranged to park at the ground floor.

The above are the normal programmes for a four-programme system usually applied to a bank of up to four lifts. For a larger bank a six-programme system comprising the following additional programmes is used.

5. HEAVIER "UP" PROGRAMME. After the morning peak or after lunch, the traffic may be in both directions but predominantly in the up direction. This requires the cars to make more stops in the up direction, necessitating more time for the up travel. Hence the Automatic Traffic Analyser dispatches cars from both terminals at automatically adjusted time intervals so that the cars are equally spaced, thus reducing passenger waiting time.

6. HEAVIER "DOWN" PROGRAMME. This requires the cars to make more stops in the down direction and the dispatch times are adjusted accordingly by the Traffic Analyser.

To maintain efficient service the above systems require high-speed door operators and effective automatic control over the opening and closing of doors. Messrs. Otis use a cold-cathode tube detector which stops and reverses car, and landing-doors when a passenger is within a few inches of either door. In the Express Traffic Sentinel, passenger movement is detected by a light-beam which operates a photo-electric receiver to control door opening and closing.

The Express Lift Company's Traffic Mark IV patented control system for use with high-speed gearless lifts has been designed to handle unlimited variations of demand for lift service. It recognizes that traffic demands and surges can be short or long, up or down, in both directions at the same time rather than a steady or definite intermittent flow of traffic.

This system therefore deals automatically with any type of traffic that may occur in a building and does not have the disadvantages of the above four- and six-programme systems. In the latter systems much time can be spent at the terminals waiting for a dispatch interval to expire, cars sometimes make useless journeys to terminal floors and although average

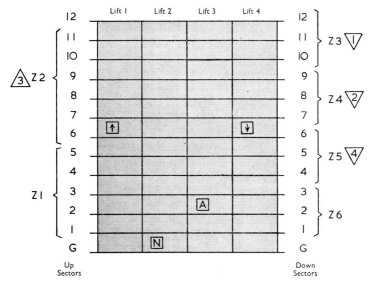

FIG. 15.4. PRINCIPLE OF EXPRESS TRAFFIC MARK IV CONTROL

(Express Lift Co.)

waiting times are low the maximum waiting time can be high.

The principle of the Express Mark IV control is illustrated in Fig. 15.4. So far as the UP and DOWN landing calls are concerned, the building is divided into a number of demand sectors. DOWN-demand sectors may have from one to three floors per sector, and UP-demand sectors from one to several floors. Fig. 15.4 shows how a 13-floor installation of four cars may be divided into six sectors: two UP sectors Z1 and Z2 and four DOWN sectors with three floors in each, Z3, Z4, Z5 and Z6. When the first call is registered within any sector, a timing device comes into operation for that sector so that at any instant it is known for how long each sector has had an unanswered call within it. This timing is measured in six predetermined periods of waiting time, designated as periods 1 to 6 in an ascending order; a typical example is 10, 15, 20, 30, 40 and 60 seconds. These periods are capable of adjustment to meet actual service requirements. As the calls in each sector are answered, the sector timing reverts to zero until the next landing call is made in that sector. Hence cars are allocated to each sector in an order of priority based on the length of time a demand has been present. This is referred to as a "sector demand of period 1 to 6." In the illustration the triangles denote the direction of travel and the numbers within the triangles denote the sector demand. As each car becomes available it is dispatched immediately to the sector with the longest waiting time, whereupon the sector demand is reset to zero. When answering a sector demand, a car clears only calls within the sector to which it has been assigned; thereafter it travels directly to the destinations arising from the sector demand. There are conditions in which a car is permitted to give service to another sector demand *en route* to its destination floors.

As cars become available they are assigned to the unallocated sector demand with the longest waiting period. In Fig. 15.4 if lift 4 was the first available car it would have been allocated to the DOWN sector Z5 with the timing period 4, and this car is shown travelling down to the highest DOWN call in this sector. If after this allocation, car 1 became available, it would be allocated to the period-3 demand in UP sector Z2, and the car is shown travelling to the lowest UP call. Car 3 is shown as an available car and will be assigned to sector Z4, when it will travel up to the highest DOWN call in this sector, where it will reverse to answer any other DOWN calls in the sector. Car 2, becoming the next UP car at the ground floor, will become available for dispatch only after all other cars have been assigned. To cover the event of more than one car being available to answer any sector demand, electronic selection is used to assign the nearest car.

The first Mark IV installation went into service in this country in 1964 on a group of four lifts serving twenty-two floors in St. George's House at the Croydon Redevelopment Centre. The first to the tenth floors are served by a 300 ft. per min. installation and the ground and eleventh to twenty-second floors by a 700 ft. per min. Mark IV system.

CHAPTER XVI

CONTROLLERS

General. The controller is usually located in the machine room, but care should be exercised in choosing its exact position in order to ensure that sufficient clearance exists between the controller and any walls, and, further, that there is no possibility of a maintenance engineer, when working on the controller, coming into contact with a moving part of the lift. The various contactors, relays, auxiliary resistances and fuses are mounted on a panel of slate, "Sindanyo," or other insulating material, the panel in turn being fixed to an angle iron framework. The control relays and the starting, accelerating, and retarding contactors, together with their associated interlocks, are electromagnetically operated, the magnet coils being supplied from the control voltage. Contacts performing certain operations require delayed actions of varying durations, and this is effected by dashpots, mechanical time lags, condenser and resistance networks, or a thermionic valve and condenser circuit. Copper-to-carbon and copper-to-copper are the contact combinations generally adopted. Blow-out coils are fitted where necessary to the contactors in order to extinguish the arc, and when copper-to-copper contacts are used these must be carefully designed to give a "rolling" contact. By this means the portions of the contact surfaces on which the circuit is made and broken, and which suffer any deterioration due to arcing, are slightly removed from the portions which actually carry the current. If the contacts are not correctly designed and the roll is insufficient, or is lost due to faulty maintenance, there is danger that the arcing will cause the contacts to weld, and if the reversing contactors are so affected serious damage may result. Copper contacts are sometimes faced with silver, silver-tungsten or silver-cadmium of a thickness up to about 0·05 in. to minimize deterioration due to arcing. The main reversing contactors carry a current depending upon the size of the motor and at the voltage of the motor supply, whilst the operating coils of the contactors, relays, and brake carry comparatively small currents, the voltage being

either that of the mains (a.c. or d.c.) or a separately produced
control voltage. D.C. magnets are more silent and generally
more reliable than a.c. magnets and it is, therefore, frequently

FIG. 16.1. CONTROLLER FOR SINGLE-SPEED LIFT
(*Wm. Wadsworth & Sons*)

the practice to supply all control circuits and the brake coil
from a direct-current voltage which, in the case of a.c. mains,
is obtained from a double-wave rectifier. The control voltage
employed is generally between 100 and 200 volts. A typical
controller for a single-speed lift is shown in Fig. 16.1, a

controller for a two-speed tandem motor lift in Fig. 16.2 and a controller for two interconnected collective control lifts in Fig. 16.3.

FIG. 16.2. TWO-SPEED CONTROLLER
(*Etchells, Congdon & Muir*)

Low Voltage Controller. The type of controller employed by the Express Lift Co. differs in many respects from those used by other British firms and a brief description is therefore warranted. Two views of a typical controller are shown in Figs. 16.7 and 16.8. The controller voltage employed has in the past been 50 volts—unusually low for a lift control voltage—which is the standard Post Office telephone exchange voltage. For more recent work however, the lift control voltage has been raised to 100—still low—to reduce troubles due to voltage drop

Fig. 16.3. Controller for Two Interconnected Collective Control Lifts
(*Express Lift Co.*)

whilst retaining the advantage of a relatively low voltage for insulation purposes. The telephone type equipment which was used has now been replaced by a heavier type designed to meet the requirements of lift service. The type of relay used is shown in Fig. 16.4 and is more robust than a telephone relay.

The contact assembly is built on two separate pile-ups of nickel silver contact springs fitted with silver contacts. Each

pile-up is clamped by two screws and is fixed to the yoke by a
third screw. The stationary contact springs are tensioned
against the moulded buffer block mounted on the armature
end of the yoke. The armature is pivoted on a pin located in an
adjustable saddle fixed to the yoke. The adjustment of this
saddle enables the armature to be correctly positioned to give
the required operating stroke. This relay may comprise between

FIG. 16.4. CONTROLLER RELAY
(*Express Lift Co.*)

two and six contacts in any combination of make and break
or, alternatively, two or four change-over contacts. The
operating time is between 30 and 50 milli-seconds and the
release time between 10 and 30 milli-seconds. The maximum
rating of the contacts is 5 amps, but for long life of the order
of ten million operations on non-inductive loads the rating is
2 amps at 50 volts, 0·7 amp. at 100 volts or 0·2 amp. at 200
volts d.c., and approximately double these current figures on
corresponding a.c. voltages. For inductive loads, spark-quench
circuits are provided to obtain reasonable contact lives. The

operation of the relay may be delayed by the use of resistances and condensers or thermionic valve and condenser circuits.

The selector for this controller has been described in Chapter XIV.

Three standard sizes of main contactors for operating from 100 volts are available for 15 h.p., 25 h.p., and 40 h.p. The main contacts are copper to carbon to eliminate as far as practicable the possibility and the consequences of contacts welding together. Auxiliary contacts for these contactors are similar to the relay contacts, i.e. nickel silver springs with silver contacts. The armatures are pivoted on the knife-edge principle. A 25 h.p. contactor is shown in Fig. 16.5. Starting contactors which have a heavy duty to perform are similar to the main contactors, but have heavy silver to silver contacts instead of copper to carbon.

Control Features. One of the most important functions of a controller is to ensure that it is impossible for the lift to be operated when any landing or car gate is open. This is usually effected by wiring the gate interlocks in series in one of the mains to the control circuits (see Chapter XI).

With automatic lifts it should not be possible to interfere with the car's motion by operating a landing button when a passenger has entered or is travelling in the car unless collective control is used. With this type of control it is possible to stop the lift car at any intermediate landing for the reception or discharge of passengers. A delayed contact associated with the car gate contact is arranged to cut out the landing buttons for about five seconds after the passenger has entered the car and closed the gates. He is thus allowed this period in which to select and operate the required car button (see Chapter XI). If the car button is not pressed during this interval, the car may be called to any floor by the operation of the relative landing button. Immediately a car button is pressed, however, the controller is arranged to cut out the landing buttons until the car has reached the end of its journey and the passenger has left the car and closed the gates. This is effected by incorporating a relay which is energized by the pressing of a car button, the relay contacts cutting out and reinserting the landing buttons at the commencement and end of each journey respectively.

To guard against the possibility of the car or counterweight encountering an obstruction, a relay is connected in the mains and is energized when a reversing contactor operates. After a period of from $1\frac{1}{2}$ to 2 complete journeys, a lagged contact (one

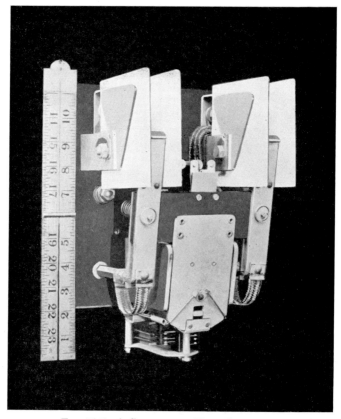

FIG. 16.5. A CONTROLLER MAIN CONTACTOR
(*Express Lift Co.*)

with delayed operation) of this relay is arranged to cut off the control supply. This contact is usually of the hand reset type, and therefore the cause of the trouble may be removed before resetting the contact. As a precaution against the main contactors welding it is sometimes the practice to arrange that this delayed contact disconnects the no-volt coil on the main circuit-breaker, and thereby cuts off both the motor and control circuits. Another method of providing this safeguard is by arranging that a condenser discharges through a neon tube or a resistance network if the controller remains energized longer than required. The discharge current operates a relay which cuts off the control supply.

A reverse-phase relay should be incorporated in the controller of a polyphase motor, the functions of this relay being to guard against damage to the motor due to low voltage, phase reversal, and phase failure. The relay operates on the induction motor principle, a disc acting as the rotor. The relay contact is opened and closed by the motion of the disc and is kept closed by the operating coils when the phases are in proper relation. When reversal occurs, the rotor turns in the opposite direction, thereby opening the relay contacts which in turn cut off the motor supply and prevent operation of the motor in the wrong direction. Low voltage reduces the relay torque and the contacts again open, thus cutting off the motor supply and preventing possible stalling, burn-out, and increased speed. When phase reversal occurs, the relay loses its torque and the contacts open and thus guard against a possible motor burn-out due to running on single phase.

It is frequently the practice to interlock the motor and control circuits so that, in the event of a failure of the motor supply, the control supply is automatically disconnected. This is achieved with a polyphase motor supply and a single-phase control supply by inserting two relays in the mains, these relays being supplied from different pairs of line wires. A contact of each relay is joined in series in the control supply so that in the event of any phase becoming disconnected the controller is cut off. With a rectifier d.c. control supply the two circuits may be interlocked by means of two opposed relay coils; one supplied from the d.c. control voltage, the other from a separately rectified d.c. voltage obtained from the polyphase motor supply. When both coils of the relay are energized, the contact closes and completes the control circuit. If either the motor supply or the control supply fails, one of the coils is disconnected and the relay contact thereby opened.

All controller wiring should be of the flame-resisting type.

The operating handle of every car switch lift should be arranged to return to the OFF position when released.

To safeguard the motor against possible damage caused by the operation of the reversing contactor before the motor has come to rest, a lagged contact is wired in the control supply to the reversing contactor coils. This contact closes a few seconds after the reversing contactor opens at the end of a journey, thus giving the motor time to come to rest before the contactor can be re-energized.

In addition to the reversing contactors being electrically interlocked, i.e. an auxiliary contact of each contactor being joined in series with the operating coil of the other, they are also mechanically interlocked as an additional safeguard against both operating together.

Other electrical interlocks have for their objects the prevention of the fast- and slow-speed contactors operating together, and ensuring that any accelerating and decelerating rheostats are inserted in correct sequence. The control circuit is sometimes interlocked with the brake hand release so that the controller is disconnected when the brake release is operated. It is thus impossible for the lift to be operated if the hand release has been overlooked after operation during testing. When the car safety gear operates, the motor is overloaded, and to prevent the possibility of damage if the fuses fail to blow, an interlock switch is frequently fitted to the gear so that when the safety gear sets, the switch cuts off the control circuit. In addition, this switch prevents the car from moving if the proper clearance does not exist and the safety jaws scrape the guides.

A further facility invariably provided on lift controllers consists of a change-over switch, the operation of which, together with the pressing of associated test buttons, permits of the lift being worked from the motor room for test purposes.

It is now recognized that accurate lift working records are an essential to efficient maintenance and service, and therefore energy meters to record the lift's electrical consumption, and counters to record the number of operations of the controller, are now being fitted as standard items of the control equipment. Operation counters are usually of the "Veeder" pattern and generally register the number of operations of the reversing or accelerating contactor. More complete data are obtained by fitting, in addition to an operation counter, a revolution counter on the main driving shaft and thus obtaining a record of the distance travelled by the car during any period.

If the floor selector is driven from the car or counterweight by a chain, tape, wire or similar means, a switch should be provided to stop the lift if the chain, tape or wire becomes slack or breaks.

Control circuits at mains voltage should be connected between phase and neutral or, on d.c. systems, between one outer and mid-point and not across phases or outer conductors. Where control circuits are supplied at a reduced voltage, the supply should be from a double-wound step-down transformer, or where a rectifier is used, one pole of such a circuit should be earthed and no single-pole switch or fuse except a testing link should be placed in that pole of the circuit.

Any pivoted joint of a contactor or relay should be short-circuited by a flexible lead if the joint carries a current.

If main contactors of opposite polarity or in different phases are adjacent to each other they should be fitted with arc shields.

All cable terminals should be suitably marked.

The incoming supply mains in the machine room should terminate either in a circuit-breaker with over-current release or in a main switch with fuses. In the latter case the lift controller should be equipped with over-current protection. For a single lift the circuit-breaker or switch should be fixed near the machine-room entrance. With more than one lift in a machine room, each circuit-breaker or switch should be conveniently situated with respect to the lift machine it controls and should be suitably labelled. The opening of the circuit-breaker or main switch should apply the machine brake.

The wiring of the control circuits should be separated from that of the main power and lighting circuits by running it in separate trunking or conduits.

CAR SWITCH CONTROLLERS

With this type of controller, floor selection and stopping are usually controlled entirely by the movement of the car switch, although with many modern high-speed controllers the stopping is performed automatically and independently of the car switch. In the latter types the car switch only performs the functions of starting and of selecting the floors at which stops are to be made, the switch being returned to the OFF position after passing the floor immediately preceding that at which a stop is required. With ordinary single- or two-speed controllers for speeds up to about 100 ft. per min. and 200 ft. per min. respectively, starting and accelerating are performed by moving the car switch in the appropriate direction to the first position. If two speeds are employed, the switch is then moved to the second position, after which the motor accelerates to its full speed. On approaching the desired floor, the switch is returned to the first position and the motor decelerates to its lower speed. Finally, the switch is returned to the OFF position and the car is brought to rest, the accuracy of the levelling depending upon the operator's judgment in returning the car switch at the correct moment.

Four-floor Car Switch Controller with "Trulevel" Machine, Manually-operated Doors, and 2-phase, 3-wire A.C. Motor. This controller, designed by the Express Lift Co., is suitable for a speed of up to about 200 ft. per min. The "Trulevel" machine is described and illustrated in Chapter XIV. Fig. 16.6 shows the wiring diagram for this controller, whilst Fig. 16.7 is a front view and Fig. 16.8 a rear view of the controller. The main motor of the "Trulevel" machine is a single speed 2-phase slip ring motor with three steps of rotor starting resistance and the small auxiliary motor is of the single speed 2-phase squirrel-cage type. The inductor system of levelling is employed and the operation of the controller is as described below. In the description N/O means a normally open contact and N/C a normally closed contact.

If all doors and other safety contacts are closed, the operation of the car switch to the UP position closes the car switch contact CSU and relay 81 operates. N/O contact 81 operates contactors 86 and 1 and the timing relays $T10$ and $T11$. N/O contact 1 prepares the circuit to the main motor for the UP direction whilst another N/O contact 1 and a N/O contact 86 prepare the circuit for the brake coil. The clutch coil is energized by N/O contact 86 and the clutch operates to isolate the level-

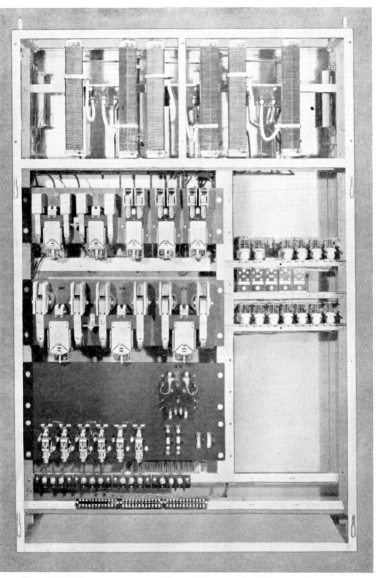

FIG. 16.7. CONTROLLER FOR FOUR FLOORS, CAR SWITCH CONTROL,
WITH "TRULEVEL" MACHINE

(*Express Lift Co.*)

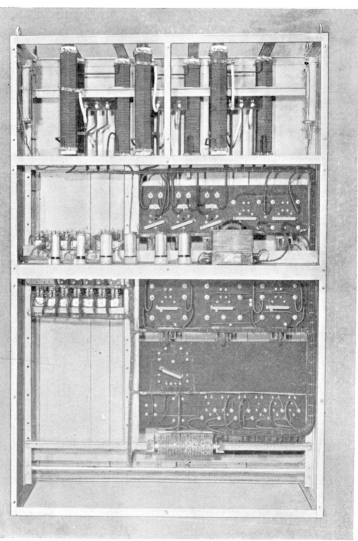

FIG. 16.8.   REAR VIEW OF CONTROLLER SHOWN IN FIG. 16.7
(*Express Lift Co.*)

ling motor. Clutch contact CL opens to enable relay CPR to operate. Another N/O contact 86 operates the timing relay $T5$. Contactor 6 now operates via N/O $T5$, N/O 1, CPR, $T10$ and $T11$. N/C contact CPR opens the circuit to contactors 91 and 92. N/O contact 6 completes the brake and motor circuit, the brake lifts and the motor commences to rotate. The brake contact BK opens and allows relay BR to operate. N/O contact BR operates contactor 9, N/O contacts of which short-circuit the first steps of rotor resistance. N/C contacts of 9 open $T10$ circuit and $T10$ times out due to its parallel condenser and resistance circuit. N/C contact $T10$ operates contactor 10 and N/O contacts 10 short circuit the second step of rotor resistance. N/C contact 10 opens $T11$ circuit and $T11$ times out. N/C contact $T11$ operates contactor 11 and N/O contact 11 short circuits the last step of rotor resistance after which the motor accelerates to full speed.

When the lift approaches the floor at which a stop is desired, the car switch is released and relay 81 is de-energized. Contactor 1 and relay 47 release and the motor is disconnected from the supply, and the brake is partially applied (the circuit to the brake coil is now via N/O contact 6, N/O contact 86 and resistance $R4$). Contactors 6 and 86 are held by N/O contact $T4$. Relay $T4$ commences to time out, this circuit being opened by N/O contact 47. As the car comes to rest, relay $T4$ times out and opens contactors 86 and 6. The release of N/O contact 86 releases the clutch coil and applies full brake. The clutch is now released and connects the slow speed motor to the supply. The release of contactor 86 energizes the levelling inductor coils $1US$ and $1DS$. With the clutch de-energized and relay CPR released, the circuit is complete to relays LU and LD via inductor contacts $1US$ and $1DS$. N/O contacts LU and LD prepare circuits to the low speed reverser contactors 91 and 92 but N/C contacts LD and LU also isolate these low speed reverser contactors. If we assume that the car stops below the level of the floor, then inductor $1DS$ will be embracing an inductor plate and the inductor contact $1DS$ will be open. Relay LD will be released and its N/C contact will then complete the circuit to the Up low speed contactor 91. N/O contacts of LU and 91 re-operate contactor 6 and N/O contacts of 6 and 91 connect the low speed motor to the supply and energize the brake coil. The car then travels Up at slow speed.

At floor level the inductor plate enters inductor $1US$ and contact $1US$ opens to release contactor 91 and relay LU. The low speed motor is disconnected and the brake applied to bring the car to rest at floor level.

If the car slides past the floor level during slowing, inductor $1DS$ moves beyond the inductor plate and therefore contact $1DS$ releases and completes the circuit to the down low speed contactor 92 via N/O contact of LD and N/C contact of LU. N/O contacts of LD and 92 operate contactor 6. The car returns to floor level and on the re-opening of inductor contact $1DS$ due to the plate entering $1DS$ inductor, relay LD and contactor 92 release.

Various protective features are incorporated in the controller. To ensure that the high speed motor is not energized until the clutch has operated to release the low speed motor, relay CPR is fitted. This operates when the clutch is fully operated, the clutch contact CL opening to remove the short circuit on CPR. If the clutch is energized but does not operate, CPR remains short circuited. When contact of 86 open circuits, the clutch relay CPR will be de-energized even if the clutch itself fails to release. N/O contact CPR in the circuit of line contactor 6 ensures that the high speed motor cannot be connected to the supply until the clutch has operated to disconnect the low speed motor from the machine. The timing relay $T5$ is operated by a N/O contact of the clutch contactor 86 and commences to time out on the release of that contactor. If the levelling operation is not completed within a predetermined time delay, relay $T5$ releases and opens the circuit to 91, and to 92 and to 6.

In Fig. 16.7 showing the front view of this controller the high speed rotor starting resistances are on the top shelf. On the next panel (left side) are from left to right, contactors 91, 9, 10 and 11, and on the next panel (left side) are, from left to right, contactors 1, 2, and 6. The bottom two panels accommodate contactor 86, the motor overload coils, and the fuses. On the right hand side are the various relays and timing condensers. The spaces above and below these relays would be occupied by additional equipment if the lift had been provided with automatic control.

AUTOMATIC CONTROLLERS

As the name implies, this type of controller is one in which the operations of starting, accelerating, and stopping the lift are performed automatically after a landing or car button has been pressed. The selection of the desired floor is effected by means of floor relays and direction switches; one relay and switch for each floor served. On pressing a car or landing button the corresponding floor relay is energized, and at the same time the main supply is connected to the motor. The floor relay is disconnected when the car reaches the desired floor and opens the direction switch.

Semi-automatic Single-speed Controller. The wiring diagram shown in Fig. 16.9 is suitable for a single-speed goods or service lift with two landings and Up, Down, and Stop buttons fitted on each landing. The supply is three-phase, and a slip-ring motor with one step of starting resistance is used. If the car is at the bottom landing, the operation of the circuit is as follows. The down terminal limit switch is open and it will be noted that pressure on either Down button (D) will have no effect as it is impossible to complete a circuit for coil B by operating a Down button. If an Up landing button (U) is pressed, the circuit for coil A is via $+$ve, the door and car contacts, stop buttons S, change-over switch, operated U button, Up terminal limit switch, coil A, interlock $B3$, rotor contactor coil R, and brake economy resistance to $-$ve. The closing of $A1$ and $A2$ energizes the brake coil and releases the brake. Contactor $R1$ closes a few seconds later and the motor accelerates to full speed corresponding to, say, 100 ft. per min. As soon as the brake has released, the brake economy switch opens and inserts the economy resistance. When the car reaches the upper floor the Up terminal limit switch fitted in the well is opened by a ramp on the car, and the car then stops due to the de-energizing of coil A and the opening of contactor $A5$.

A circuit for coil B is similarly provided on the operation of a Down button.

The change-over switch permits of the operation of the car from the motor room by moving the switch to test and depressing a test button.

Contact $P1$ is lagged for a period of from $1\frac{1}{2}$ to 2 times of a complete journey and by cutting off the control circuit after

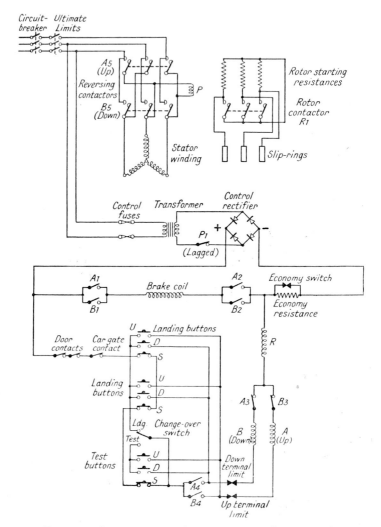

FIG. 16.9. SEMI-AUTOMATIC CONTROLLER FOR SERVICE LIFT

this period so prevents the motor from being connected to the mains in the event of a brake coil failure.

Fully-automatic Single-speed Controller, Four Floors with Prelocks and Inductor Stopping. The diagrams in Fig. 16.10 and Fig. 16.11 show the principle of the Express Lift Co.'s con-

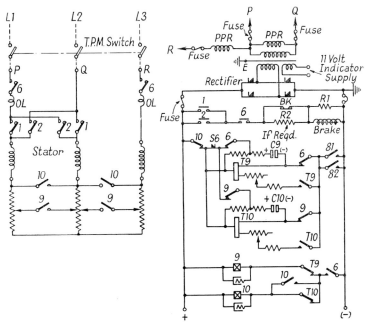

FIG. 16.10. POWER CIRCUIT FOR AUTOMATIC SINGLE-SPEED CONTROLLER, FOUR FLOORS, WITH PRELOCKS AND INDUCTOR STOPPING

troller for a speed of about 100 ft. per min. employing a squirrel-cage motor with starting resistance. The selector and inductor described in Chapter XIV are both used in this installation. In the description below, N/O means normally open and N/C normally closed.

If the lift is standing at an intermediate floor, both relays 23 and 24 are operated and their contacts prepare the circuits for the UP direction relay 81 and the DOWN direction relay 82. Assume that the car is standing at the second floor and

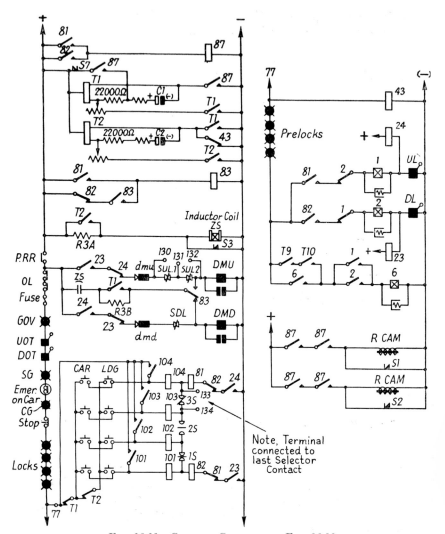

Fig. 16.11. Control Circuit for Fig. 16.10

then selector contacts $1S$ and $2S$ will be open and if the doors are
closed relay 43 will be operated. If a passenger opens the doors
and enters the car, relay 43 releases, N/C contact 43 closes
and operates relay $T2$ and N/C contact. $T2$ then opens the
landing button feed. On closing the door, relay 43 re-operates
and N/C contact 43 opens and cuts off relay $T2$ which commences to time out after about 5 seconds. This is effected by
the condenser and resistance circuit in parallel with the relay.
If the passenger operates the 3rd floor push button, relay 103
and 81 operate and self hold via N/O contact 103. N/O contact
81 prepares the circuit for contactor 1 and a second N/O
contact 81 operates relay 87. The retiring cam, relay $T1$ and
relay 83 now all operate. N/C contact $T1$ cuts off the button
feed, whilst N/O contact $T1$ shorts the selector reducing
resistance $R3B$. N/O contact $T1$ operates relay $T2$ and N/O
contact $T2$ shorts the ZS inductor reducing resistance $R3A$.
The retiring cam locks the doors and prelocks the circuit to
contactor 1. The main contacts of contactor 1 prepare the
motor windings for the UP direction and N/O contact 1
prepares the brake circuit. N/O contact 81 operates relays $T9$
and $T10$ and timing condensers $C9$ and $C10$ then charge. N/C
contacts $T9$ and $T10$ open circuits contactors 9 and 10 which
have not yet operated, whilst N/O contacts $T9$ and $T10$
complete the circuit to contactor 6 which operates. The brake
now lifts and the motor accelerates. The brake economy
resistance $R2$ and switch BK are fitted if required. N/C
contact 6 opens the timing circuit of $T9$ which times out and
then N/C contact $T9$ operates contactor 9. The main contacts
of contactor 9 short out the first step of starting resistance.
Similarly the main contacts of contactor 10 short out the second
step of starting resistance later, after $T10$ times out.

On leaving the second floor, inductor contact ZS closes,
the selector steps to position 4 and contact $1S$ closes. The
operation of the selector is described in Chapter XIV. On
approaching the 3rd floor, contact ZS again closes, selector
steps to position 5 and contact $3S$ opens. Relays 103, 81 and
87 all release. N/O contact 81 releases contactor 1 and contacts of 1 open the motor circuit and release the brake. N/O
contact 1 releases contactor 6. The car comes to rest at the
floor, the retiring cam advances and opens the prelocks, $T1$

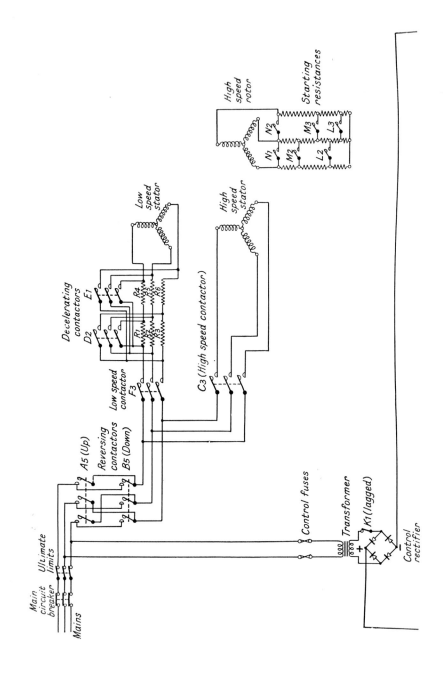

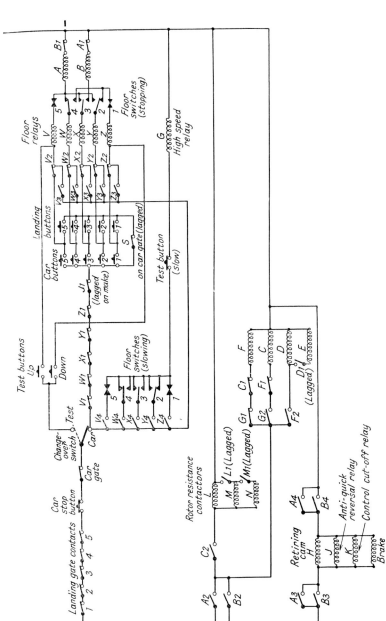

Fig. 16.12. Two-speed Automatic Controller with Tandem Motor

times out to insert the selector reducing resistance, and restores the push button feed. $T2$ starts timing out, but when the passenger opens the doors, relay 43 releases and N/C contact 43 maintains $T2$. When the passenger leaves the car and closes the doors, relay 43 re-operates to open $T2$. After 5 seconds $T2$ times out to restore the landing push button feed and the lift is now ready to accept landing calls.

At the terminal floors the stopping is controlled by the limit switches UL and DL which open the circuit to contactors 1 and 2 and relays 23 and 24.

Operation in the Down direction is similar when relay 82 operates.

Fully-automatic Two-speed Controller with Tandem Motor. The diagram shown in Fig. 16.12 is for an automatic controller with the car operating between five floors. A two-speed tandem motor is employed and the controller is suitable for car speeds up to about 250 ft. per min. Starting and full-speed running are performed on the high-speed slip-ring section, whilst the slow running speed of, say, one-sixth the fast speed is obtained from the squirrel-cage section.

If the car is at the bottom landing and the top landing button No. 5 is pressed to bring the car to that floor, the circuit for the Up contactor coil A is via landing gate contacts, car stop button, car gate contact, change-over switch, $V1, W1, X1, Y1, Z1, J1, S$, landing button No. 5, $V2$, floor relay V, No. 5 stopping switch, coil A, and interlock $B1$. Contact $A1$ on opening ensures that B cannot operate; $A3$ and $A4$ on closing energize the brake coil and release the brake. Coils H, J, and K are also fed by contacts $A3$ and $A4$. The opening of $V2$ and closing of $V3$ transfers the feed of coils V and A from the change-over switch direct to floor relay V, thus cutting out the buttons and preventing interference when the car is in motion. The closing of $V4$ provides a parallel circuit for the high-speed relay G via $V4$, slowing switch No. 5, and slow test button. Contact $G1$ opens and guards against the low-speed contactor $F3$ operating, whilst $G2$ closes and provides a circuit for C via $A2$, $G2$, and $F1$. Reversing contactor $A5$ has operated, and with the energizing of coil C contactor $C3$ closes and the mains are then connected direct to the high-speed stator, thus causing the motor to rotate. The closing of $C2$ provides a circuit for the first rotor contactor coil L, whilst $L2$ and $L3$ cut out the first step of rotor resistance. After a brief interval $L1$, which is lagged, closes and energizes M, thus causing $M2$ and $M3$ to cut out the second step of rotor resistance. Similarly, $M1$ provides a circuit for N so that $N1$ and $N2$ cut out the last step of resistance and the lift attains its full speed of, say, 250 ft. per min.

When the car approaches floor No. 5, slowing switch No. 5 of the floor selector is opened automatically. This cuts off the supply from coil G, and the opening of $G2$ disconnects the high-speed coil C, with the result that contactor $C3$ opens. The closing of $G1$ energizes F, and contactor $F3$ closes, thus connecting the mains to the low-speed stator via buffer resistances $R1$–6. On closing, shortly afterwards, $F2$ feeds coil D, and contactor $D2$ short-circuits the first steps $R1$, $R2$, and $R3$ of the buffer resistance. $D1$ is slightly lagged, and on closing energizes coil E, and thus the second steps of buffer resistance $R4$, $R5$, and $R6$ are short-circuited. Alternatively, these resistances may be short-circuited by a single magnet as shown in Fig. 6.14. The lift is now running at the slow levelling speed of, say, 40 ft. per min. On approaching still nearer to floor No. 5, the stopping switch No. 5 is opened and thus disconnects coil A, which in turn causes contactor $A5$ to open. The supply is now cut off from the slow-speed section and the opening of $A3$ and $A4$ causes the brake to operate and the car to level at the floor.

Operation in the down direction is similar to that described above, except that coil B is energized and contactor $B5$ operated instead of $A5$.

The up and down test buttons enable the car to be operated from the motor room either at fast or slow speed, the latter if the slow test button is also operated.

Various protective and non-interference devices may be incorporated in the controller. In the diagram shown, H is the retiring cam coil and is energized when either the UP or DOWN contactor operates, and by withdrawing the cam on the car beyond reach of the gate lock striker prevents a landing gate from being "snatched" open as the car passes. The coil releases the cam when the reversing contactor opens. Contact $J1$ on the main control feed is lagged on make for a period of a few seconds, sufficiently long to enable the motor to come to rest after the reversing contactor has opened, and this prevents the motor from being reversed before it has stopped rotating. The control cut-off contact $K1$ disconnects the control supply if the motor is connected for a period of more than about $1\frac{1}{2}$ times a full journey. Associated with the car gate contact is a contact S, lagged for about five seconds, and this disconnects the landing buttons for this period after a passenger has entered the car. The ultimate limits cut off both motor and control circuits in the event of the car over-travelling the terminal landings.

Down Collective Control with Single-speed Motor, Rheostatic Starting (Express Lift Co. Ltd.)

1. GENERAL DESCRIPTION

Pushes are provided in the car corresponding to each floor served. A single call push is fitted at each landing. Registration of a landing call is indicated by illumination of the push.

All calls which are registered from either the car or landings are stored in the system. Car calls are answered successively by the car which reverses at the highest call.

Landing calls are answered successively by the car when it is travelling in the DOWN direction only. The car does not stop for landing calls when it is travelling in the UP direction, with the exception of the highest landing call which is answered only if there are no car calls ahead of this landing call.

After all calls have been answered, the car automatically returns to the parking floor and remains there with the doors closed.

The circuit diagram is shown in Fig. 16.13 and the controller in Fig. 16.14.

2. SEQUENCE OF OPERATIONS

Car Calls. Car call relays $1C$, $2C$, etc., are operated by the corresponding pushes in the car via winding B–A, a $N/C\ G$ contact and relay CP. These relays self-hold via their own N/O contacts, a $N/O\ 48$ contact, $N/O\ T5A$, $N/O\ FC$ and $N/C\ MAA$.

A car call relay is cancelled by the opening of a $N/O\ 48$ contact when the car stops.

The $N/C\ G$ contact in series with each car push prevents a car call from being registered for a floor while the car is in the zone of the floor. Without this feature, the registration of the car call would cause the car to stop away from floor level if

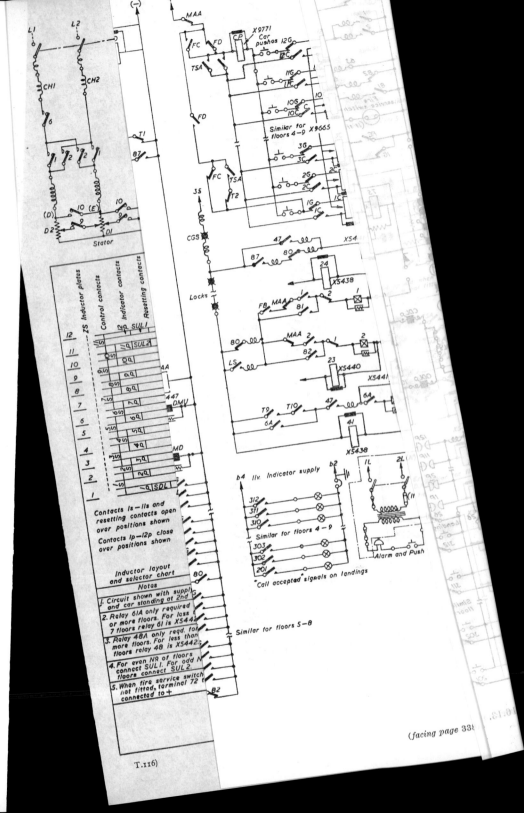

the call was registered while the car was travelling away from the floor and while it was still in the floor zone.

Landing Calls. Landing call relays 201, 302, 303, etc., are operated by the corresponding landing pushes via winding $B–A$, N/O FC and the N/C G contact or a N/O 61 contact. These relays hold via their own N/O contacts, winding $B–A$ and a N/C G contactor a N/O 61 contact. They are also held via winding $D–C$, their own contacts and N/O 26.

The N/O 61 contact in series with winding $B–A$ is for a similar purpose to that described above. (See next section for description of 61 and 61A.)

A landing call relay is cancelled by the opening of N/O 26 in series with winding $D–C$ and by the opening of N/O 61 or N/O 61A in series with winding $B–A$.

Registration of a landing call is indicated by illumination of the push.

Relays 61 and 61A. These relays are operated for the Up direction via N/C 2 and N/O 26. They are released for the Down direction by the opening of N/C 2.

Contacts of these relays in the negative feed to winding

FIG. 16.14. CONTROLLER FOR DOWN COLLECTIVE CONTROL
(*Express Lift Co.*)

$B–A$ of the landing call relays enable a call relay to be operated for opening the doors at the floor at which the lift is standing. When the doors have fully opened, N/O 26 releases 61 and 61A and so cancels the call relay.

A second function of the contacts of the above relays is to enable landing calls to be registered when the lift is travelling in the Up direction, even though the lift may be in the zone of the floor for which the call is registered.

When the lift is travelling Down, 61 and 61A are released, and a landing call cannot be registered for a floor while the lift is in the floor zone.

Selector Circuit and G *Relays.* For a description of the operation of the Selector, see Chapter XIV.

With inductor ZS energized, the selector is stopped by the closing of N/O LS contacts each time the car passes a ZS plate in the well (LS is operated by ZS inductor contacts).

The positions over which the various contacts open and close are shown on the Selector Chart.

Selector contacts 1P, 2P, etc., close to operate successively relays 1G, 2G, etc., as the car passes through the corresponding floor zones. (For example, 1G will be operated while the car is at the first floor.)

The operation of contacts 1S, 2S, etc., is described later.

Directional Set-Up. The direction in which the car will travel is determined by the operation of relays 81 (Up direction) and 82 (Down direction). Contacts of these relays control the main reversing contactors 1 and 2.

Relays 81 and 82 are controlled by contacts of the call relays in conjunction with the selector S contacts. Selector contacts 1S, 2S, etc., are arranged in a chain and a contact is open on each side of the feed to the chain, corresponding to the position of the car. For example, with the car at the second floor, 1S and 2S will be open.

The feeds to the chain are via N/O contacts of the call relays. Hence a call above the car will operate 81 and a call below the car will operate 82.

Relays 81 and 82 are electrically interlocked, so that the operation of one prevents the operation of the other. Therefore, with a number of calls stored in the system, either 81 or 82 will remain operated, and the car will continue its direction of travel until it has answered all calls ahead of it.

For example, if the car has been travelling Up (81 operated) and calls exist below, on answering the highest call 81 will be released, thus preparing the circuit to 82 via N/C 81.

82 operates when $T1$ times out. The circuit to 82 is via the call relay contacts, the chain of N/C selector S contacts, N/O 23, N/C 81 and N/C $T1$. The function of $T1$ is to prevent sudden reversal of the car.

N/O 82 operates 87.

N/O 87 closes to hold 82.

Relay 83 is operated by N/O 81 for the Up direction and self-holds via its own N/O contact and N/C 82. 83 is released for the Down direction by the opening of N/C 82.

C/O 83 selects either DMU or DMD to step the selector for the correct direction.

3. RELAY 13

Relay 13 is operated via winding B–A, N/O FB, N/O $T2$ and N/O $6A$ when the car is running.

When the car has stopped, relay 13 is timed out by N/O $T2$.

N/O 13 releases 80.

N/C 13 operates 20 to close the doors.

N/C 80 prepares the circuit to 43 via N/O 87 and CGS contacts.

N/C 80 prepares the circuit to 1 or 2.

Registration of a car call will cancel the time delay of 13. The operation of a car push operates relay CP so long as pressure is maintained on the push.

N/O CP energizes the neutralizing winding C–D of relay 13 via N/C $6A$ and N/O 13; and 13 releases.

The early release of 13 enables the car to start more quickly.

4. DOOR OPERATION

Door Opening. The car door and landing doors are connected by a solenoid-operated retractable vane. The solenoid which operates the vane is operated by relay 43.

N/O 47 releases 43 when the car stops.

N/O 43's open to release the retiring cam which ejects the vane to unlock the landing doors and to connect them with the car door.

The opening of the landing door lock contacts releases relay 41.

The opening of the doors is controlled by contactor 21.

When the car is approaching a floor at which it will stop, relay 80 operates.

N/O 80 operates 100 via N/O 26 and N/O FB.

100 holds via its own N/O contact, N/O FB and N/O 26.

N/O 100 operates 21 via $DGOL$'s, limit ODL, N/C 43, N/C 41, N/C 20 and N/C 47.

N/O 21's close to energize the door gear motor, and the doors commence to open. CGS contacts open.

N/O 21's operate $T2$.

N/O $T2$ operates 13 via winding B–A and N/O 100.

When the doors have fully opened, limit ODL opens to release 26 and contactor 21.

N/O 26 releases 100.

N/O 21's open to de-energize the door gear motor.

N/O 21's open to commence the timing out of $T2$.

The doors may be opened by pressure on the OPEN DOORS push in the car. This push operates relay 100 which opens the doors as described above.

The doors may be opened at the floor at which the lift is standing by pressing the landing push at that floor. This operates the corresponding landing call relay, which self-holds.

Contacts of the call relay operate 80 via a N/O G contact and N/C 1.

N/O 80 operates 100 which opens the doors as described above. When the doors have fully opened, N/O 26 and N/O 61 or N/O 61A release the call relay.

Door Closing. The closing of the doors is controlled by contactor 20.

When relay 13 times out, N/C 13 operates 20 via $DGOL$'s, N/O FD, N/O $T5A$, N/C 100, N/C 21 and N/C 47. (See section 3.)

N/O 20's close to energize the door gear motor, and the doors commence to close. When the doors have fully closed, limit CDL opens to release 20.

N/O 20's open to de-energize the door gear motor.

The doors may be reopened during closing by pressure on the safety edges or by operation of the OPEN DOORS push.

Relay 100 operates.

N/C 100 opens to release 20.

N/O 20's open to de-energize the door gear motor to stop the doors.

N/C 20 with N/O 100 operate 21 to reopen the doors.

Door Test Switches. Switches are provided in the car to enable the doors to be tested under power. The TEST–RESET switch is placed in the TEST position to cut off the normal feed to 81 and 82, and to operate MA and MAA. N/C MA isolates the landing call relays and N/C MAA isolates the car call relays.

The doors may now be tested by operating the OPEN–CLOSE switch. Relay 100 is operated to open the doors and 87 is operated to close the doors.

With relays MA and MAA operated, the lift is on MAINTEN-ANCE control and is operated from the UP and DOWN pushes on top of the car.

Door Protection Timer. When the doors commence to open, the limit CDL closes, thus making the supply to the $T5$ circuit.

$T5$ will operate and will remain operated during the normal door open-close cycle. When the doors are reclosed, limit CDL will break the circuit to $T5$. Under this condition $T5$ will remain operated.

If, for some reason, the doors do not fully close (e.g. the safety edge continually meeting with some obstruction), $T5$ will time out after the normal open-close time has elapsed.

N/O $T5$ will release $T5A$.

N/O $T5A$ will open in the circuit to 20.

N/C $T5A$ will maintain 20 via N/O 87.

The safety edges will again operate 100 to open the doors and release relay 26.

N/O contacts of 26 and $T5A$ open to release the existing car calls and landing calls.

81 or 82 releases, and N/O 81 or N/O 82 releases 87.

When $T5$ has timed out, an attempt may be made to close the doors by operating a landing push, or they may be closed by a car push.

The circuit to 20 is via N/O 87 and N/C $T5A$.

5. POWER SECTION

Starting. Starting is initiated automatically by the release of relay 13 (see section 3) provided that calls exist in the system.

The sequence is shown for the Up direction.

N/C 13 operates 20 to close the doors, as described elsewhere.

81 and 87 will be operated.

N/O 87 operates $T9$ and $T10$.

N/O 13 releases 80 (see end of this section).

N/C 80 prepares the circuit to 43 via N/O 87.

N/C 80 prepares the circuit to 1 via N/O 81, N/C 2 and limit UL.

When the doors have closed and locked, the lock contacts complete the circuit to 1, which operates.

1 self-holds via its own N/O contact, N/C MAA, N/O FB and N/C 80.

N/O 1 operates 47.

N/O 47's operate $T1$.

N/O $T1$ operates $T2$.

N/O 47 maintains 43.

N/O 47 operates $6A$ via N/O $T9$ and N/O $T10$.

N/O $6A$ operates 6.

N/O $6A$ holds $6A$ and 6.

N/O 1's and N/O 6's close to energize the main motor.

N/O 1 and N/O 6 close to lift the brake and the car starts with full resistance in circuit.

N/O 6 operates 48 and $48A$.

Acceleration. N/C 6's open to commence the timing out of $T9$.

$T9$ times out and N/C $T9$ operates 9 via N/O 6.

N/O 9's close to move the star point to $D2$, $E2$, $F2$, and the motor accelerates.

N/C 9's open to commence the timing out of $T10$.

$T10$ times out and N/C $T10$ operates 10 via N/O 6.

N/O 10's close to move the star point to D, E, F, and the motor accelerates to full speed.

Stopping. Stopping is initiated by the operation of relay 80.

Relay 80 operates to stop the car in answer to a car call, irrespective of the direction of travel. Assume that a car call for the 3rd floor has been registered:

When the lift enters the 3rd floor zone, relay $3G$ operates.

N/O $3G$ operates 80 via N/O $3C$ and N/O 13.

80 self-holds via its own N/O contact and N/O 13.

Relay LS operates as the car comes on to the ZS plate which precedes the 3rd floor (when $3G$ operates).

N/O LS holds across N/C 80 to maintain 1.

As the car comes off the ZS plate, relay LS releases and so releases 1.

N/O 1 releases 47.

N/O 47 releases 6.

N/O 1's and N/O 6's open to de-energize the motor.

N/O 1 and N/O 6 open to apply the brake which stops the car.

The car stops for landing calls when it is travelling in the DOWN direction only.

Relay 80 is operated by N/O G contacts via N/O contacts of the landing call relays and N/C 1.

The car will stop for a landing call when it is travelling UP provided there are no calls ahead of it. When this occurs, relay 80 is operated by N/C 81 via N/O 83 and N/O 13.

The car then reverses to answer the existing landing calls which were by-passed on the UP journey.

80 is released each time by the timing out of 13, in preparation for the next journey.

6. PARKING FEATURE

A parking call for the main floor is registered by the timing out of $T2$, each time the car answers the first call away from the main floor.

N/C $T2$ operates $1C$ via winding B–A, N/O FD, N/O $T5A$ and N/C $1G$ contacts.

When there are no further calls to be answered, the car will start for the main floor, and when it arrives will park with doors closed.

Relay $1C$ is cancelled by the opening of N/O $6A$ contact when the car stops.

7. FIRE SERVICE SWITCH

Relay FC is operated via the Fire Service switch.

N/O FC operates FD.

N/O FC operates FB.

FD holds via its own N/O contact and N/C $1G$.

Placing the Fire Service switch in the OPEN position enables a person to gain control of the lift in an emergency.

Relay FC releases and N/O FC releases FB.

N/O FC contacts open to release existing car calls and landing calls.

Assume that the lift is standing at a floor and the doors are opening.

N/O FB opens to release 100.

N/O FB releases 13.

N/O 100 releases 21 to stop the doors.

N/C 100 with N/C 13 close to operate 20, which immediately closes the doors.

N/C FC operates $1C$ to start the car for the main floor.

When the car enters the first-floor zone, N/C $1G$ releases FD.

N/C FD reoperates FB.

N/O FD opens to release $1C$ and to prevent the registration of a parking call while the lift is on Fire Service.

N/O FD opens in the circuit to 20.

N/O FB reoperates 13 via N/O $T2$.

N/O 13 operates 80 via N/C 82 and N/C 1.

N/O 80 operates 100.

N/O 100 prepares 21 for opening the doors.

When the car stops the doors open (41 and 43 release).

The doors are prevented from closing automatically by the inclusion of N/O FD in series with N/C 13 in the 20 circuit.

N/C FD restores the feed to the car pushes via N/O $T5A$.

The lift is now completely under the control of a person in the car.

The registration of a car call will operate 81.

N/O 81 operates 87.

N/O 87 operates 20 via N/O CP, to close the doors.

Note. Constant pressure must be maintained on the car push until the doors have fully closed. Premature release of the push will release 20 to stop the doors, due to the release of CP. When the car arrives at the required floor, it will remain there with the doors open until a further call is registered.

If the Fire Service switch is opened while the lift is travelling Down, it will proceed directly to the main floor.

Assume that the Fire Service switch is opened while the lift is travelling Up:

N/O FB opens to release 1.

N/O 1 releases 47.

N/O 47 releases 6.

N/O 1's and N/O 6's open to de-energize the motor.

N/O 1 and N/O 6 open to apply the brake, which stops the car.

Relay $1C$ is operated as described above.

When $T1$ times out, N/C $T1$ operates 82 to initiate starting for the Down direction.

If the Fire Service switch is opened while the lift is parked at the main floor with the doors closed, the doors open as follows—

N/O FC releases FD (N/C $1G$ is open in hold circuit).

N/C FD in series with N/C 87 operates 21 to open the doors. (The doors cannot be opened by a landing push because the pushes are isolated by N/O FC.)

If the Fire Service switch is opened during closing of the doors, the doors will continue to close.

N/O 20 maintains contactor 20 until the doors have fully closed.

The car will then proceed to the parking floor as described earlier.

Note. Chokes and capacitors are fitted to various parts of the circuit to minimize interference to radio and television reception.

Variable-voltage Dual Directional Collective Controller. The following description relates to a typical Otis controller of this type, the circuit diagram of which is in Fig. 16.15. The motor generator set, lift motor, and gearing are shown in Fig. 16.16 and the controller in Fig. 16.17.

1. Names of Field Apparatus and Operating Equipment

ADL	. .	Attendant Down Light
ASC	. .	Auxiliary Stop Contacts
ATS	. .	Attendant Cut-out Switch
AUL	. .	Attendant Up Light
B	. .	Brake
BS	. .	Brake-operated Switch
BU	. .	Buzzer
BTS	. .	Broken-tape Switch
CAC	. .	Car Contacts
CAR	. .	Car Reset Brush
CB	. .	Car Buttons
$CDDL$	.	Down Car Direction Light
CM	. .	Retiring Cam Magnet
$CUDL$	.	Up Car Direction Light
D	. .	Down Inspection Button
DAS	. .	Down Auxiliary Stop Brush

FIG. 16.17. VARIABLE-VOLTAGE
DUAL COLLECTIVE CONTROLLER
(*By courtesy of the Postmaster-General*)

FIG. 16.16. VARIABLE-VOLTAGE GEARED LIFT DRIVE
(*By courtesy of the Postmaster-General*)

DCS	. .	Down Car Stop Brush
DHC	. .	Down Hall Contacts
DHR	. .	Down Hall Reset Brush
DHS	. .	Down Hall Stop Brush
DS	. .	Door Contacts
DV	. .	M.G. Set Driving Motor
EEC	. .	Emergency Exit Contact
ES	. .	Emergency Stop Switch
FDC	. .	Floor Directional Cam
FH	. .	Floor Brush Switch
GA	. .	Generator Armature
GIP	. .	Generator Interpole Field
GLF	. .	Generator Levelling Field
GS	. .	Gate Contact
GSEF	. .	Generator Self-excited Field
GSF	. .	Generator Series Field
GSH	. .	Generator Series Field Shunt
GW	. .	Gate Contact Switch
HB	. .	Hall Buttons
HDDL	.	Down Hall Direction Light
HPI	. .	Hall Position Indicator
HUDL	.	Up Hall Direction Light
INS	. .	Inspection Knife Switch
KS	. .	Knife Switch
1LS	. .	Down Final Limit Switch
2LS	. .	Up Final Limit Switch
LV	. .	Levelling Switch
MA	. .	Elevator Motor Armature
MCB	. .	Main Circuit Breaker
MF	. .	Elevator Motor Shunt Field
MIP	. .	Elevator Motor Interpole Field
NSB	. .	Non-Stop Button
OS	. .	Governor Overspeed Switch
PLT	. .	Pilot Light
IRF, etc.	.	Rectifiers
SES	. .	Service Emergency Stop Switch
SKS1	. .	Service Knife Switch (Motor)
SKS2	. .	Service Knife Switch (Controller)
SOS	. .	Safety-operated Switch
SS	. .	Car-stopping Switch

SSK	. .	Signal Cut-out Knife Switch
TKS	. .	Test Button Knife Switch
TRF	. .	Rectifier Transformer
TRS	. .	Signal Transformer
U	. .	Up Inspection Button
UAS	. .	Up Auxiliary Stop Brush
UCS	. .	Up Car Stop Brush
UHC	. .	Up Hall Contacts
UHR	. .	Up Hall Reset Brush
UHS	. .	Up Hall Stop Brush

2. NAMES OF SWITCHES AND RELAYS ON CONTROLLER AND RELAY PANEL

1*AE*	. .	1st Attendant Switch
2*AE*	. .	2nd Attendant Switch
ATC	. .	Attendant Control Relay
ATD	. .	Down Attendant Switch
ATU	. .	Up Attendant Switch
BL	. .	Buzzer Switch
C	. .	Potential Switch
1*C–TC*	.	Car-stop Floor Relays
D	. .	Down Direction Switch
2*D–TD*	.	Down Floor Relays
DF	. .	Direction Field Relay
DTB	. .	Down Test Button
DX	. .	Down Auxiliary Direction Switch
ER	. .	Excitation Switch
ERT	. .	Excitation Time Switch
1*E*	. .	1st Slow-speed Switch
2*E*	. .	2nd Slow-speed Switch
3*E*	. .	3rd Slow-speed Switch
GL	. .	Levelling Slow-speed Switch
H	. .	Field and Brake Switch
HX	. .	Auxiliary Field and Brake Switch
J	. .	Reverse Phase Relay
K	. .	Starting Switch
KR	. .	Running Switch
LD	. .	Down Light Switch
LE	. .	Levelling Fast-speed Switch
LU	. .	Up Light Switch

MC	. .	Minimum-current Shunt Field Relay
NT	. .	Hall Time Switch
$1P, 2P, 3P,$		
and $4P$	.	Overload Switches
Q	. .	Load Switch
S	. .	Starting Switch
SF	. .	Series Field Switch
ST	. .	Stopping Switch
STX	. .	Auxiliary Stopping Switch
SUD	. .	Suicide Delay Switch
TS	. .	Double-journey Relay
U	. .	Up Direction Switch
$1U$, etc.	.	Up Floor Relays
UTB	. .	Up Test Button
UX	. .	Up Auxiliary Direction Switch
VR	. .	Voltage Relay
XD	. .	Down-direction Switch Relay
XKR	. .	Running Switch Relay
XQ	. .	Load Switch Relay
XU	. .	Up-direction Switch Relay

3. SEQUENCE OF OPERATION

Assume that—

Attendant switch ATC is closed to AUTO.

Main circuit breaker MCB and service knife switches $SKS1$ and 2 are closed.

Transformer TRF and rectifier $1RF$ are energized. Rectifier $1RF$ supplies continuous d.c. power to operating circuit terminals $C14$ to $HL1$.

Reverse phase relay J is energized and its contact closed.

Car is parked at the bottom landing with doors closed.

Motor generator is shut down.

All mechanical switches are in their normal position as shown on diagram.

An intending passenger wishes to travel from the bottom landing to the third landing.

Car Call Operation and Starting Motor Generator

1. Passenger opens landing and car doors, enters car and registers a third-landing call by pressing the appropriate car

button. Assume that a top floor down hall call is also registered. Switches LU and XU pull in.

2. Switches ERT, ER and K pull in.

Rectifier $2RF$ is energized to supply d.c. power to the main operating circuit terminals MF and $HL1$.

3. M.G. driving motor DV starts up on star connexion and generator armature GA comes up to speed.

4. Shunt field MF builds up to minimum standing value and relay MC pulls in.

5. Switch XKR pulls in and switch K drops out after a slight time delay.

Switches KR and C pull in.

M.G. driving motor DV transfer from star to delta connexion.

6. Meanwhile, the doors have been closed and gate contact switch GW pulls in. Contacts ADS and GS are closed.

7. Switch NT drops out after a time delay and switch S pulls in.

8. Coil LV is energized to lift the contacts off the cam, and contacts $LV2$ and $LV6$ make.

9. Coil CM is energized to retire the lock cam. Contacts DS are closed.

10. Switches UX, U and H pull in.

Switches HX, SUD and SF pull in.

Generator levelling field GLF builds up, and switch DF pulls in.

Motor field MF builds up to full strength.

Brake B lifts.

Motor armature MA starts to turn and car starts moving up.

11. Switches $1E$, $2E$, $3E$, and GL pull in.

Switch SF drops out.

Generator field $GSEF$ is connected across the generator and starts building up.

Switch VR pulls in.

Generator armature GA voltage builds up to rated value and armature MA accelerates up to speed.

Car runs full contract speed up.

Slow-down and Stopping

12. Crosshead panel contact UCS touches third-floor bar contact CAC, and switch ST pulls in.

At the same time crosshead contact UAS touches third-floor bar contact ASC.

Switch STX pulls in.

Switch ST drops out, and floor relay $3C$ resets.

13. Crosshead contact UAS rides off bar contact ASC, and switches S and STX drop out.

14. Levelling switch LV releases.

Switch $1E$ drops out, and switch $2E$ starts drop-out timing.

Resistance $GF1$ is inserted in series with generator field $GSEF$, and the car slows down first step.

15. Switch $2E$ drops out to insert series resistance $GF2$ and shunt resistance $GF5$ in generator field $GSEF$.

Car slows down second step.

16. Levelling switch contacts No. 1, No. 3, and No. 4 are bridged together by levelling cam, and switch LE pulls in.

17. Levelling contact $LV2$ breaks.

Switches UX and HX drop out.

Switch $3E$ drops out and switch SF pulls in.

Resistance $GF5$ is reduced and field $GSEF$ is disconnected.

Switch NT starts drop-out timing.

Car slows down to fast-speed levelling speed.

18. Coil CM is de-energized, lock cam advances and contacts DS open.

19. Levelling switch cam rides off contact $LV4$, and switch LE drops out.

Switch GL starts drop-out timing.

Car slows down to slow-speed levelling speed.

20. Levelling switch cam rides off contact $LV3$, and switches U and H drop out.

Switch GL drops out, and switch SUD follows after slight time delay.

Switch SF drops out.

Brake sets and car comes to a stop at second floor.

21. Passenger opens doors, leaves car and recloses doors.

Hall Call Operation

22. Top floor hall call TD is registered and set.

Switches LU and XU remain in.

23. Switch NT drops out.

24. Switch S pulls in.

Coil LV is energized to lift levelling contacts off cam.

25. Steps 9 to 11 are now repeated.

Slow-down and Stopping

26. Spring contact (FH 1–2) on top-floor bar breaks. Switches LU and XU drop out.

27. Crosshead contact UAS touches top-floor bar contact ASC, and switch STX pulls in.

28. Steps 13 through 18 now follow in order.

29. Steps 19 through 21 now follow in order.

4. NOTES ON SPECIAL SWITCHES AND RELAYS

1. Load relay XQ and load switch Q operate to compensate for loss of speed due to load.

2. Voltage relay VR is set to operate at about 50 per cent armature voltage to insure that XQ operates at a specified load only.

3. Switch GL drops out with switch H to boost car back to the floor on a relevel operation. It also boosts car into floor when lifting load, and pulls the car out of a stall if necessary.

4. Relay DF must operate for fast-speed operation. Main field GF cannot be connected across armature GA without first setting up direction field.

5. On attendant operation, switches $1AE$ and $2AE$ operate to make operation subject to attendant.

They also permit operator to select direction of operation.

6. Switches ATU and ATD operate for up and down operation on both attendant and inspection operation.

Car operates at slow speed only when on inspection operation.

Variable-voltage Duplex Collective Controller. The following is a description of the Otis Elevator Company's 21 UCL controller applied to an installation of two passenger lifts of contract load 2 000 lb and contract speed 300 ft. per min. The circuit diagrams are shown in Figs. 16.18, 16.19 and 16.20.

1. CONTROL SYSTEM

Duplex Collective, with stopping switches on
6850CA Selector
6850H Relay Panel
6970A Door Operator Resistance Timed

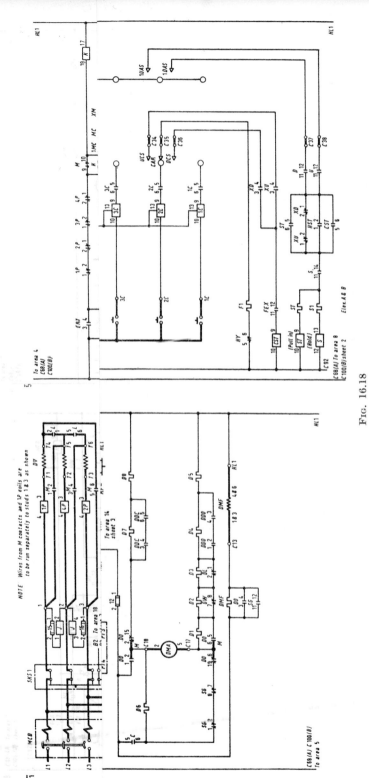

(facing page 354)

FIG. 16.18

VARIABLE-VOLTAGE DUPLEX COLLECTIVE CONTROLLER
(sheet 1)
(Otis Elevator Co.)

(T.116)

Symbol	Apparatus	Location on Diagram
2LS	Up Safety Limit Switch	14
3LS	Down Final Limit Switch	14
4LS	Up Final Limit Switch	14
LNS	Load Non-stop Switch	10
LV	Levelling Switch	15, 16
MA	Elevator Motor Armature	19
MCB	Main Circuit Breaker	1
MF	Elevator Motor Shunt Field	14
MIP	Elevator Motor Interpole Field	19
OS	Governor Overspeed Switch	14
PIB	Position Indicator Brush	2
PIC	Position Indicator Contacts	2
SCS	Signal Cutout Switch	2
1SGC	Safety Shoe Contact No. 1	11
2SGS	Safety Contact No. 2	11
1SL	Down Selector Slowdown Limit	16
2SL	Up Selector Slowdown Limit	16
3SL	Down Selector Direction Limit	15
4SL	Up Selector Direction Limit	15
SKS1	4 Pole Switch Fuse	1
SKS2	3 Pole Switch Fuse	2
SKS3	2 Pole Switch Fuse	2
SOS	Safety Operated Switch	14
TES	Top of Car Emergency Stop Switch	14
TTL	Tell-tale Lights	2
U	Car Up Test Button	10
1UAS	1st Up Auxiliary Stop Brush	7
2UAS	2nd Up Auxiliary Stop Brush	16
UBB	Up Call Below Brush	9
UBC	Up Call Below Contacts	9
UCS	Up Car Stop Brush	7
UHC	Up Hall Contacts	6
UHR	Up Hall Reset Brush	6
UHS	Up Hall Stop Brush	6

3. Names of Switches and Relays on Controller and Relay Panel

ACT	Acceleration Time Switch	16
BKS	Button Knife Switch	10
C	Potential Switch	14
1C-TC	Car-stop Floor Relays	7
CK	Car Call Switch	7
CND	Condenser Damping Switch	17
CST	Car Stopping Switch Relay	7
CX	Auxiliary Potential Switch	15
D	Down Reversing and Levelling Switch	15

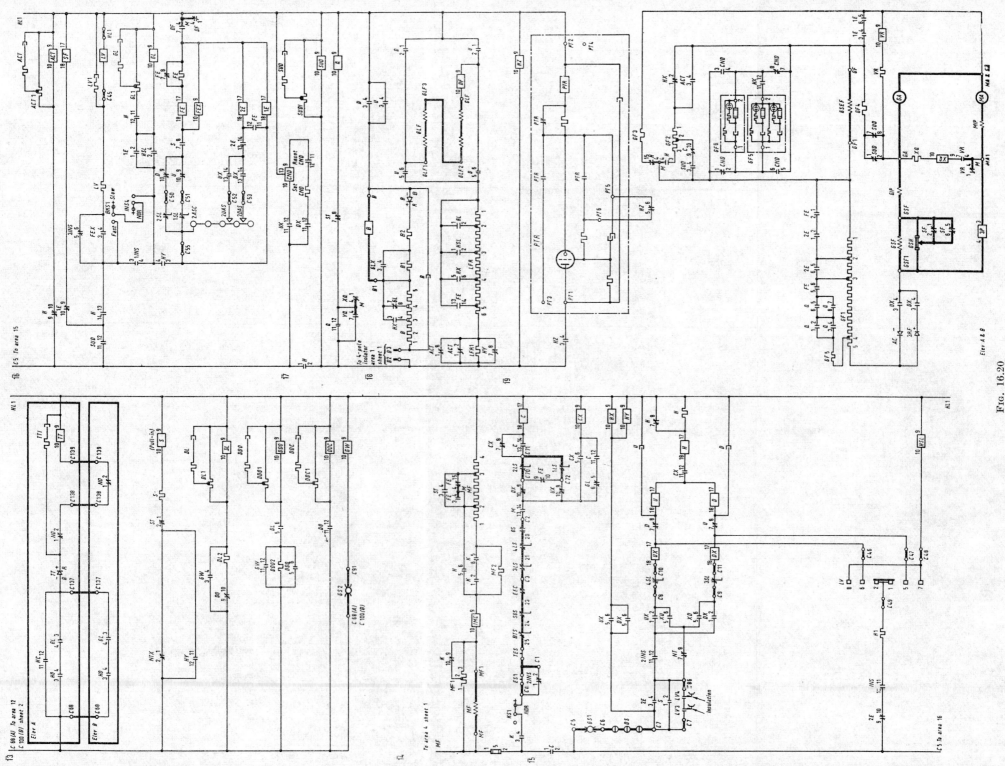

FIG. 16.20

VARIABLE-VOLTAGE DUPLEX COLLECTIVE CONTROLLER
(sheet 3)

(Otis Elevator Co.)

(facing page 356)

(T.116)

CONTROLLERS

357

Symbol	Apparatus	Location on Diagram
2D-TD	Down Floor Relays	6
DDC	Door Close Speed Switch	13
DDO	Door Open Speed Switch	13
DF	Direction Field Relay	18
DO	Door Open and Close Switch	11
DL	Door Limit Switch	13
DTB	Down Test Button	10
DX	Auxiliary Down Reversing Switch	15
1E	1st Speed Switch	16
2E	2nd Speed Switch	16
EKS	Earth Knife Switch	4
ERT	Excitation Time Switch	5
FC	Selection Switch	8
FCX	Auxiliary Selection Switch	8
FE	Full-speed Switch	16
FEX	Auxiliary Full-speed Switch	16
GL	Timed Levelling Speed Switch	16
GLX	Auxiliary Timed Levelling Speed Switch	5
GSH	Generator Shunt	19
H	Field and Brake Switch	15
*HC	Hall Call Switch	9
HO	Home Station Switch	8
HSL	Fast-speed Levelling Switch	15
HST	Hall Stopping Switch Relay	6
HX	1st Auxiliary Field and Brake Switch	15
HY	2nd Auxiliary Field and Brake Switch	15
HZ	3rd Field and Brake Switch	19
INS	Inspection Knife Switch	5, 10, 15, 16
1INS	1st Inspection Relay	10
2INS	2nd Inspection Relay	10
3INS	3rd Inspection Relay	10
J	Reverse Phase Relay	1
K	Running Switch	5
KS	Knife Switch	4, 14
L	Star Switch	5
LD	Down Light Switch	10
LDX	Down Auxiliary Light Switch	10
LU	Up Light Switch	10
LUX	Up Auxiliary Switch	10
M	Delta Switch	5
1MC	1st Minimum Current Shunt Field Relay	14
NO	Non-stop Switch	10
NT	Door Time Switch	8
NTX	Auxiliary Door Time Switch	11
1P, 2P, 4P	M.G. Driving Motor Protective Relays	1
3P	Generator Protective Relay	19

* Denotes Elevator A only.

Symbol	Apparatus	Location on Diagram
PC	Passed Call Switch	10
PTR	Protective Time Relay	19
Q	Load Switch	17
RF, etc.	Rectifiers	3, 6, 19
*RV	Rectifier Voltage Switch	6
S	Starting Switch	7, 13
SF	Series Field Switch	16
SG	Safety Shoe Switch	11
ST	Stopping Switch	6, 7
SUD	Suicide Delay Switch	17
TG	Transfer Switch	5
TRF	Rectifier Transformer	2
†TRS	Signal Transformer	2
*TT	Transfer Timing Relay	13
U	Up Reversing and Levelling Switch	15
†1U, etc.	Up Floor Relays	6
UB	Up Call Below Switch	9
UTB	Up Test Button	10
UX	Auxiliary Up Reversing Switch	15
VR	Voltage Relay	19
XD	Down Direction Switch Relay	10
XM	Motor-generator Set Running Switch Relay	5
XQ	Load Switch Relay	19
XU	Up Direction Switch Relay	10
	Fuse 4	4
	Fuse 5	14
	†Fuse 8	6
	†Fuses 9, 10	2
	Fuse 12	4
	Fuses 15, 16	1

* Denotes Elevator A only.
† Denotes Elevator B only.

4. SEQUENCE OF OPERATION

Note. Numbers in parentheses adjacent to various symbols in text denote the position of the symbol on the diagram.

A. A more detailed description of Duplex operation is at *E* on page 362. It will assist the reader to assume the following points. Elevator A is at the main (1st) landing (ground floor) with its doors closed and M.G. set shut down.

Elevator B is at the 2nd landing with its doors closed and M.G. set shut down.

A passenger wishes to travel from the first landing (ground floor) to the third landing.

Assume that Elevator B is the scheduled car.

Switches FC (8) on Elevator B are pulled in.

Switch HO (8) on Elevator A is pulled in, Elevator A is home car.

See further notes on Duplex operation on page 362.

Main circuit breaker MCB (1) is closed.

Reverse phase relay J is energized and contact is closed.

Transformer TRF (2) and rectifier RF (2) are energized, and continuous d.c. power is supplied to terminals MF and $HL1$.

All mechanical switches are in their "normal" position as shown on the diagram.

B. *Starting Motor Generator Set and Re-opening Doors*

1. Passenger presses the UP hall button.

Floor relay $1U$ (6) sets.

Switch HST (6) pulls in.

Switch NT (8) pulls in.

Switch NTX (11) pulls in.

2. Switch ERT (5) pulls in.

Switches K (5) and then L (5) pull in.

M.G. set driving motor DV (1) starts up on star connexion and generator armature GA (19) comes up to speed.

3. Shunt field MF (14) builds up to a minimum standing value and relay $1MC$ (14) pulls in.

4. Switch XM (5) pulls in and switch L (5) drops out after a short delay.

Switch $1U$ (6) is reset.

Switch HST drops out.

5. Switches M (5), (14) CX (15) and ACT (16) pull in.

Switch DO (11) pulls in.

M.G. driving motor DV (1) transfers from star to delta connexion.

Doors start to open.

Switch NT (8) starts drop-out timing.

C. *Car Call Operation*

6. Passenger registers 3rd landing car call.

Relay $3C$ (7) sets and switch CK (7) pulls in and drops out to remove timing from switch NT (8).

Assume that at this point a 5th landing hall call is also registered.
Switches XU (10), LU (10) and LUX (10) pull in.

7. When timing is removed from switch NT (8) it drops out immediately.

Switch NTX (11) drops out.

Switch DO (11) drops out.

Doors start to close.

8. Switch S (13) pulls in on its inner winding.

Doors close fully.

Starting and Acceleration

9. Contacts GS (13, 15) and DS (15) close.

Switches UX (15), U (15) H (15) and HZ pull in.

Relay PTR (17) is energized and commences timing.

Switch CND (17) sets.

Generator levelling field GLF (18) builds up and switch DF (18) pulls in. (See special note on operation of DF at end of text, page 364).

Switch SUD (17) pulls in. Switch SF (16) pulls in.

Switches HX (15), HY (15) pull in.

Switches FE (16), FEX (16) $1E$ (16) GL (16) and GLX (5) pull in followed by switch $2E$ (16). Switch SF (16) drops out.

Coil LV (16) is energized to lift levelling contacts of the cam and contacts $LV3$ (15) and $LV4$ (15) make.

Motor field (14) builds up to full strength.

Brake B (18) lifts.

Motor armature MA (19) starts to turn and car starts moving up.

10. Generator field $GSEF$ (19) is connected across the generator and starts building up.

Switch VR (19) pulls in.

Generator GA (19) voltage builds up to rated value and armature MA (19) accelerates up to speed.

Car runs at full contract speed up.

Slow-down and Stopping

11. Crosshead panel contact UCS (17) touches 3rd landing floor bar contact CAC (7) and switch CST (7) pulls in.

At the same time crosshead brush contact UAS (7) touches 3rd landing floor bar contact ASC (7).

Switch ST (Pull in) (7) pulls in.

Switch S (Hold) (7) is energized.
Switch S (Pull in) (13) is de-energized.

12. Crosshead brush contact UAS (7) rides off floor bar contact.
ASC (7) and switch S (Hold) (7) drops out.

13. Switches FE (16) and FEX (16) drop out, and switch $2E$ (16) starts drop-out timing.
Part of resistance $GF1$ (19) is inserted in series with generator field $GSEF$ (19) and the car slows down first step.
Coil LV (16) is de-energized.
Levelling switch contacts fall into path of next cam.

14. Switch $2E$ (16) drops out to insert a further section of $GF1$ (19)
Resistance in series with $GSEF$ (19).
Car slows down second step.

15. Levelling switch contacts $LV2$, (15) $LV6$ (15) and $LV8$ (15) are bridged together by the levelling cam and switch HSL (15) pulls in.
Switch GL (16) pulls in.

16. Levelling contact $LV4$ (15) breaks.
Switches UX (15) HX (15) and HY (15) drop out.
Switch $1E$ (16) drops out and switch SG (16) pulls in.
Field $GSEF$ (19) is disconnected.
Car slows down to fast levelling speed.

17. Levelling switch cam rides off $LV8$ (15) and switch HSL (15) drops out.
Switch GL (16) starts drop-out timing.
Car slows down to slow levelling speed.

18. Door zone contacts $DZ1$ (11) followed by $DZ2$ (11) close.
Switch DO (11) pulls in.
Doors start to open.

19. Levelling switch cam rides off contact $LV6$ (15) and switches U (15), H (15) and HZ (19) drop out.
Switches GL (16) and GLX (5) drop out and switch SUD (17) drops out after a short delay.
Switch SF (16) drops out.
Brake sets and car comes to rest at the fourth landing.

20. Switch NT (8) starts drop-out timing.
Doors fully open.

D. *Hall Call Operation*

21. A $5U$ (6) hall call is registered and set.

Switches XU, (10) LU (10) and LUX (10) remain in.

22. Switches NT (8) and NTX (11) drop out.

23. Switch DO (11) drops out.

Switch S (13) pulls in.

Doors start to close.

24. Steps 8 to 10 are repeated.

Slowdown and Stopping

25. Floor brush switch FH (10) breaks and switches XU (10), LU (10) and LUX (10) drop out.

26. Crosshead brush contact UAS (7) touches fifth landing floor bar contact ASC (7) and switch ST (7) pulls in.

Switch S (Hold) (13) is energized.

Switch S (Pull in) (13) is de-energized.

27. Steps 12 to 17 now follow.

When crosshead brush contact UHR (6) touches 5th landing bar contact $5UHC$ (6) switch HST (6) pulls in to reset fifth landing hall call relay $5U$.

28. Steps 18 to 20 now follow.

E. *Duplex Operation*

On Duplex installations two cars designated A and B operate in conjunction with each other, and in such a manner that the registered hall calls will be answered in a reasonable time. To accomplish this, each controller has seven additional switches (FC, FCX, HO, PC, TG and UB) and the controller for car A has three additional relays (TT, RV and HC) which function commonly to both cars. In addition there is only one set of UP and DOWN hall call relays to which both cars respond.

Each individual elevator car in a Duplex installation is always in one or another of the following three situations (unless it is out of service)—

(i) It may be idle at the home landing. In this case it is called a HOME car.

(ii) It may be idle at a landing other than the home landing, in which case it is called a FREE car. Where both cars are idle at the home station, the one which has been there longest is also called a FREE car.

(iii) It may be responding to calls, either from landing buttons or from the buttons on its own operating panel, in which case it is called a BUSY car.

On normal Duplex operation, one car, the HOME car, remains at the home landing (main floor) while the other, the FREE or BUSY car, remains somewhere in the hoistway. All car calls above the main floor are answered by the BUSY or FREE car, except under certain conditions as explained under features. Each car answers its own car calls in the proper order.

The features of Duplex operation are as follows—

1. Automatic return of a car to home landing, after hall calls are answered, to become a Home car.
Selection switches *FCX* (8) and *FC* (8) drop out on first car to open up direction circuits, and they will establish a holding circuit for *FCX* (8) and *FC* (8) on other car.

2. Automatic start-up of HOME car in the event that an UP or DOWN call is registered above the down-travelling BUSY car.
Passed call switch or *PC* (10) pulls in on controller for BUSY car and completes the circuit for switches *FCX* (8) and *FC* (8) for the HOME car, making it a BUSY car.

3. Automatic start-up of the HOME car in the event that an UP hall call is registered below the up-travelling BUSY car.
UP call below switch *UB* (8) on the BUSY car drops out to complete a circuit for the selection switches *FCX* (8) and *FC* (8) for the HOME car, thus making it a BUSY car.

4. Transfer of calls from one car to another in the event of power failure, the opening of the emergency stop switch or changing to inspection operating.
Transfer switch *TG* (5) drops out on machine having failure, thus dropping out selection switches *FCX* (8) and *FC* (8) to transfer calls.
On a power failure, rectifier voltage switch *RV* (6) drops out also and ties in hall button relay circuit with power supply from other machine.

5. Response of the HOME car to any hall call within 30 to 40 seconds, after registration should the other car fail to respond

for reasons other than those listed in the preceding paragraph (4). Hall call relay HC (9) drops out when a hall call is registered and starts condenser timing on relay TT (13). If neither car starts (contact HO 3–4) within 30 to 40 seconds, relay TT (13) drops out to set up direction circuit for HOME car, thus transferring calls.

If knife switch $KS1$ (14) is open and the other car does not cancel all hall calls within 30 to 40 seconds, relay TT (13) drops out to set up, direction circuit for HOME car thus transferring calls.

6. Home station switch HO (8) operates when car comes into home landing. When a BUSY car arrives at home landing on a car call it changes the HOME car to a FREE or BUSY car and the arriving car becomes the HOME car.

It also serves to control door re-open so that when both HOME and FREE cars are at home landing only the FREE car will open its doors in response to a hall button at that floor but when only the HOME car is at home station its doors will respond to hall buttons at that floor.

F. *Inspection Operation*

(*a*) Inspection Control from the Car

Inspection Initiation key switch CIS (10) in the Car Operating Panel is thrown to INSPECTION position. Switches $1INS$, $2INS$ and $3INS$ (10) drop out. Direction switches XU (10) and XD (10) are now subject to the buttons D (10) and U (10) in the Car Operating Panel. Car operates at levelling speed.

(*b*) Inspection Control from the Controller

Knife switch INS is thrown to INSPECTION position. Switches $1INS$, $2INS$ and $3INS$ (10) drop out. Direction switches XD (10) and XU (10) are now subject to test buttons DTB (10) and UTB (10) in the controller. In addition the car may be run at full or levelling speed; this feature is controlled by knife switch $BKS1$ (16).

5. FUNCTION OF SPECIAL SWITCHES AND RELAYS

1. Relay DF (18) must operate for fast-speed operation of the car. It is set to operate at minimum generator levelling

shunt field current. This ensures a proper build-up of generator self-excited field for direction of car travel.

2. Switch GL (16) is adjusted to drop out in one second on timing. It serves to boost the car into the floor when lifting load, and pulls the car out of a stall when necessary.
Timing is removed on a re-level operation so that the switch may serve to boost the car back to floor.

3. Relay $1MC$ (14) operates on minimum lift motor shunt field current to permit transfer of M.G. drive motor DV (1) from star to delta operation.

4. Switch SUD (17) is delayed $\frac{1}{4}$ second to permit dynamic braking to be fully effective until the brake is set and the car stopped.
If the timing of this switch increased appreciably the suicide protection will be lost and generator voltage may built up to cause overload switch $3P$ (19) to operate.

5. Relay VR (19) is set to operate at approximately 50 per cent armature voltage to insure that relay XQ (19) operates at a specified load only and that it is not affected by starting currents.

6. Load relay XQ (19) is set to just operate "empty-car-down" and serves to complete the circuit for switch Q (7) which compensates for loss of speed due to load.

7. Condenser damping switch CND (17) connects capacitors $GF5$ (19) and $GF6$ (19) appropriately to smooth out speed changes during slow down.

8. Relay PTR (17) permits use of the lift for enough time to allow the lift to make one full up-and-down journey of the shaft at contract speed.

6. 6970A DOOR OPERATOR

This operator has a speed-controlled d.c. motor. Condenser timed relays accelerate and decelerate the operator motor by shorting out or inserting resistances in series with the door motor armature. The door is moved positively by rotation of the motor in either direction and the door motor armature remains energized while the door is in the fully opened position and the fully closed position, to maintain torque.
The armature is de-energized only when the motor generator set shuts down.

The direction of rotation and/or torque is determined by a single switch, *DO*, energized when the door is open or opening and de-energized when the door is closing or closed.

Switch settings should be as follows—

Switch *DL* should be set to drop out $\frac{1}{2}$ second after switch *DW* is de-energized.

Switch *DDO* should be set to drop out $1\frac{1}{2}$ seconds after *DL* is de-energized and *DDC* $2\frac{1}{2}$ seconds after switch *DO* is de-energized. The gate contact *GS* in the *DW* circuit should be set to open as soon as the interlock opens.

Resistance Settings for Door Operation

Door Opening

D5—set to give good maximum opening speed.

D2 and *D3*—determine initial speed at which door starts to open. Their combined resistance should be as low as possible and still permit interlock to open without noise.

D2 will be shorted out as soon as interlock opens.

D3 will be shorted out 1–2 seconds later for full speed from the door motor.

Adjust time on *DL* to give smooth transition to full speed.

D4—adjust as high as possible without making door motion too slow for the last part of its travel. It must also be low enough so that the door is held positively open.

DDO should drop out when door is three-quarters open.

Door Closing

D8—adjust to provide satisfactory maximum closing speed.

D7—adjust to slow the door for the last one-quarter of travel. *D7* should be as high as possible but must be low enough to hold the door positively closed.

DDC should drop out when door is three-quarters closed.

DMF-weak motor field is used in closing direction only, so that the force necessary to prevent the door from closing is kept down to about 30 lb. Retain this setting for the initial adjustments.

Should the stalled force exceed 30 lb it may be necessary to increase this resistance.

Door Re-opening

Adjust resistance $DL2$ so that DL relay will be delayed 1–2 seconds on pulling in through DO 9/10 contact.

Adjust resistance $DDO2$ so that the DDO relay will be delayed 1–2 seconds pulling in.

These delays prevent full torque from being applied if a reversal occurs near the fully open position which would result in the door slamming and perhaps cause mechanical damage.

Reversal at any point after DL and DDO have been energized will start at less than full torque.

Note that the dropping out time of these relays will be considerably less on reversal than on an original opening since the condenser cannot charge fully with resistors $DDO2$ and $DL2$ in series with switch coils.

The values of resistors $D1$ and $D6$ are not critical.

When a door reversal occurs, even before the DO making contacts close, current will flow through $D6$ resistor, DMA and $D1$ resistor to give an immediate reverse current through the armature and smooth out the reversal.

In the door-closing operation $D1$ and $D6$ resistors perform the function of keeping the door-closing speed approximately constant regardless of changes in frictional forces opposing the operation of the door.

Express Traffic Mark IV Controller. The main features of this patented controller are described in Chapter XV and the following are the principles of operation of the control circuits of a twelve-landing three-car installation.

Fig. 16.21 shows the landing call registration circuit in which push-buttons $1UP$ to $11UP$ are provided at landings 1–11 and are pressed by passengers intending to travel upwards. Push buttons $2DP$–$12DP$ are provided at landings 2–12 and are pressed by passengers intending to travel downwards. Operation of any of these buttons energizes the respective one of the relay coils $1U$–$11U$ and $2D$–$12D$. These relays are self-holding via n/o contacts, $1U$, $2U$, etc.

The landing call buttons and associated relays are segregated into six groups or zones. Zone 1 is $3UP$–$6UP$, zone 2 is $7UP$–$10UP$, zone 3 is $11UP$, $11DP$ and $12DP$, zone 4 is $7DP$–$10DP$, zone 5 is $3DP$–$6DP$ and zone 6 is $1UP$, $2UP$ and $2DP$. Hence the operation of one of the push-buttons in any zone energizes one of the zone relays $LCZ1–LCZ6$ to indicate the registration of a landing call in this zone.

As the ground floor, which is floor No. 2 in the figure (floor

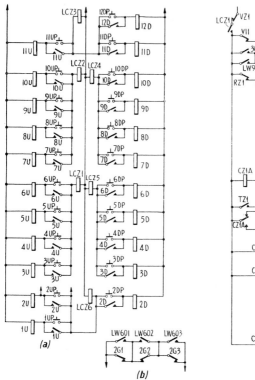

(a)

(b)

Fig. 16.21. Express Traffic
Mark IV: Landing Call
Registration Circuit
(*Express Lift Co.*)

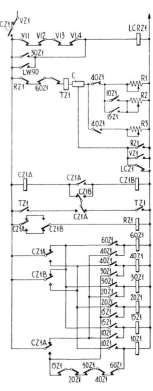

Fig. 16.22. Express Traffic
Mark IV: Zone Timing
Circuit
(*Express Lift Co.*)

No. 1 is the basement) is likely to be the busiest, it is convenient to provide a priority service for this floor. This is provided by the parking call circuit (Fig. 16.21 (*b*)) which is connected across the push-button $2UP$ as indicated by the arrow heads. Thus a landing call is automatically inserted for the ground floor unless a car is already parked there when one set of the contacts

$2G1$ to $2G3$, each of which is associated with a respective one of the cars, is open to de-energize relay $2U$ and also $LCZ6$ if there are no other landing calls in this zone. When the car parked at the ground floor is loaded to say 60 per cent the call is again automatically inserted by closure of the respective one of sets of contacts $LW601$ to $LW603$ which by-passes the corresponding set of open G contacts.

Fig. 16.22 shows the zone timing circuit for zone 1 only, the others being similar. This zone timing circuit indicates how long a zone has waited for a car to answer its call; thus with the operation of zone relay $LCZ1$ n/o contacts $LCZ1$ are closed to energize relays $LCRZ1$ and $TZ1$, the latter being energized via a timing relay unit which is controlled by three pre-set potentiometers to give nominal time delays of 5, 10 and 20 seconds.

The timing relay is a thermionic valve controlled capacitance resistance timing circuit, the three pre-set potentiometers $R1$, $R2$ and $R3$ constituting the resistance. The thermionic valve and capacitors have been shown in block form at C. The values of $R1$, $R2$ and $R3$ are set such that when $R1$ is connected alone in series with relay coil $TZ1$ a time interval of 10 sec. is provided. When $R1$ and $R2$ are connected in parallel a time interval of 5 sec. is provided and when $R3$ alone is connected in series with relay coil $TZ1$ a time interval of 20 sec. is provided. With contacts $10Z1$, $15Z1$ and $40Z1$, which control the connexion of the potentiometers $R1$, $R2$ and $R3$, initially in the position shown, $R1$ alone is connected in series with relay coil $TZ1$ and the first time interval is therefore 10 sec.

After the first 10-second interval from closure of contacts $LCZ1$, relay $TZ1$ is energized and closes its sets of n/o contacts $TZ1$. The relay self-holds via one set of $TZ1$ contacts. Closure of another set causes energization of relay $CZ1A$ and at the same time provides a short-circuit across relay coil $CZ1B$ to prevent this relay from operating. Energization of relay $CZ1A$ causes switching of the $CZ1A$ contacts from the position shown. This causes energization of relay $10Z1$ which self-holds via n/o contacts $10Z1$ and $CZ1A$. Energization of relay $CZ1A$ also causes energization of relay $RZ1$ which in turn opens n/o contacts $RZ1$ in the circuit of relay coil $TZ1$ to de-energize this relay. Operation of relay $10Z1$ also causes closing of n/o contacts $10Z1$ in the timing circuit to connect potentiometers

$R1$ and $R2$ in parallel so that the next timing interval will be 5 sec. When relay $TZ1$ becomes de-energized the relay coil $CZ1B$ is no longer short-circuited and so is operated to switch contacts $CZ1B$, so that relay $RZ1$ is de-energized to close its n/c contacts in the circuit of relay $TZ1$. Hence the next time-interval starts with $CZ1A$ and $CZ1B$ both operated.

After the 5-second interval, relay $TZ1$ is again operated to close its n/o contacts $TZ1$ and as relay $CZ1B$ has switched its contacts from the position shown, relay coil $CZ1A$ is short-circuited and is thus de-energized to return its contacts to the position shown. Relay $10Z1$ is still self-holding through contacts $CZ1B$, and thus relay coil $15Z1$ is energized through contacts $CZ1A$ and $10Z1$. Operation of $15Z1$ causes opening of n/c contacts $15Z1$ in the circuit of relay $10Z1$. Relay $CZ1B$ being energized, de-energization of $CZ1A$ also causes energization of $RZ1$ which de-energizes relay $TZ1$ as before to de-energize $CZ1B$ and hence relay $10Z1$. Thus the circuit at this stage has relay $15Z1$ operated, and holding through contacts $CZ1A$ and $CZ1B$ in parallel, both relays $CZ1A$ and $CZ1B$ being in the de-energized state as they were initially. Thereafter the circuit repeats the above operations in which $CZ1A$ and $CZ1B$ are operated in turn, thus working progressively through the relays $20Z1$–$60Z1$.

Operation of relay $15Z1$ causes closure of n/o contacts $15Z1$ in series with potentiometer $R2$; thus the next time interval will also be 5 sec. When $15Z1$ is de-energized, potentiometer $R1$ alone is connected in series with relay $TZ1$ so that $30Z1$ and $40Z1$ will be operated after 10-second intervals. When $40Z1$ is operated, potentiometer $R3$ alone is connected in series with $TZ1$ and the time interval before $60Z1$ is operated, is therefore 20 sec. Thus considering the total time from the instant when a call is inserted in zone 1, $10Z1$, $15Z1$, $20Z1$, $30Z1$, $40Z1$ and $60Z1$ are operated after 10, 15, 20, 30, 40 and 60 sec. respectively. The time relay continues to time up to the maximum time unless all calls have been answered in the zone, when contacts $LCZ1$ open to reset the timer to zero, or until a car is selected to answer calls in this zone, when contacts $VZ1$ open.

The timing circuits for the other zones are similar, and where it is necessary in the description to refer to contacts of relays whose coils are in the timing circuits for the other zones, they will be suffixed by a number corresponding to the zone number. Thus the contacts for zone 2 corresponding to contacts $10Z1$ for zone 1 will be $10Z2$ and for zone 6 will be $10Z6$.

The zone selection circuit is shown in Fig. 16.23. When a car becomes free its relay FC operates; thus for car No. 1 relay $FC1$ operates. This causes closure of n/o contacts $FC1$–$FC3$ depending on which car becomes free, and provided one of the zone relays has been operated to close one or more sets of n/o contacts $LCRZ1$–$LCRZ6$, relay CFC is energized. This causes n/o contacts CFC to close and via the gating circuit shown, operates the one of the relays $Z1$ to $Z6$ for the zone with the longest time interval. Hence if the state of the timing circuit is such that say relay $60Z1$ has operated and for the other zones $30Z2$ and $15Z3$–$15Z6$ have operated, relay $Z1$ will be operated via n/o contacts $60Z1$ and $LCRZ1$. Also n/c $60Z1$ will be opened to cut out the rest of the gating circuit. Similarly, if $30Z3$ is the highest-order relay to be operated, relay $Z3$ will be operated via n/o $30Z3$ and $LCRZ3$, and n/c contacts $30Z3$ will be opened to cut out the rest of the gating circuit below $30Z3$. Thus the zone with the longest waiting time will be allocated the car by operation of the appropriate one of the relays $Z1$ to $Z6$.

If no zone has had a 10-second waiting time, the car will be dispatched to a zone with outstanding calls, the appropriate one of relays $Z1$ to $Z6$ being energized via n/o contacts $LCZ1$ to $LCZ6$. Because of the n/c Z contacts above the Z relay coils, the zones are given a priority order, under these circumstances zone 6 having the highest priority.

Figs. 16.24 and 16.25 show the car selection circuit which determines the free car to be allocated to answer one or more calls in a selected zone. The principle is that if there is only one free car, this car is dispatched to the selected zone and if there is more than one free car, that car nearest to the first call to be answered in the selected zone will be dispatched, i.e. the car nearest to the lowest UP call in the selected zone or to the highest DOWN call in the selected zone.

The car selection is achieved by cold-cathode trigger tubes $CC1$ to $CC6$ arranged in pairs so that if $CC1$ and $CC2$ fire ($CC2$ firing as a result of $CC1$ firing), relay $N1$ is operated to select

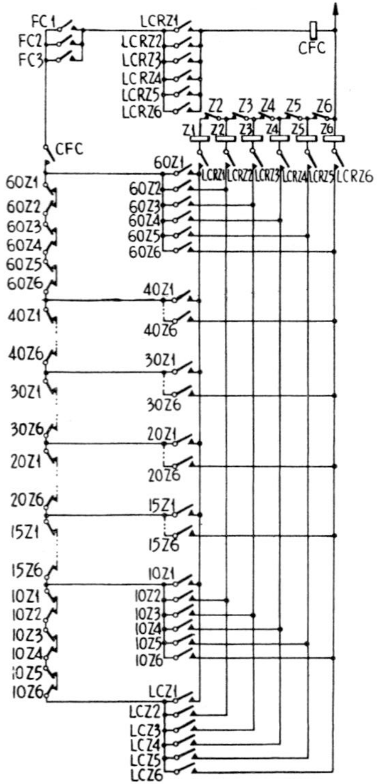

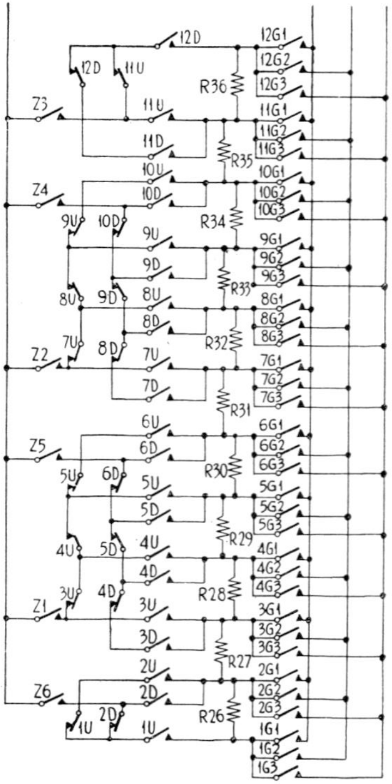

FIG. 16.25. EXPRESS TRAFFIC
MARK IV: CAR SELECTION
CIRCUIT
(Express Lift Co.)

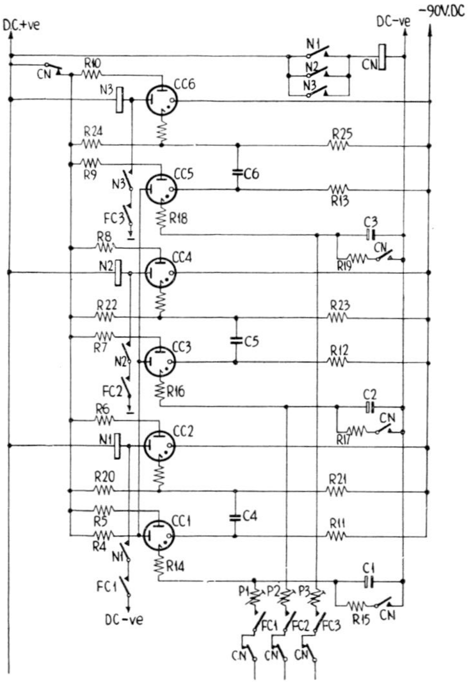

FIG. 16.24. EXPRESS TRAFFIC MARK IV:
CAR SELECTION CIRCUIT
(Express Lift Co.)

FIG. 16.23. EXPRESS TRAFFIC
MARK IV: ZONE SELECTION
CIRCUIT
(Express Lift Co.)

car No. 1 to be dispatched; similarly if $CC3$ and $CC4$ fire, car No. 2 is selected by operation of relay $N2$, and if $CC5$ and $CC6$ fire, car No. 3 is selected by operation of relay $N3$.

The main anodes of cold-cathode tubes $CC1$, $CC3$ and $CC5$ are interconnected and have a common anode load $R4$ through which they are connected to D.C. $+$ve via n/c contacts CN. Hence if any one of cold-cathode tubes $CC1$, $CC3$ and $CC5$ fire, the anode voltage on all of them is reduced to prevent the others firing. The main anodes of tubes $CC2$, $CC4$ and $CC6$ are connected to D.C. $+$ve through the coils of the relays $N1$, $N2$ and $N3$ respectively. The auxiliary anodes of tubes $CC1$ to $CC6$ are connected via resistors $R5$ to $R10$ respectively and n/c contacts CN to D.C. $+$ve. The cathodes of tubes $CC1$, $CC3$ and $CC5$ are connected to -90 V D.C. (which is more negative than D.C. $-$ve) through resistors $R11$, $R12$ and $R13$ respectively, and the cathodes of tubes $CC2$, $CC4$ and $CC6$ are connected to -90 V D.C.

The trigger electrode of tube $CC1$ is connected to D.C. $-$ve via resistor $R14$ and capacitor $C1$ which is provided with a discharge circuit through resistor $R15$ and n/o contacts CN. The trigger electrodes of tubes $CC3$ and $CC5$ are similarly connected via resistors $R16$–$R19$, capacitors $C2$ and $C3$ and n/o contacts CN. The trigger electrode of $CC2$ is connected to the junction of resistors $R20$ and $R21$ which are connected in series between D.C. $+$ve and -90 V D.C. through n/c contacts CN. Similarly the trigger electrodes of tubes $CC4$ and $CC6$ are connected to the junctions of resistors $R22$, $R23$ and $R24$, $R25$ respectively. Capacitors $C4$, $C5$ and $C6$ interconnect the cathodes of tubes $CC1$, $CC3$ and $CC5$ and the trigger electrodes of tubes $CC2$, $CC4$ and $CC6$ respectively.

Control of the firing of the tubes is shown in Fig. 16.25 in which various amounts of resistance are inserted in series with capacitors $C1$, $C2$ and $C3$ across the D.C. $+$ve and D.C. $-$ve to control the rate of charging of the capacitor. The amount of resistance inserted is related to the distance between the position of the car and the first call to be answered in the selected zone. Thus the car to be selected, i.e. the one of the relays $N1$ to $N3$ to be operated, will be that associated with the one of capacitors $C1$ to $C3$ having the least resistance in series with it as the voltage on the capacitors $C1$, $C2$ and $C3$

determines the voltage on the trigger electrodes of the associated tubes $CC1$, $CC3$ and $CC5$. The value of the resistance is selected by controlling the number of resistance sectors $R26$ to $R36$ connected in series with each of the capacitors $C1$, $C2$ and $C3$. This control is effected by n/o contacts $Z1$ to $Z6$ associated with the zones, n/o and n/c contacts $1U$ to $11U$ and $2D$ to $12D$ which are associated with the landing call relays $1U$ to $11U$ and $2D$ to $12D$ (Fig. 16.21) and contacts $1G$ (1), (2) and (3) to $12G$ (1), (2) and (3). Hence if zone 1 is selected with a call at $4U$ and the free car is No. 3 which is at the eighth floor, n/o $Z1$, n/o $4U$ and n/o $8G$ (3) are closed and n/c $4U$ are opened.

To obtain the full car selection circuit, Fig. 16.24 has to be placed above Fig. 16.25, the connexion being through n/o contacts $FC1$, $FC2$ and $FC3$ which are closed when cars Nos. 1, 2 and 3 respectively are available for dispatch to a zone and n/c CN contacts which are opened to reset the system after a car has been selected. Potentiometers $P1$, $P2$ and $P3$ provide a priority order of operation of tubes $CC1$, $CC3$ and $CC5$. If the same value of resistance from the sectors $R26$ to $R36$ is included in series with more than one of the capacitors $C1$, $C2$ and $C3$, the different settings of the potentiometers $P1$, $P2$ and $P3$ determine the one to be operated.

If we consider that there is an UP call from the third landing in the selected zone, i.e. zone 1 with car 1 free at the fifth floor and car 2 free at the eighth floor, n/o contacts $Z1$, $3U$, $FC1$, $FC3$, $5G(1)$ and $8G(2)$ will close and n/c $3U$ will open. Hence the resistor sectors connected in series with capacitor $C1$ are $R28$ and $R29$ and in series with capacitor $C2$ is $R28$ to $R32$ and tube $CC1$ will fire. Firing of $CC1$ will, because of the common anode load $R4$, prevent firing of tubes $CC3$ and $CC5$. Firing of tube $CC1$ will cause an increase in its cathode voltage and because of the coupling capacitor $C4$ this increase will be reflected on the trigger electrode of tube $CC2$. This will cause it to fire, and hence relay $N1$ to be operated to select car No. 1, i.e. the car nearer to the landing call. Closure of one set of n/o contacts $N1$ causes the relay to self-hold via n/o contacts $FC1$ and at the same time to cut off tube $CC2$. Closure of another set of n/o $N1$ contacts in Fig. 16.26 causes car No. 1 to be dispatched. Closure of a further set of n/o $N1$ contact causes operation of relay CN. This causes opening of its n/c contacts to break the anode circuit of the tubes causing, in this case, tube $CC1$ to cut off and to break the upper and lower parts of the circuit thus disconnecting the capacitor $C1$ from the supply. Operation of CN also causes closing of its n/o contacts to effect discharge of capacitor $C1$ through resistor $R15$. Hence when

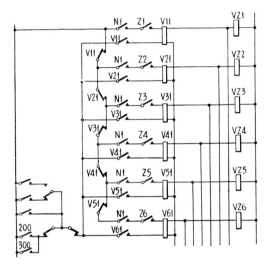

Fig. 16.26. Express Traffic Mark IV: Car Dispatch Circuit

(*Express Lift Co.*)

car No. 1 is no longer a free car so that its n/o contacts $FC1$ re-open to de-energize relay $N1$, the circuit is returned to its initial state ready for the next selection.

From Fig. 16.24 it will be appreciated that if cars 1 and 2 were free at the twelfth and fourth floors respectively and up landing calls were registered in selected zone 2 from the ninth and seventh floors, car 2, i.e. the car nearest to the lowest Up call will be selected, although both cars are three floors from one of the calls. The reason is that the charging rate of capacitor $C1$ is controlled by five sectors, $R32$ to $R36$ (due to the call at the seventh floor, n/c contacts $7U$ open to prevent sections $R34$ and $R36$ alone controlling the charging of $C1$) whilst the charging rate of $C2$ is controlled by three sectors $R29$ to $R31$ and therefore tube $CC3$ will fire.

Fig. 16.26 shows the car dispatch circuit for car No. 1, the circuits for the other two cars being similar. A car is dispatched by the individual landing call relays 200 for calls upwards and 300 for calls downwards. Assume that car No. 1 is selected as the next free car. Owing to a landing call existing n/o contacts 200 or 300, depending on the direction of travel, will be closed. If zone 1 has been selected for this car, n/o contacts $Z1$ are closed and n/o contacts $N1$ are closed by energization of relay $N1$ (Fig. 16.25) to energize relay $V11$. The car will then be dispatched to answer calls in the zone selected. The V relays are fitted with "pick-up" and "hold" windings and whilst the pick-up windings are energized, the respective one of the series-connected relays $VZ1–VZ6$ is energized to reset the zone timing circuit to zero. In this example it resets by the opening of n/c contacts VZ in the timing circuit for zone 1 (Fig. 16.22). When a car is dispatched to a zone it will travel to the highest Down call or the lowest Up call and then proceed to answer other calls in the zone working on a "collector" system, i.e. a car will answer calls in the direction of travel, reverse at the highest call and answer calls in the opposite direction. Assume that the car is at the eighth floor and is dispatched to zone 1 which has calls $3U$, $4U$, $5U$ registered. The car will travel from the eighth to the third floor without showing any indicators. When a car arrives at the third floor, an arrow above the car entrance illuminates to show that the car will leave in the Up direction. The car then travels to answer calls $4U$ and $5U$, each time illuminating the Up arrow. When the car leaves the fifth floor it will have only car calls registered; thus when the car stops in response to the last car call it becomes "free" and no directional arrows are illuminated.

Another method by which calls may be answered is to extend the system so that a car, when allocated to a zone, answers calls within that zone and then continues to answer all calls ahead of it. For example, if a car is allocated to zone 1 which has calls $3U$, $4U$ and $5U$ registered, the car after answering these calls, will continue to answer all Up calls above $5U$.

Alternatively, the system may be arranged so that cars

when answering Up call zones, respond to the individual zones only, but when travelling to Down call zones answer all calls ahead of the car after answering all calls in that zone.

Yet another method allows these additional calls to be inserted only if they have been accepted for longer than a

Fig. 16.27. Controllers for Four Gearless Lifts
(*Express Lift Co.*)

predetermined time interval, say 30 seconds, and no other car is available to answer them. If calls are inserted later, the car will answer them, provided that they are ahead of the car when they are inserted.

By selection of the landings in each zone and individual adjustment of each zone timer this Mark IV is flexible and can satisfy a variety of requirements.

Fig. 16.27 shows part of the machine room of a bank of four Express Traffic Mark IV gearless lifts. The first panel is the individual automatic control section housing the switch-gear for car location, monitoring and call acceptance. The central portion shows the electromechanical floor selectors. The second panel is the power section for one car and contains the switchgear and relays for the control of the a.c. driving motor and generator field of the Ward-Leonard motor generator set and of the gearless lift motor. The ground-floor lobby supervisory panel for this installation is shown in Chapter XII.

Controllers with Transistor Static Switching Elements. The development of the transistor has been rapid since its introduction in 1948 and although it has some important limitations it has been used successfully for many applications where thermionic valves and electromagnetic switchgear have been employed. One of its properties is that it can be used efficiently as a switching device having no contacts requiring adjustment and cleaning. It has a long life, needs little space and has a high degree of reliability. These advantages have prompted lift manufacturers to examine the possibility of their replacing electromagnetic switches in performing some of the functions in lift control. Mullard Equipment Ltd. was one of the first firms to adapt their "Norbit" transistor for this purpose, and a number of lifts have been installed by several manufacturers incorporating substantial sections of "Norbit" static switches in their controllers. It is probable that they will prove economical because of their reliability and low maintenance cost.

THE TRANSISTOR. The transistor consists of three parts, the emitter, the base and the collector as shown in Fig. 16.28. If a wafer of germanium which has been "doped" with a suitable impurity is heated in contact with two pellets of indium to a temperature just below the melting point of indium, the germanium dissolves some of the indium at the areas of contact. As a result, a rectifier is formed at each of the two indium–germanium junctions. The doped germanium wafer is termed "n" type germanium and the solution of indium in germanium is termed "p" type germanium. The resultant transistor is therefore two rectifiers with a common cathode connexion called the "base." The connexions to the indium pellets are called the "emitter" and the "collector." The electrical equivalent of the transistor is shown in Fig. 16.29. If both diodes are reverse-biased, the transistor is non-conducting and is said to be "off." When one diode is forward-biased and the other reverse-biased, interaction occurs between the two diodes because of their proximity and the current in the reverse diode, instead of being a very small leakage current, is

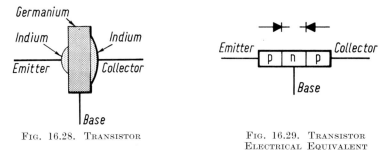

FIG. 16.28. TRANSISTOR

FIG. 16.29. TRANSISTOR ELECTRICAL EQUIVALENT

proportional to, and nearly equal to, the current in the forward-biased diode. The transistor then conducts and is said to bottom or be in the "on" condition. The fundamental current relationships in the transistor are shown in Fig. 16.30.

TRANSISTOR USED AS A SWITCH. Although the transistor

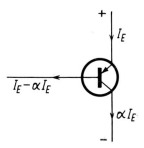

FIG. 16.30. TRANSISTOR CURRENT RELATIONSHIP

can be used for several different purposes in a variety of circuits we are interested here in its use as a switch to replace electromagnetic switches in lift controllers. From the above it is seen that the transistor can be in the conducting or insulating condition depending on the voltage connected to its base. If the base of the transistor is positive to the emitter, the emitter–base junction is reverse-biased and the transistor is in the "off" state with only a very small leakage current flowing in the collector circuit. This leakage current is of the order of 1 μA for a germanium transistor and 0·01 μA for a silicon transistor. Hence the transistor, when "off," simulates very closely an open switch. If the base is biased negative to the emitter potential so that sufficient (negative) base current flows, the transistor is turned "on" and conducts and simulates

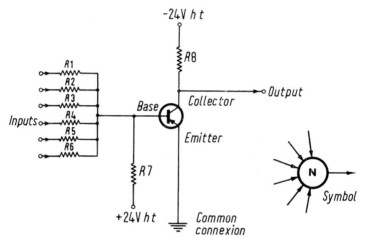

FIG. 16.31. TRANSISTOR-RESISTOR NOR CIRCUIT

a closed switch with only a very small voltage between the collector and emitter. With collector currents up to 1 A the voltage between the collector and the emitter may not exceed 0·2 V in the "on" state and for small collector currents may not exceed 1 mV.

THE NORBIT STATIC SWITCHING ELEMENT. The transistor–resistor NOR circuit used by Messrs. Mullard on lift controllers is shown in Fig. 16.31 and from this it is possible to build extensive switching systems using only combinations of this one type of unit. Six input resistors are connected to the base, with a bias resistor $R7$ which is connected to the +24 V bias line. The load resistor $R8$ is connected to the collector and is returned to —24 V. A negative output is obtained from this circuit whenever none of the inputs 1, 2, 3, 4, 5 or 6 is positive, this form of operation being referred to as a "NOR" function.

If a "1" is a negative voltage exceeding 7 volts and a "0" is a voltage between 0 and -0.2 V, a "1" signal present on any of the six inputs causes the transistor to conduct. The transistor bottoms and the output voltage is a "0." This function is "OR" (similar to switches in parallel with each other) followed by polarity inversion due to the transistor.

If all six inputs are at "0" level the transistor is cut off as

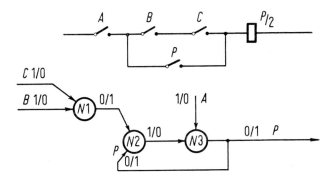

FIG. 16.32. NOR SWITCHING CIRCUIT

the base is positive, due to the $+24$ V bias line, the bias resistor, and the input resistors returned to 0 V. The output voltage is then negative, its value depending on the load. This function is "AND" (similar to switches in series with each other) followed by polarity inversion.

As an illustration of how NOR elements can replace switches in a lift controller, Fig. 16.32 shows a simple switching circuit and the NOR equivalent. Relay P is energized if contacts A, B and C are closed, and contact P then holds relay P energized regardless of B or C until contact A opens. The NOR element $N1$ replaces the B and C series contacts and functions as an AND gate. If B and C are "0" the output is "1" and this is fed to the second element $N2$ which replaces contact P and functions as an OR gate. Element $N3$ replaces contact A and functions as an AND gate being fed by the output of $N2$. When the inputs of $N3$ are "0" the output is "1" and this is fed back to $N2$ so maintaining the output regardless of B or C. If A becomes "1" the output becomes "0" and the circuit resets.

Simplex Collective Static Switch Controller. The following is a description of the Mullard Equipment Ltd. (M.E.L.) controller using their Norbit static switches. It will be appreciated that these functions and the equipment employed are typical and can be modified considerably depending upon the customer's requirements.

Fig. 16.33 is a schematic diagram of two floors of a simplex collective control installation, with one terminal floor, one intermediate floor and other common central control circuits. The various blocks are described briefly below and this is followed later by a more detailed account of circuit operation.

1. *Call Memories.* This consists of a memory circuit to register and retain the car and landing calls. The input to each memory is from a normally closed contact on the car or landing call button.

2. *Call Cancellation.* This cancels the call memory when the call has been answered, and consists of three AND gates, resetting the up, down and car memories.

3. *Call Direction.* The determination of the direction in which the lift must move to answer a call is based on the principle of call blocking. A call into a floor module causes a call α (call lift up) and a call β (call lift down) to be sent via other modules (α through those above and β through those below). The presence of the lift in a zone will block either the α-call (if the lift is above the call) or the β-call (if the lift is below the call) thus transmitting the correct call to the central control. The blocking is done by the lift zone signal obtained from either an internal selector or an external selector. Assuming that the UP button is pressed on a floor below the lift and there are no calls below this floor, the α-call is blocked by the zone signal (lift position) above the call, the β-call results in the lift being called down.

4. *Slowing Circuits.* These consist of three AND gates—up slow, down slow and car slow. The outputs of these gates are combined and trigger a slowing memory in the SLOWS-AND-DOORS module. The UP, DOWN and CAR-CALL signals are taken from the call memories.

5. *Doors (Call).* These circuits provide a DOORS-TO-OPEN

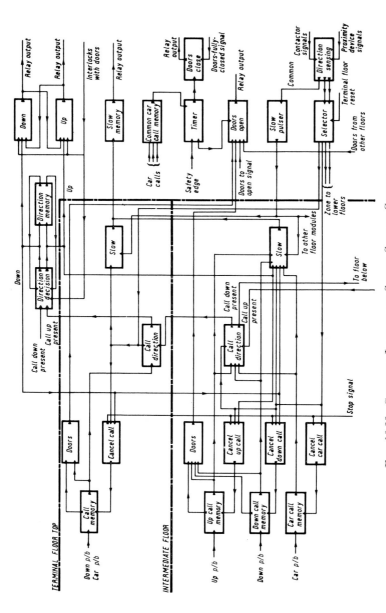

Fig. 16.33. General Layout of Static Switch Controller

(*Mullard Equipment Ltd.*)

signal when a landing call is made at the floor at which the lift is standing. The signal is taken to the SLOWS-AND-DOORS module.

6. *Central Control.* This takes the control information on the α- and β-lines, and in conjunction with the direction memory, controls the two pilot relay outputs (UP and DOWN) and the direction information fed back into the floor modules. This direction information is taken down to all floor modules to provide the signals mentioned in (2) and (4) above. The incoming α and β information are from the floor modules.

7. *Doors (Open and Close).* A DOORS-TO-OPEN memory is provided with drive to a pilot relay. This circuit can be driven from the DOORS circuits on the floor modules, by signals from external apparatus, e.g. a contact from a levelling circuit or via a door safety-edge or light-beam circuit.

A DOORS-TO-CLOSE memory is provided with drive to a pilot relay. This relay is interlocked with the DOORS-TO-OPEN relay so that they cannot operate together. The circuit can be driven from a common car-call circuit, from attendant push-buttons, or from an automatic timing circuit. The timing circuit is arranged to start its operation when the doors are fully open, and a DOORS-TO-CLOSE signal is given after a preset time delay. Car calls will over-ride this time delay so that doors start to close immediately.

The operation of the DOORS-TO-CLOSE circuit is subject to safety-edge or light-beam interruption. This signal will inhibit the DOORS-TO-CLOSE operation and automatically start the DOORS-TO-OPEN cycle.

8. *Slowing Memory.* This circuit has inputs from all floor circuits and drives a pilot relay. It resets when the lift has stopped (derived from both UP and DOWN contactors on the lift equipment).

9. *Selector.* This can be external to M.E.L. equipment or can be to M.E.L. design.

The requirement from external equipment is—

Zone signal for each floor; a contact would be provided on each floor. It is important that the zone signal should change *before* slowing region is reached.

The changeover should be as quick as possible. It is

desirable, but not essential, that a small overlap of zones should occur.

If the M.E.L. static switching selector is used, this consists of a Norbit ring counter. This can be driven in forward and reverse directions from signals derived from photo-electric devices or other types of proximity device.

Alternatively a both-ways uniselector may be used as a counter operated from the proximity detectors.

The UP and DOWN contactor signals mentioned previously are used to define direction, and pulses from a single proximity device are adequate if a suitable vane is used. (A vane about 1 ft. long and with entry and exit of the proximity device at the slowing points).

Note. A single proximity device can only be used if there is no cross-over of the slowing point, i.e. if the UP slow point for a given floor is above the DOWN slow point for the floor below. If cross-over exists, it is necessary to use two sets of vanes and proximity devices, one for UP, and one for DOWN.

Entry into the vane is used to cause the zone to change. Exit from the vane causes a slow pulse (used to interrogate the slowing circuits on floor circuits).

A reset must be provided at a terminal floor.

This is normally only required on switching-on or after power failure. But it is convenient to provide the facility as a permanent feature, the setting of the counter being confirmed every time the lift goes to the terminal floor.

On re-establishing power after a failure, circuits in the power supply cause the control equipment to provide a general reset to all functions and to cause the lift to travel to the terminal floor where reset of the counter occurs.

Inputs

Bypass signal	Attendant bypass feature
Direction signals	Main contactors
Proximity signals	Contact or electrical if suitable for Norbits
Reset at terminal floor	Contact at terminal floor.

Outputs

Zone signals to M.E.L. floor circuits.

Zone signals via power amplifiers for lamp circuits (2·8 W 24 V under run at 1·5 W). Up to 14 lamps.

10. *Slow Pulser.* This is self-contained with selector if provided. If selector is external, a signal is required from the external circuits. This must occur after the zone changeover and can be a contact closure or proximity-device signal. Duration required can be extremely short (a few milliseconds) or may last till stop signal is given. (This is not desirable as it could cause late slowing if a button is pressed during this period.)

The following are more detailed descriptions of the modules for an intermediate floor, a terminal floor, the central control and for door operating, all of which can be modified to cater for various requirements. The diagrams do not show input filter circuits and it should be remembered that all signals from external contacts and relays are connected via input filters which integrate the signal and have a voltage threshold.

INTERMEDIATE FLOOR LOGIC

1. *Call Memories.* Fig. 16.34 shows a simplified logic diagram of a floor module.

L.P.O. unit 1 and Norbit 2 form a memory circuit with power available to light a 24 V 2·8 W lamp, under-run at 1·5 W. An emitter follower is used on the output of 2 due to the drive required for L.P.O. 1. This memory operates from a push-button signal. (The NOR "1" is required to "set" the memory.) L.P.O. 5 and Norbit 6 form an identical circuit from the DOWN push button. Norbits 3 and 4 form a memory without indication, operating from the car push-button. Inputs for reset are provided on Norbits 2, 4 and 6 for resetting on switching-on.

2. *Call Cancellation.* Norbits 16, 17 and 18 make up the call cancellation circuit. Each Norbit is an AND gate, one for UP calls, one for DOWN calls and the last for car calls.

The UP circuit has three inputs—

(i) Direction ($D = 0 =$ either idle or UP)

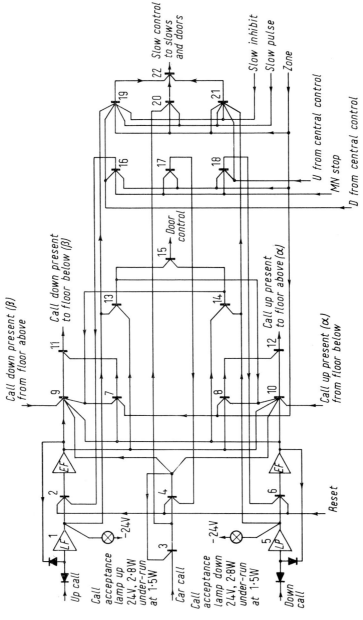

Fig. 16.34. Intermediate Floor Logic of Static Switch Controller

(*Mullard Equipment Ltd.*)

(ii) Lift has stopped (STOP $= 0 =$ both contactors are out)

(iii) Zone ($Z = 0 =$ lift is in zone associated with circuits)

The down circuit also has three inputs—

(i) Direction ($U = 0 =$ either idle or DOWN)

(ii) Lift has stopped

(iii) Zone

The car circuit has two inputs—

(i) Lift has stopped

(ii) Zone

Direction has no influence on cancellation of car calls.

It is only necessary to ensure that the necessary signals of correct polarity are available at the correct time, i.e. direction signals are available from central control. "Lift has stopped" signal is available from lift control apparatus. Zone signal either from lift control apparatus or from counter determining lift position.

A call will be cancelled if the lift has stopped at the call floor and the lift is programmed to proceed in the direction required by the call, except in the case of car call, where no direction preference is needed.

3. *Call Direction.* This is made up of Norbits, 7, 8, 9, 10, 11 and 12. Consider first Norbits 9, 10, 11 and 12.

The determination of call direction is based on the principle of call blocking. A call at a floor module causes a call α (call lift up) and a call β (call lift down) to be sent via other modules, (α through those above and β through those below).

The presence of the lift in a zone will block either the α call, (if lift is above call), or the β call (if lift is below call). This will result in the correct call being transmitted to the central control. *The zone blocking* is achieved by forcing a "1" on to Norbits 11 and 12 whilst the lift is in the particular zone.

Consider now calls inserted whilst the lift is in the zone associated with the calls. In the simple case we considered above, all calls are blocked by the zone signal. Therefore Norbits 7 and 8 are introduced. The blocking of 11 is a function of: Zone $= 0$ (i.e. lift is in zone) and DOWN call $= 0$ (i.e. no DOWN call). This means that if there is no DOWN call on the circuit the β line will be blocked in that zone. If, however, a DOWN call is present in that zone the β line will not block. Similarly Norbit 12 is blocked through the function

$$\text{Zone} = 0$$
$$\text{and Up call} = 0$$

The car call is not required to unblock these paths.

A call made this way on to the α- and β-lines is known as an "injected" call to distinguish it from the normal call mechanism. The action of an injected call is important when considering a call which necessitates a lift reversal at that point. This condition will of course only be acted upon if no other call requiring no reversal is present on the system.

We will consider a particular case to clarify the action.

Assume: The UP push-button is pressed on a floor below the lift. There are no calls below this floor. The α-call is blocked by the lift above the call, the β-call results in the lift being called down.

It will be seen later that direction preferences in the slowing circuits would prevent slowing of the lift when the direction is down and the call is UP.

The zone block is imposed before the slowing pulse. The zone block prevents the β-call from reaching the central control. However, the presence of the UP call prevents the zone block acting on the α-line and therefore an α-call is propagated up to the central control. It will be seen later that this call, as it is not present with a β-call, can change over the direction memory and will therefore result in a slow signal from the slowing circuits.

It is therefore evident from the above, that the direction information on the U and D lines is "Direction in which lift is programmed" and not "Direction in which lift is travelling."

The requirements of this circuit to some extent dictate the operation of selector and slowing circuits.

(i) Zone signal must occur either before or coincident with the slowing signal.

(ii) There must be time enough for the injected call to propagate to the central control, direction memory to change over, new direction information to be presented to slowing circuits before the slow pulse ends.

4. *Slowing Circuits.* Norbits 19, 20, 21 and 22 form the slowing circuits.

19 is an AND gate with the following inputs—

    (i) Up call = 0 when a call is present
    (ii) Direction D = 0 when idle or Up
    (iii) Zone = 0 when lift is in appropriate zone
    (iv) Inhibit = 0. Bypass not operated
    (v) Slow pulse = 0 (for duration of pulse)

20 is an AND gate with the following inputs—

    (i) Car call = 0 when call is present
    (ii) Zone = 0 when lift is in appropriate zone
    (iii) Slow pulse = 0 (for duration of pulse)

21 is an AND gate with the following inputs—

    (i) Down call = 0 when call is present
    (ii) Direction U = 0 when idle or Down
    (iii) Zone = 0 when lift is appropriate zone
    (iv) Inhibit = 0 when bypass not operated
    (v) Slow pulse = 0 (for duration of pulse)

The three outputs are combined in an OR gate 22 before leaving this module.

5. *Doors (Call).* Norbits 13 and 14 are AND gates with the following inputs—

Norbit 13

    Call Up present (0)—call button pressed
    Reset on call up present (0)—this is taken from the second output of the call memory.

Norbit 14

    Call Down present (0)—call button pressed
    Reset on call down present—this input is taken from the second output of the call memory.

These circuits ensure that when a landing call is inserted at the floor at which the lift is standing, the command is given for the doors to open. The two command signals are applied to the OR gate 15 before leaving the module. The combination in the floor module is for convenience and easing the wiring problem, and is, as in the slowing circuit, a matter of choice.

It should be noted that the outputs of 13 and 14 are used to inject a call into the α-and β-lines, to give direction preference immediately the landing button is pressed. This can cause maloperation if a button is deliberately held, and it is advisable to fit a further circuit to prevent this condition. An *RC* circuit consisting of a diode, a resistance and a capacitor is connected from the UP push-button circuit to two paralleled inputs of 13. On pressing the UP button the output of 13 becomes a "1" as previously described, but after a few milliseconds the *RC* circuit changes and causes 13 output to be a "0." The pulse output of 13 sets the doors to open memory and also changes over the direction memory (if necessary). A similar circuit can be connected to 14 from the DOWN push-button circuit. In addition the RESET signal should be connected to a further input on 13 and 14 so that the two Norbits are completely out of circuit when RESET is applied (e.g. on test of goods control).

TERMINAL FLOOR LOGIC

Fig. 16.35 shows the terminal floor logic required.

L.P.O. (1) and Norbit 2 form a call memory with lamp drive available. It can be set from either *DN* landing button or car push-button.

L.P.O. (3) and Norbit 4 perform the same function for the lower floor landing and car call.

Norbits 5, 6, 7, 8, 9 and 10 combine the functions of call direction and reset. It is convenient to be able to apply an inhibit signal to 9 and 10 to prevent all calls reaching the central control under, for example, test conditions.

Norbits 6 and 7 perform the function of reset, on zone and STOP signals being 0. Norbits 5 and 9 transmit the α call (call up) and Norbits 8 and 10 transmit the β call (call down).

In the top zone and the lift stopped α calls are blocked by the output of 6.

In the bottom zone and the lift stopped β calls are blocked by the output of 7.

The Norbits 11 and 12 provide DOORS-TO-OPEN signals as described in the description of an intermediate floor module.

No slowing circuits are provided, although these are provided

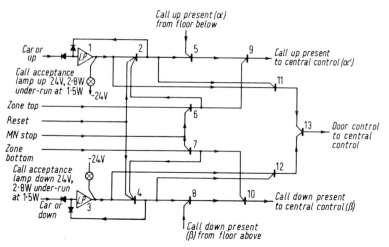

FIG. 16.35. TERMINAL FLOOR LOGIC OF STATIC SWITCH
CONTROLLER

(*Mullard Equipment Ltd.*)

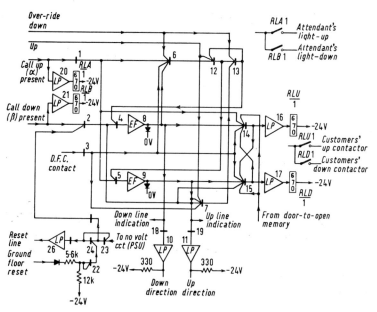

FIG. 16.36. CENTRAL CONTROL LOGIC OF STATIC SWITCH
CONTROLLER

(*Mullard Equipment Ltd.*)

elsewhere to give a slow signal whenever the lift enters either the top or bottom zones.

It is recommended that the lift manufacturer should always provide "back-up" slowing contacts at terminal floors if a counter type of selector is used.

It should be noted that if a type of control requiring inhibit of landing calls is required, the combination of landing and car calls in the way described is not possible.

CENTRAL CONTROL LOGIC (SIMPLEX)

Fig. 16.36 shows the central control for a simplex. Norbits 12 and 13 comprise the direction memory. Norbits 1 to 7 are concerned with decision of direction, whilst emitter followers 8 and 9 and L.P.O.s 10 and 11 provide drive to floor modules and to amplifiers to drive lamps. Norbits 18 and 19 are inverters. Norbits 14 and 15 and L.P.O.s 16 and 17 operate the pilot relays for the UP and DOWN contactors.

When the direction memory is in the set state, i.e. output of Norbit 13 is 1 the memory is in the up condition.

When in the non-set state, i.e. output of Norbit 13 is 0 the memory is in the down condition.

If a call up is present on the system, a "1" signal appears on the α line. If a call down is present on the system a "1" signal appears on the β line.

It is possible to write a table summarizing the action of the control. Each case will then be considered and the action described.

Case	α	β	Previous Direction Memory State	Resulting Direction Memory State	Direction Information
1	0	0	Up	Up	Not Up Not Down
2	0	0	Down	Down	Not Up Not Down
3	0	1	Down	Down	Down
4	0	1	Up	Down	Down
5	1	0	Up	Up	Up
6	1	0	Down	Up	Up
7	1	1	Up	Up	Up
8	1	1	Down	Down	Down

The following explanation assumes over-rides are 0 and DFC (DOORS-FULLY-CLOSED signal) is 1.

Case 1: *No calls*

Direction memory remains in condition on answering last call.

U line $= 0$ due to output of Norbit 1 ($/\bar{\alpha}$)

D line $= 0$ due to output of Norbit 2 ($/\bar{\beta}$)

Case 2: (*a*) *Call down, no call up, memory in "down" state from last call*

Output of Norbit 2 ($/\bar{\beta}$) is 0. Both inputs to 4 are therefore 0 as the memory is down. The D line is therefore 1.

Both inputs to 5 are 1 from Norbit 1 ($/\bar{\alpha}$) and from the memory which is down. The U line is therefore 0.

The memory state remains unchanged as the output of Norbit 7 is 1 and Norbit 6 a 0.

(*b*) *Call down, no call up, memory in "up" state from last call*

Norbit 6 has a 1 input from Norbit 1 ($/\bar{\alpha}$) Norbit 5 has also a 1 from Norbit 1 causing a 0 on to Norbit 7. The other input to Norbit 7 is from 2 ($/\bar{\beta}$ which is also 0). The output of 7 is therefore 1. The memory changes over to down state.

The conditions are now as for 2 (*a*).

Case 3: (*a*) *Call up present, no call down and memory in "up" state*

The action is similar to that in case 2 (*a*) the roles of the α and β lines being interchanged.

The U line $= 1$
The D line $= 0$

(*b*) *Call up present, no call down and memory in "down" state.*

The action is similar to that in case 2 (*b*) with the roles of α and β lines interchanged, with final state as for 3 (*b*).

Case 4: (*a*) *Calls on both lines, memory "up"*

Norbit 4 has a 1 input from the memory. Therefore the inputs to Norbit 6 are both 0 and the input set of the memory is as for up.

Norbit 5 has both inputs 0, one from the memory, one from Norbit 1 ($/\bar{\alpha}$). One input to Norbit 7 is therefore 1 and the reset to the memory is 0.

The *U* line is 1
The *D* line is 0

The memory is therefore unchanged.

(*b*) *Calls on both lines, memory "down"*

Norbit 4 has both inputs 0, one from the memory, one from Norbit 2 ($/\bar{\beta}$). One input to Norbit 6 is therefore 1 and set to memory is 0.

Norbit 5 has a 1 input from the memory. Therefore the inputs to Norbit 7 are both 0 and the reset input of the memory is as for down.

The memory is therefore unchanged.

The *U* line is 0
The *D* line is 1

Norbit 3 inverts the signal obtained from the *DFC* contact (Doors-fully-closed). This signal is 1 for the doors fully closed and 0 for all other conditions.

The output of Norbit 3 is therefore 1 when the doors are not fully closed. This signal freezes the direction memory. Direction memory information is preserved whilst the doors are open. In order to make the direction over-ride from attendant push buttons effective, it is necessary to apply the required signal to the direction memory and to inhibit any drive from either Norbit 6 or 7. Therefore, to set the memory to Up, drive is applied to Norbit 12 and Norbit 7 is inhibited, and to set the memory to Down, drive is applied to Norbit 13 and Norbit 6 is inhibited.

Norbit 14 is an AND gate with the following inputs—

1. Call Up present ($/\bar{\alpha} = 0$).

2. Doors are to open—this will be a 1 immediately the doors to open signal is given.

3. Direction ($D = 0$) IDLE or UP.
4. Doors fully closed. Output of Norbit $3 = 0$.
5. Norbit 15 is not operated. Output of Norbit $15 = 0$.

When all these conditions are 0, the output of Norbit14 will be 1 and will drive the UP pilot relay via the L.P.O. 16.

Norbit 15 is an AND gate with the following inputs—

1. Call DOWN present ($/\bar{\beta} = 0$).
2. Doors to open in this will be a 1 immediately the doors to open signal is given.
3. Direction ($U = 0$) IDLE or DOWN.
4. Doors fully closed. Output of Norbit $3 = 0$.
5. Norbit 14 is not operated. Output of Norbit $14 = 0$.

When all these conditions are 0 the output of Norbit 15 will be 1 and will drive the DOWN pilot relay via the L.P.O. 17.

Input 2 is taken from the DOORS-TO-OPEN memory and ensures that a call at the floor at which the lift is standing, immediately inhibits the pilot relays. Norbits 23 and 24 form a reset memory set from a "no volt circuit" in the power supply. This memory resets from the ground-floor contact via inverter 22. (This assumes a $+24$ signal as the reset signal.) This is so if a static type of selector is used where ground floor reset applies a $+24$ signal to collectors of a ring counter through diodes and resistors. The reset signal may in other cases be a negative signal and be applied to 24 without the inverter 22.

The output of 23 is a 0 while memory 23, 24 is set and this signal is inverted and applied to 2 as a call down. On switching on, a pulse sets the memory 23, 24 and therefore a CALL-DOWN signal is applied. This causes the lift to move down to the lowest terminal floor where it slows in response to the terminal floor slowing contacts and then the ground floor reset signal is applied, resetting memory 23 and 24. The lift is then ready for service. It is important to note that as the ground floor reset signal is used to reset the selector (if a counter type is used), this signal should be removed when the lift departs from the terminal floor in the opposite direction, before the proximity detector pulses the selector.

Norbit 27 has two inputs, $/\alpha$ and the DOWN signal. These are both 0 if there is an α call and the direction signal is either UP or IDLE. The output of 27 drives RLA through $LP20$. RLA can be used for attendant indication, or generator starting circuits. It should be noted that this signal differs from the UP signal at the output of 11, as the output of 27 will be 1 if there is an up call and the lift is idle. The output of $LP11$ would not register a 1 under this condition if the direction memory was set to DOWN and the doors were open.

The output of 11 would be a 1 when the doors closed as the direction memory can then change over.

DOORS (OPEN AND CLOSE) AND SLOWING

The basic requirements of the doors-operating circuits are—

1. The doors will start to open just before floor level. This can be achieved by taking a signal from normal levelling circuits on the lift control.

2. On normal operation the doors will stay open for a period adjustable from 0 to 10 seconds, the time starting from the instant the doors are fully open. After this period, the doors should automatically close.

3. If, on a person entering the lift, a safety device is operated (either a light beam or a safety edge on the door), the doors, if closing, will re-open and the time delay will reset.

4. If a car call is made during the timing period, the timing inhibit will be over-ridden and the doors will immediately close. Should the safety device operate during the closing, the doors will re-open and the timing will be re-instated. The doors would therefore close after the time delay. Should another car call be inserted the time delay will again be over-ridden and the doors will immediately start to close.

5. On attendant operation the doors when opened will remain open, unless the attendant UP or DOWN push buttons are operated.

If these buttons are released before the doors are fully closed, the doors will re-open. The operation is again subject to the safety device, and if this is operated the doors will re-open. But when the doors have reached the fully open position and the push is still operated, the doors will immediately close.

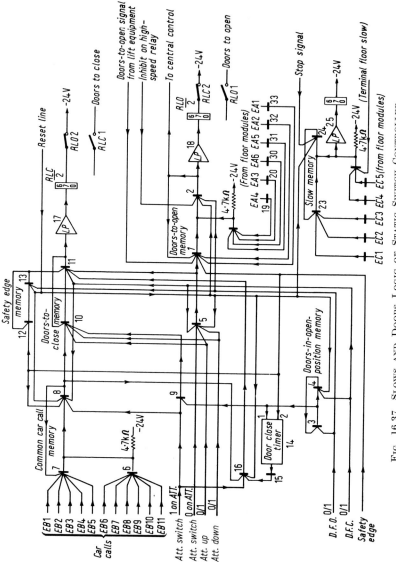

FIG. 16.37. SLOWS AND DOORS LOGIC OF STATIC SWITCH CONTROLLER

(Mullard Equipment Ltd.)

Fig. 16.37 shows a circuit which will meet the above requirements.

Norbits 1 and 2 form a DOORS-TO-OPEN memory, driving a relay through an L.P.O. (18).

Norbits 10 and 11 form a DOORS-TO-CLOSE memory, driving a relay through an L.P.O. (17). The DOORS-OPEN relay (DOR) and the DOORS-CLOSE relay (DCR) are interlocked on their contacts to prevent simultaneous operation.

We will describe the operation from the instant a slowing signal is generated.

The slowing signal is used usually to release a fast speed relay in the lift control circuits. A contact on this relay is used to inhibit the DOORS-TO-OPEN. When the lift is at high speed, the fast speed relay is operated. On the FSR releasing, the inhibit is released.

A levelling contact is opened at the instant the levelling commences. This sets the D.O. memory, operating the DOR and the doors start to open. When the doors are fully open, a contact (DOORS FULLY OPEN, DFO) opens and a reset is applied to the D.O. memory.

This contact also sets a DOORS-IN-OPEN-POSITION memory (Norbits 3 and 4). Operation of this memory applies a 0 to Norbit 9 and to input 1 of timer 14. The second input to Norbit 9 is held on a 0 by the attendant switch in the normal condition. The output of 9 therefore applies a 1 to the DOORS-TO-CLOSE memory. This memory is inhibited by the 1 on the timer output through Norbits 15 and 16. Other inputs to Norbit 16 are 0. The application of a 0 to the timer input 1 starts the timer, and, after the time delay the output of the timer will go to 0, releasing the inhibit on the D.C. memory through Norbits 15 and 16. This memory can now set and operates the DCR through LPO 18. The DCR can operate only after the DOR has been released, but the D.C. memory has set and will maintain if the DOR has not completed its release. The doors will start to close, and when fully closed the D.C. memory will reset from a DOORS-FULLY-CLOSED (DFC) contact opening. This contact also resets the DOORS-IN-OPEN-POSITION memory on the input to Norbit 4.

If, however, the safety edge (SE) is operated a 1 is applied to the input 2 of timer 14 for the duration of the operation. This input 2 resets the timer and for the duration of the pulse the timer output is 0. It is therefore necessary to apply the pulse direct to the D.C. memory as an inhibit through a memory, Norbits 12 and 13, which resets from the *DFO* contact. On the pulse ending, the timer output goes to 1 and the inhibit transfers to the output of Norbit 16. The safety edge pulse is also applied to the D.O. memory and sets this. The inhibiting of the D.C. memory releases *DCR*, and on release of *DCR* the D.O. memory drives the *DOR* and the doors re-open as described previously. The doors will remain open until the timer output releases the inhibit on the D.C. memory. At the end of the timing period the D.C. memory will again set as the output of Norbit 9 is still a 1. The sequence will then occur as previously described.

Note. The safety edge pulse must also be applied direct to Norbit 11 as the safety edge may still be operated when the doors are fully open.

If a car button is pressed, the common car call memory will set, applying a 1 to the input of the D.C. memory, and removing the inhibit by applying a 1 to Norbit 16. The doors will therefore start to close immediately. The operation is subject to the safety edge again, as the S.E. contact resets the common car calls memory, imposes an inhibit on the D.C. memory, and resets the timer. When the S.E. contact is released the timer output goes to 1 and continues the inhibit through Norbits 15 and 16. When the doors are fully open the safety edge memory resets, but the timer maintains the inhibit until the time delay has elapsed. The doors will re-close after the time delay as the output of Norbit 9 is still a 1. Alternatively, operation of a car push again will repeat the sequence described above. It should be noted here that the car call memory drive to the D.C. memory is not strictly necessary as a 1 is maintained from Norbit 9. Also the safety edge memory is not necessary for normal operation, but is needed for attendant operation.

Under attendant operation an input to Norbit 9 is a 1 and the output of this Norbit is 0. The common car call memory is also inhibited. The doors to close memory cannot operate automatically. Norbit 16 output is also held at 0 by the attendant switch on a third input. Norbit 5 has five inputs, the first being 0 on attendant control. The second and third from the attendant pushes Au and Ad are 0 when pushes are not operated. The fourth input is from the DFC contact and is 0 except when the doors are fully closed. The fifth input is from reset. The output of Norbit 5 is therefore 1 in the conditions above and this 1 inhibits the D.C. memory and applies a 1 to operate the D.O. memory. On operating an attendant push, Au or Ad, the output of Norbit 5 is 0, releasing the inhibit on the D.C. memory, and the output from the push is also applied to set the D.C. memory. The doors therefore start to close. If the push is released, the D.C. memory is inhibited and the D.O. memory sets via the output of Norbit 5.

If the S.E. contact operates, the S.E. memory operates and inhibits the D.C. memory. The D.O. memory is also set and the doors open. When the doors are fully open the S.E. memory resets from the DFO contact and if the attendant button is still pressed, or is re-pressed, the sequence will repeat.

Norbits 19 and 20, 30, 31, 32, 33, 34 accept signals from the common circuits in the floor modules for doors opening commands.

Norbits 21, 22, 35, 37 and 38 accept slow signals from the common circuits in floor modules as described in section 4. A slow signal sets the memory consisting of Norbits 23 and 24 and operates the slow Relay (SR), via LPO 25. This memory can be reset when the lift has stopped (Stop Signal). The reset signal is applied to all memories and to Norbits 5 and 16, ensuring the outputs of these are 0 during reset.

The above circuits should be considered only as a guide to the door and slowing circuits. The particular functions required can be provided by modifications applied to this basic circuit.

The Norbits used as inverters at the input to the slowing and doors memories can often be used to inhibit these functions when required. It is sometimes better to drive the slow relay via an inverter from the outside of the slow memory, rather than the inside, as a reset on the second Norbit does not inhibit the operation of the inside of the memory.

It is sometimes desirable to add a circuit following the safety edge input, preventing the operation if the lift is moving (inverse of STOP).

POWER SUPPLIES

Standard brick-built power supply units are used to provide a stabilized -24 volt supply at $3\cdot5$ A and a stabilized $+24$ volt supply at 1 A. (Note that the -24 line has the load of call acceptance lamps and the worst case condition of all lamps on must be considered.)

A further -100 V unstabilized supply provides up to 1 A for all push-button and switch contact circuits.

If a static selector is used, the output power amplifiers for position lamps should be driven from a separate unstabilized -24 V supply. It may be necessary to provide drive to a position lamp on each floor and in the car, as well as power to direction arrows at each floor and in the car.

If a uniselector-type selector is used, it is advisable to provide a separate unstabilized line for operation of the uniselector coils, but lamp circuits can be driven from an a.c. supply via the bank contacts.

M.E.L. Duplex Controller. This has been developed for two adjacent lifts and can be extended to a larger number of lifts in adjacent wells. The central control and slows and doors logic are similar to that for a simplex system. The selector can be mechanical, electro-mechanical or static. At each intermediate floor there is an UP and a DOWN button, these being common to the two wells and a single button at each terminal floor.

The control for car calls is the same as for simplex but in responding to landing calls a decision must be made to determine which lift will answer the call. In simplex operation calls are answered so as to cause the minimum reversals of the lifts and in duplex the lift requiring the minimum reversals answers the call. To make the decision with duplex it is necessary to know the direction of travel of each lift and the position of each lift with respect to the call. The priority system for UP and DOWN calls is as follows—

Up Landing Calls	*Direction*	*Lift Position*
1st Priority	Up	Below
2nd Priority	Idle	Below
3rd Priority	Not up	Above
Down Landing Calls		
1st Priority	Down	Above
2nd Priority	Idle	Above
3rd Priority	Not down	Below

The first priority does not require reversal of the lift, the second priority applies to a stationary lift, placed so that when started to answer the call it will become a first Priority call. The third priority is where the lift is approaching from

FIG. 16.38. DUPLEX STATIC SWITCH CONTROLLER
(*Marryat and Scott*)

the opposite direction to that desired and one reversal is required. An idle lift is included in the direction condition "not up" or "not down."

Priority gates are included in the logic circuits of both wells and are inserted between the call memory and the call direction circuit. The priority circuits are self-organizing by cross-inhibits between the two sets of priority gates. A higher priority will always inhibit lower priorities in the second well.

It is necessary for the lift position to be available on the duplex controller as this determines the priority of calls. A memory on each floor, triggered from pulses derived from the proximity selector used in slowing, gives this information. When set, the information is that the lift is below or at the floor. When not set, the information is that the lift is above the floor. If two lifts are proceeding in the same direction almost together, only one should answer the call, and cross-inhibits used on the slowing gates, ensure that only one lift stops at the required floor. Fig. 16.38 shows a Marryat and Scott Duplex Static switch controller for two geared variable-voltage 300 ft. per min. lifts. The electromagnetic portion of the controller is on the right and the M.E.L. Norbit equipment on the left.

CHAPTER XVII

LIFTS FOR SPECIAL PURPOSES

MANY lifts are installed in buildings other than office blocks and because of the rather unusual nature of the traffic, sometimes demand special consideration. The main features of some of these types of lift are mentioned below.

(a) **Lifts for Multi-storey Flats.** The Ministry of Housing and Local Government* recommends that a lift should be provided in buildings in which the entrance to any dwelling is on the fourth storey or above and that two lifts should normally be provided in buildings of more than six storeys, to minimize inconvenience during household removals, breakdowns, servicing or re-roping. In buildings with more than one lift, the installation should be arranged so that if one lift is out of action the other will serve the same dwellings.

Access to the lifts at the ground floor must be direct and the space in front of the lifts should be large enough for the movement of passengers, delivery of tradesmen's goods and the manoeuvring of prams and furniture. The grouping of lifts is advisable as this will allow the lift controls to be interconnected. The positions of the well and machine room and their insulation from dwellings should be carefully considered to reduce noise and if possible no habitable room should adjoin the well or machine room. The lift well is totally enclosed and the lift machines should be insulated from the floor by pads of compressed cork or rubber to reduce vibration. Noise is also reduced by using a floor selector in the machine room or inductors instead of slowing and stopping switches in the well.

The British Standard Code of Practice requires each block of flats to have a "fire lift" with a platform area not less than $15\frac{1}{2}$ sq. ft. and with a contract load of not less than 1 200 lb. This means an eight-person car. Some authorities insist on lifts larger than eight-person so that stretchers can be accommodated, and in high buildings 10- or 13-person cars may be necessary to give an adequate standard of service. The lifts

* *The Selection and Planning of Passenger Lifts in High Flats*, Ministry of Housing and Local Government Design Bulletin No. 3, Part 2.

must be large enough to carry prams and most furniture, and a deep lift is therefore needed, the recommended dimensions being 5 ft. 2 in. deep by 3 ft. 9 in. wide. The car is constructed of steel or hardwood and faced with formica, which is durable, not easily defaced and obtainable in a variety of colours and patterns. Corners and angles should be kept to a minimum to facilitate cleaning. Another very satisfactory car lining for this purpose is made of sheets of aluminium-manganese alloy. The surface of this material, fluted or ribbed vertically, requires little maintenance, and disfiguration by scratching or writing is not easy. This material is also used for the door surfaces and the landing architraves. To reduce malicious damage to a minimum the fixing screws for any equipment should be inaccessible from the landings or the car.

Single sliding fire-resisting power-operated doors are most suitable for flats. They are simple, minimize maintenance costs and are less costly to install. The opening size for the doors is not less than 2 ft. 9 in. wide by 6 ft. 8 in. high, which has been found sufficient for the largest prams. The doors should have vision panels of $\frac{1}{4}$ in. thick plain wired glass. Door closing is sometimes effected by a simple door closer of the falling-weight type instead of power operation. Another safety feature in common use is a concealed switch in the leading edges of the car and landing doors, which if operated by the door meeting an obstruction automatically reverses the door operator motor and opens the door. The door-bottom tracks are of the self-cleaning type to minimize maintenance and possible failure of the door operator. The clearance between the front edge of the car floor and the landing cill should not exceed half an inch to avoid the risk of trapping the heels of ladies' shoes and the clearance between each landing door and architrave should not exceed $\frac{1}{8}$ in. To minimize operation by children, the push-buttons are fitted at a height of about 5 ft. 6 in. above the floor, and if a large number of floors is served they may be mounted horizontally instead of vertically so as not to extend more than six feet above the floor. In some cases Yale locks are fitted instead of push-buttons, the tenants being supplied with keys.

The minimum speed of travel in high blocks is determined by the fire service requirements and the grade of service needed. A fire lift should be capable of reaching the top floor from ground level non-stop in one minute. If the floor-to-floor height is 8 ft. 6 in. this requirement can be met with a speed of 100 ft. per min. for 10–12 storeys, 150 ft. per min. for 13–17 storeys and 200 ft. per min. for 18–21 storeys. Higher speeds than these may be necessary in high buildings to give the required grade of service.

Where two lifts serve the same floor they should be close together so that the controls can be interconnected, calls registered thus being answered by the nearest available car. In blocks of flats most journeys either start or finish at the ground floor. Calls originating on floors above the ground therefore will generally be for downward journeys. Hence the most effective form of control is "highest call down collective." This provides for the registration of all up calls made by passengers entering the car at the ground floor and the answering of these calls in floor sequence. The lift then answers the highest call first and all other landing calls on the downward journey. If there is likely to be a large number of journeys between floors, full collective control is used although this is more costly. Some blocks of flats have two lifts, each of which stops at alternate floors above the ground and this results in a saving of about £150–£250 per floor but it reduces the grade of service.

In office buildings there are peak demands at the beginning and end of the working day and also probably at lunch time. This means that as many as 75 per cent of the building's population above the first floor will have to be transported in 30 minutes. In flats, peak demands are not usually as high. People start and finish work at different times, children go to school at different schools and women shop at different times. The peak will vary for different buildings. For example, it is likely to be higher in a block of bed-sitting rooms or one-bedroom flats where almost everyone goes out to work than in a block of family flats. Traffic counts at a 17-storey block of London flats showed a peak of only 3·75 per cent of the building population above the first floor using the lifts in a period of five minutes. In the U.S.A., lift requirements are calculated on the basis of the movement in five minutes of 7 per cent of the building population on the first floor and above of luxury flats and of 5 per cent of this population in municipal flats. It seems desirable therefore, to base the standard of service on the movement in five minutes of 6 per cent of the population above the first floor. In designing the type of installation

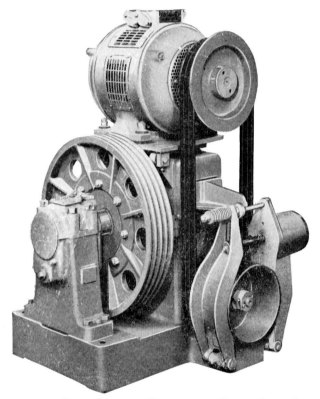

Fig. 17.1. Small Winding Machine for Single Speed Lift
(*Express Lift Co.*)

required to meet this demand a figure of 90 seconds is taken as the acceptable waiting interval, which is much more than the acceptable interval for office buildings.

A small belt-driven machine with a single-speed induction motor, suitable for 100 ft. per min. lifts in flats is shown in Fig. 17.1. The life of this vee belt drive is quite satisfactory, and it has some advantages over the usual direct couple motor drive. It provides a small and compact winding machine and eliminates the necessity to align carefully and correctly the motor sheave and the worm shaft, which is a highly-skilled operation. Further, the motor can readily be removed for repair or replacement.

As an additional safeguard the control circuit voltage is kept as low as practicable and is usually 100 volts d.c. Thus, in the event of malicious damage to any control wiring, the risk of a serious electric shock is diminished.

(*b*) **Hospital Bed Lifts.** The general principles of design are similar to those for passenger lifts but for this purpose the requirements are a deep car to accommodate beds, relatively low speeds and accurate floor levelling. Recommendations for hospital lifts are contained in B.S. Code of Practice C.P. 407.101 and B.S. 2655 Part 3.

Because a deep and relatively narrow car is required the best well arrangement is with the counterweight at the side of the car and the car entrance on one of the narrow sides. If a second entrance is needed this can be provided at the back of the car. Adequate space should be available at the landings for easy manoeuvring of bed trolleys.

Two-speed solid sliding doors are recommended for car and landing entrances but where space is restricted, multi-leaf landing doors may be used as they occupy less space than two-speed sliding doors. In these circumstances it may also be necessary to fit a collapsible gate to the car. Landing doors should be arranged for automatic opening and closing, or for manual opening and automatic closing, and the car door should preferably be automatically opened and closed.

Bed lift cars should have flush panels and the finish should be suitable for withstanding frequent washing. A handrail is often required around the inside of the car, projecting not more than $1\frac{1}{2}$ in.

The most suitable type of control is usually automatic with a call button at each landing and a full set of buttons together with an emergency stop button in the car. For long travels collective control is sometimes used. Accurate floor levelling is necessary for the wheeling in and out of beds or stretcher trolleys and automatic levelling should therefore be provided.

A speed of 100 ft. per min. is adequate for most bed lifts although 150 or 200 ft. per min. may be used for the longer travels.

To avoid the risk of overloading when full of passengers it is recommended that bed lifts should be rated as for passenger lifts of the same car floor area.

It is desirable to take all possible steps in the choice and installation of the lift equipment to reduce noise to a minimum.

The table on page 411 gives recommended dimensions for hospital bed passengers lifts, rated as for normal passenger lifts. They are intended for hospitals where the loading on visiting days can be as great as that for passenger lifts. The dimensions are for two-panel side opening doors but to provide a wider entry two-speed centre opening doors may be used, in which case the landing entrances will be increased by about 3 in. and the lift well sizes may also be slightly modified. If counterweight safety gear is required, the lift well sizes may also be increased. A typical bed lift is shown in Fig. 17.2.

(c) **Lifts in Ships.** As these are subjected to the rolling and pitching of the ship, special treatment must be given to many items of the control and to other equipment. It is desirable that the centre of gravity of the ship should be as low as possible, and therefore the ship designer prefers the lift machine to be installed at the bottom of the well. On the other hand, the lift engineer is well aware of the advantages of having the machine above the well, this feature being referred to in an earlier chapter. Considerable discussion is necessary between both parties before one or other of these positions is finally agreed. As a compromise the machine room is sometimes located at an intermediate position.

The counterweight does not travel down to solid ground as in a shore installation, and it is therefore necessary to fit safety gear to the counterweight as well as the car.

The travelling cables need special protection to prevent their swaying with the ship's motion and fouling projections in the lift well. It is usual to enclose these in canvas sheaths and house them in a special flush trunk inside the steel well structure. Consequently the well clearances required are somewhat greater than those needed for land lifts.

It is necessary to ensure that the movement of the doors is

RECOMMENDED DIMENSIONS FOR HOSPITAL BED/PASSENGER LIFTS

Load		Speed ft. per min.	Lift Well		Mach. Room Floor Area	Top Landing to Mach. Room Floor	Mach. Room Floor to Underside of Lifting Beam	Pit Depth	Landing Entrance Height	Landing Entrance Width	Inside Car		Car Platform	
Persons	lb.		Width	Depth							Width	Depth	Width	Depth
			ft. in.	ft. in.	sq. ft.	ft. in.	ft. in.	ft. in.	ft. in.	ft. in.	ft. in.	ft. in.	ft. in.	ft. in.
23	3 500	100	7 5	9 3	130	14 6	7 0	4 6	7 0	3 10	5 1	7 9	5 4	8 4
23	3 500	150	7 5	9 3	130	15 0	7 0	4 6	7 0	3 10	5 1	7 9	5 4	8 4
23	3 500	200	7 5	9 3	130	16 0	7 0	5 3	7 0	3 10	5 1	7 9	5 4	8 4
27	4 000	100	7 9	9 7	130	14 6	7 0	4 6	7 0	4 0	5 5	8 1	5 8	8 8
27	4 000	150	7 9	9 7	130	15 6	7 0	4 6	7 0	4 0	5 5	8 1	5 8	8 8
27	4 000	200	7 9	9 7	130	16 0	7 0	5 3	7 0	4 0	5 5	8 1	5 8	8 8
						For special applications only								
30	4 500	100	8 1	10 3	140	15 6	7 0	4 3	7 0	4 3	5 7	8 9	5 10	9 4
30	4 500	150	8 1	10 3	140	16 6	7 0	5 0	7 0	4 3	5 7	8 9	5 10	9 4
30	4 500	200	8 1	10 3	140	17 0	7 0	5 6	7 0	4 3	5 7	8 9	5 10	9 4

not erratic because of the ship's rolling, and this is achieved by using power-operated centre-opening doors, the two panels moving in opposite directions and being self-balancing.

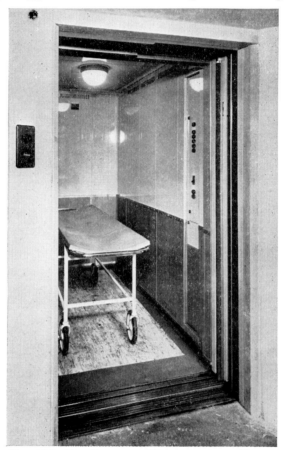

FIG. 17.2. BED LIFT WITH POWER-OPERATED DOORS
(*Express Lift Co.*)

The electricity supply in ships is usually d.c., which is not common to-day in land installations, and the equipment is therefore somewhat different from the lift manufacturer's standard equipment.

The speed of ships' lifts is usually either 100 or 200 ft. per minute, and the capacity of the passenger lifts eight or ten persons. A ship's passenger lift is shown in Fig. 17.3.

(d) **Private Residence Lifts.** In large private residences a lift similar in most respects to that provided in the smaller office buildings is usually suitable. Special landing entrance,

FIG. 17.3. PASSENGER LIFT ON T.S.S. "OCEAN MONARCH"
(*Express Lift Co.*)

door and car finishes, however, are often required. The Shepard Home Lift shown in Fig. 17.4 is a small 350 lb. capacity electric lift of unique design manufactured by Messrs. Hammond & Champness and is of particular value for invalids and old persons. The outside dimensions of the car are 3 ft. by 3 ft., and it will accommodate a wheel chair. For single phase a.c. supply a $\frac{1}{2}$ horse-power capacitor motor drives the car through a worm gear, electromagnetic brake and two simplex roller chains operating over sprocket wheels at a contract speed of 15 ft. per minute. Because of the small power

available, frictional resistances are kept to a minimum and
ball bearings used wherever possible. The car shoes consist of
four roller-bearing bakelite wheels which travel within the two

FIG. 17.4. SHEPARD HOME LIFT
(*Hammond & Champness*)

dry special shaped steel channel guides. The small winding
machine, which is 12 in. in height, is housed behind the
transome panel above the lift entrance and is provided with a
handle for winding in an emergency. Serrated cam safety

gear is fitted to the car and operates if either suspending chain breaks. A counterweight is not provided, the free ends of the chains passing over guide pulleys and thence to light supports on the winding machine framework. Control is automatic with an UP and a DOWN button at each landing and in the car. The car entrance is protected by a collapsible gate which, being constructed of polished hardwood pickets and plastic lattice bars, is very easy to operate. The enclosure on the upper floor is protected by a solid door with an electro-magnetic interlock. This lift is a self-contained unit, the guides are fitted to a wall and no enclosure is required on the lower floor. If the car meets an obstruction during descent, the false movable underfloor is compressed and thus operates two switches which cut off the power supply and automatically stop the car.

CHAPTER XVIII

MAINTENANCE AND TESTING

IT is essential that a lift should receive proper periodical maintenance in order to obtain good service and to prevent the possibility of accidents to passengers. The inspections and tests should be performed by a qualified engineer, thoroughly acquainted with the mechanical and electrical details of the lift he is called upon to maintain. It is not sufficient to leave a lift in the hands of the "electrician," who is probably more acquainted with renewing fuses and removing faults from lighting and power circuits. If the services of a qualified lift mechanic are not available, it is recommended that the periodical maintenance be undertaken by a firm of lift engineers, preferably the makers of the lift. Almost all the reputable lift manufacturers have schemes whereby they carry out periodical inspections on lifts, for a comparatively small charge, and after each inspection submit to the owner a complete test report showing the condition of the lift. In fact, most firms would prefer this method of maintaining their own lift. Usually these standard inspections are carried out each month, during which time any necessary oiling is performed, adjustments made and, in addition, recommendations are made regarding any item of equipment, such as ropes, which require renewal. By this means the owner is informed of the condition of his lift or lifts every month and he is thus given an opportunity of remedying any defects before damage is done to the equipment or accidents occur to passengers.

Maintenance Costs. When the owner is not able to carry out his own maintenance work a maintenance contract is usually entered into with the lift manufacturer. These contracts are generally of three types, the cost varying with the particular type of contract, the type of installation, the number of floors served and the intensity of traffic.

(*a*) The cheapest form of contract provides for a monthly inspection which includes cleaning, lubrication, adjustment and the provision of a report on the general condition of the lift. This costs about £40 to £50 per annum per lift.

(*b*) A more expensive form of contract includes the above and in addition, regular visits (determined by the nature of the equipment) for adjustments and repair, and also visits to rectify causes of stoppages occurring between regular visits. The cost of such a contract is about £60 to £80 per annum per lift.

(*c*) A fully comprehensive contract includes, in addition to the above, all work considered necessary to maintain the lifts in good working order. This covers the repair or replacement of all items subject to wear including motors, controllers, gears, electric cables, lifting ropes, signalling buttons and doors. In fact, this is complete maintenance. This type of contract costs about £90 to £120 per annum per lift.

All the above costs are for simple installations with single-speed motors and a speed of 100 ft. per min. If two-speed motors and speeds up to 200 ft. per min. are employed, the comprehensive contract would cost about £130 to £160 per annum per lift. At a speed of 300 ft. per min. the equipment would probably be geared variable voltage for which the comprehensive contract would be about £250 per annum for a twelve-floor lift. With a higher speed of 500 ft. per min., gearless variable voltage equipment would be used, and the cost of a comprehensive maintenance contract for a twelve-floor lift would be approximately £350 per annum.

Insurance. The annual charge for lift insurance, which usually includes two inspections and the provision of inspection reports, depends on the type of lift and the sum insured. An annual premium of about £8 would be required for a 100 ft. per min. six-floor passenger lift insured for £1,000 and a premium of about £14 for a 300 ft. per min. twelve-floor passenger lift insured for £2,000.

Inspections. The methods adopted by lift engineers for detecting faults, and the order of procedure in carrying out these tests vary, but the main items requiring inspection and means of curing troubles will be discussed. It is, of course, not necessary to perform all the tests detailed below during each visit. To carry out a thorough inspection the mechanic must have the assistance of a mate.

(1) INSPECTIONS MADE IN THE MACHINE ROOM

Switches and Fuses. The motor room is probably the best and most usual place in which to commence an inspection, and before any testing of electrical parts is done the main switches for both motor and control circuits should be switched off. Although an electric shock from a low voltage may not normally be dangerous, it may cause one to jump suddenly and come in contact with a moving part of the equipment which might result in serious consequences. The size of the motor fuses and the setting of any circuit-breakers should be checked against the nameplate rating of the motor. The carrying currents of the motor fuses will vary from twice to five times full load, depending upon the type of motor. The control circuit current will generally not exceed about five amperes.

Platform. Examine the platform over the well to ensure that it is strong and rigid and securely fastened.

Motor. On examining the motor it will be noted that the end of the shaft is squared, the object being to allow of the insertion of a winding handwheel so that the car may be moved, in cases of emergency, or after overrunning a terminal landing and operating the final limit switch. Before the motor can be turned by hand, however, it is necessary to release the brake. This is done either by turning the brake hand release lever which is fitted to some brakes, or by wedging the shoes apart with a piece of wood. If the need for using the winding hand wheel arises, care must be taken afterwards to remove it and either reset the brake hand release or remove the wooden wedge. On some lifts an interlocking contact is fitted to the brake hand release which, by disconnecting the control circuit, prevents the restarting of the lift until the release has been returned to its normal position. The commutator or slip-rings and brushes should be examined to make sure that the latter are making good contact, and are renewed before undue wear takes place. If plain bearings with ring lubricator are fitted, it should be noted whether the oil is at the correct level and whether any leakage is occurring at the drain plug, tap, or oil gauge. Ball or roller bearings are packed with grease by the makers, and replenishment is only necessary about every six months, when a little pure petroleum jelly should be used. Excessive wear of the bearings can often be detected by a distinct knock, and results in undue strain being placed on the coupling and the worm gear bearings.

Brake. The next items are the brake and the coupling upon which the brake usually operates. The coupling should be examined to ensure that the bolts are tight and that the keys joining the shafts on each side are not loose. If the motor winding wheel is inserted, and turned in each direction, looseness of the key on the motor shaft can be detected. Similarly, the key on the worm shaft may be observed, but in this case the brake must be released. Particular attention should be given to the adjustment of the brake, as an incorrectly adjusted brake will cause faulty floor levelling. When the design permits, each half should be adjusted independently so that the clearance of each shoe is the same. The clearance between the brake linings should be as small as possible, and will be found in practice to be between $\frac{1}{100}$ in. and $\frac{1}{64}$ in., if correctly adjusted. Next adjust the solenoid plunger so that the maximum possible pull is obtained. This is the case when the reluctance of the magnetic circuit is as small as possible and the maximum number of lines of force is present in the air gap. Under these conditions the brake should just operate when the magnetic circuit closes. The spring pressure should then be adjusted until smooth and rapid stopping are obtained; the actual stopping distance will vary with the car speed. After the brake has set, the car travel should not exceed about four inches for each 100 ft. per min. of running speed. The operation of any emergency brake gear should also be examined.

Gearing. The gearing is adjacent to the brake and will probably suggest itself as the next item for inspection. Examine the level of the oil in the gearbox which, if the gear is of the under-type, should be sufficiently high just to cover the worm. If the over-type gear is employed, the level should be about four inches above the bottom of the worm-wheel. The gear oil recommended by the makers should be used. As in the case of all gearing, however, the mixing of oils in the gearcase is not recommended. The oil should be free from metal particles and should not be semi-solid, gummy or have an offensive odour. When vegetable oils become rancid most of the lubricating qualities have been lost. The gearcase should be examined for leaks and tightened where necessary, and if the gland is showing signs of leakage, the gland nuts should be tightened up in rotation. Wear of the gear teeth can be detected by removing the gearbox cover and turning the motor by hand. Another method of detecting wear entails loading the car until the lift is balanced, and then, by overbalancing one side, a movement of the worm-wheel will indicate excessive wear of the teeth. The fixing of the worm-wheel rim to its centre should be carefully examined. A double thrust race is usually fitted at the outer end of the worm shaft to accommodate the worm thrust in both directions of travel, and an inspection of this thrust race is necessary to ensure that excessive wear has not taken place. If undue wear is present, the thrust is likely to be transferred to the motor bearings, which may overheat and cause a breakdown. Wear may be compensated for by removing the thrust and making adjustment.

Sheaves and Pulleys. An examination may next be made of the sheave or drum and any diverting pulleys to detect whether any looseness on the shafts is present due to keys working loose. Play may be detected by observing the line of contact of the sheave and its shaft when the lift starts or stops, and the "bubbling" of a little oil placed around the line of contact will clearly show any relative movement between the sheave or drum and the shaft. If looseness is present, the key should be driven in tightly, but it may be necessary to fit a new key. With a traction drive, the sheave grooves should be examined for wear, and if ridges are present in the grooves these may, in time, cause rope slip which will result in rope wear. Rope slip may be detected by chalking the rope and sheave, and after the lift has made a few journeys, again examining the marks, which should still coincide. The sheave should be examined carefully to ensure that all ropes seat to the same depth in the grooves. In addition to worn grooves, excessive use of lubricant on the ropes may be responsible for rope slip. In the former case the remedy is removing the sheave and having the grooves re-cut, whilst in the latter, removal of the excessive oil with a paraffined rag will cure the trouble. If the ropes "bottom" in the grooves, slip may again result and it is then necessary to fit larger ropes. If there is any doubt as to whether the traction is adequate, a contract load test should be made. It should also be noted whether the ropes lead on and off the sheave without binding on the sides of the grooves. The presence of a crack in the sheave, drum, or diverting pulleys may be detected by hammer testing. Test the bearing bolts for tightness.

Controller. The controller may now be inspected, and in

carrying this out more good will probably result from a very careful visual inspection than anything else that may be done. Loose, disconnected, or short-circuited wires are frequent sources of trouble, whilst if stranded wire is used on any contactors, this should be examined for broken strands which may cause a breakdown in the near future. Badly burned contacts should be cleaned and if necessary adjusted. The contacts are copper-to-carbon or copper-to-copper, and in the latter case, if cleaning is necessary, care must be taken to ensure that any "roll" incorporated in the design is not reduced or destroyed. If this is done, welding of the contactors may result. When an appreciable amount of cleaning or filing of contacts is necessary, the profile of the contacts should be tested with a template having the same curvature as a new contact, before the contacts are replaced. The mechanical interlock between the reversing contactors should be examined if such an interlock is fitted. The pins on which the various contactors operate should be perfectly free and should occasionally be given a drop of thin oil. Careful inspection of any controller dashpots is necessary to ensure that they provide the time lags intended. It is important to use only the type of dashpot oil recommended by the makers, or the settings will become inaccurate. An electrical insulation test should be taken on motor and control circuits at a voltage of 500 volts d.c. and when taken from the main switch with all coils in circuit should not be less than $1M\Omega$.

Floor Selector. Now that the brake has been correctly adjusted, an inspection may be made of the floor selector gear if the lift is automatically controlled, and any necessary adjustments made. The operation of each striker arm and its associated switch should be examined to see whether the best possible levelling is obtained at each floor. Commencing with the up direction switch contacts, the lift should be called up from the bottom floor to the next floor, with no load in the car. The lift should level slightly high, say $\frac{3}{4}$ in. After measuring this distance, call the car up to the next floor and take a similar measurement of the distance that the lift stops above this floor. After this has been done, repeat the tests, but with full load in the car, and in these cases the lift should stop a small distance below each floor. If necessary, the striker arms on the selector should be adjusted so that the stopping distances above the floors, when light, and the stopping distances below the floors, when carrying full load, are all equal. This adjustment will provide the best average levelling in the up direction for all loads if the counterweight is equal to the weight of the car plus 50 per cent full load. Hence, with 50 per cent full load, accurate levelling in the up direction should be obtained. If, however, the counterweight is, say, equal to car plus 40 per cent full load, it will be necessary for the car load during the load levelling test to be equal to 80 per cent full load instead of full load. This will ensure that correct levelling is obtained in the up direction with 40 per cent full load in the car. The strikers operating the down direction switches must be adjusted in a similar manner to that for the up switches, but with the car travelling in the down direction. During these down direction tests it will be noted that the empty car will again level high and the loaded car low. When a two-speed motor is employed, the floor selector will have slowing switches fitted for each direction in addition to stopping switches. The slowing switch strikers should be adjusted so that the lift speed is reduced to the low-speed value before the stopping switches are operated and the brake applied.

If stopping and levelling are performed by direction switches in the well and a cam is fitted on the car (see Chapter XIV), instead of a floor selector in the motor room, it will be necessary, in order to check the levelling, to take measurements at the floors after stopping the lift at each floor when travelling up empty. Adjustments should be made to the positions of the switches (except the ground floor switch) so that the stopping distances above each floor are equal. Similarly, the position of the ground floor switch should be adjusted until the stopping distances below each floor are equal when the car travels down loaded. If the actual distance in the well between the up and down stopping points at each landing is now made equal to say, $1\frac{1}{2}$ in., i.e. $\frac{3}{4}$ in. above the landing and $\frac{3}{4}$ in. below the landing, then levelling to within $\frac{3}{4}$ in. will be obtained. For example, if the up empty levels were all $1\frac{1}{2}$ in. high, and the down loaded levels all $\frac{1}{2}$ in. low, the top horn on the cam must be raised by $\frac{3}{4}$ in. and the bottom horn raised $\frac{1}{4}$ in. When the switches and cam have been adjusted, the only reason for inaccurate floor levelling will be that the brake is out of adjustment. It will probably be found more convenient to test these direction switches, when fitted, after the motor room inspection has been completed.

Final Terminal Stopping Switch. This should occasionally be tested by holding in the appropriate controller contactors and allowing the car to operate the switch after over-travelling each terminal landing in turn. The final limit switch is usually operated by a striker on the car engaging with a stop on the limit switch operating rope, one stop being fitted for each terminal landing. The overrun, after passing the terminal floor and before the limit switch operates, may be adjusted by altering the position of the rope stop. Two paint marks on the operating rope or pieces of adhesive tape are often used to indicate that, when the marks are opposite each other, the rope stops are in their correct positions in the well.

Governor. The overspeed governor must be lubricated where necessary and kept clean. During inspection the weights should be operated by hand to see that, firstly, the control cut-off switch operates, and secondly, the governor gripping jaws are released and grip the governor rope.

Motor Generator Set. With variable voltage equipment, an inspection of the motor generator set is necessary, particularly to ascertain the condition of the commutator and bushgear.

Ropes. Before leaving the motor room, a careful visual examination should be made of those portions of the lifting ropes which pass over the sheave or drum during a complete journey of the lift. Any "needling" of the ropes, i.e. broken wires, should be carefully noted; the detection of these "needles" being greatly facilitated by the aid of a small mirror used for the underside of the ropes or by a wad of cotton waste held lightly against the ropes. If preformed ropes are used, much greater care will be necessary in detecting broken wires as they retain their original positions in the strands even when fractured. The presence of a few broken wires does not indicate that the rope should be immediately renewed, as the factor of safety is usually about 10, but rather that the ropes should be kept under careful observation during subsequent visits. It is very difficult to quote any rule for determining when ropes should be renewed as cases have occurred of ropes failing when no broken wires have been visible, and others when the rope has given long service after a comparatively large number of wires have been broken. The decision to renew a rope rests largely upon the engineer's experience and is usually governed by the number of adjacent broken wires. Undue rope stretch is another indication of approaching rope failure.

When rope renewal is considered necessary, careful measurements must be made to ascertain the length and size of rope required. Due allowance must be made for wear and stretch of the old ropes in determining the correct rope size, and usually the measured diameter of the old ropes will lie between two standard sizes. The larger of the two sizes will invariably be the correct size of rope required, the specified size being that of the diameter of the circumscribed circle. Rope fastenings are made either by sockets, splicing, or the use of bulldog grips, and in some circumstances it may be preferable for the splices to be made before the rope is delivered, whilst in others these could more conveniently be made during fixing, providing a qualified splicer is available. During re-roping, care must be taken to bind the ends of the ropes before cutting (unless preformed ropes are used) to prevent unravelling. The actual method adopted for re-roping will depend upon the type of lift and other local conditions, but it is usually possible to position the car at the top landing so that the car top is accessible from that landing, and the bottom accessible from the landing below. When in this position, the counterweight is supported by a wooden prop in the well bottom. The car is next raised a few inches by means of lifting tackle and a sling around the car crosshead and all is in readiness for removal of the old ropes and fixing of the new ones. The worn ropes may conveniently be used for towing the new ropes into position by temporarily fixing the former to the latter. Oil should not be used on the lifting ropes unless the lift is in a damp location, when a little may be applied to prevent rusting of the ropes.

(2) INSPECTIONS MADE FROM THE LANDINGS

Well Enclosure. If the well is of the open type, examine the enclosure at all floors to ascertain that it extends from floor to floor and, if mesh screens are used, that they are of the required mesh and are properly fastened.

Call Buttons. Call the car to each landing by pressing the landing buttons in turn and if the lift is arranged for dual control and the inspector's mate rides in the car during these tests, the operation of the car call indicator and position indicator can be observed at the same time.

Landing Indicators. These can be inspected for satisfactory working when the landing buttons are tested.

Landing Gates or Doors. The lock of each landing gate or door should be tested by ascertaining that it is impossible to open any gate or door by pulling or lifting, or to stop the lift by breaking the electric interlock circuit, unless the car is at that landing.

It should also not be possible to move the car away from any landing with the gate or door at that landing open and if this is confirmed at each landing, then the landing door electric interlocks are breaking contact satisfactorily.

If gates are fitted, it is now a convenient time to note whether they can be opened or closed readily and if not, a drop of oil on each picket pin and on the overhead supporting roller pins, unless the latter are of the ball bearing type, will ensure easy operation. See that door hangers and tracks are clear and adequately lubricated.

Retiring Cam. This can be tested by maintaining a steady opening pressure on each landing door in turn in an endeavour to open it as the car passes the particular landing. It should, of course, be impossible to "snatch" open the door in this manner when the car is passing.

Landing Gate or Door Emergency Key. If this facility is provided, ascertain that the key is available to the maintenance engineer and that it will enable any landing gate or door to be opened for inspection or emergency purposes irrespective of the position of the car.

(3) INSPECTIONS MADE FROM INSIDE THE CAR

Car Floor Switch. Some old passenger lifts have a switch fitted under the car floor, the switch being operated by pressure on the movable floor of the car. This switch is wired in parallel with the car gate or door contact and when the floor is in the UP position, i.e. no passengers in the car, the floor switch contacts are made. Hence it is possible for the empty car to be called to another landing with the car gate or door open.

When a passenger enters the car, however, the floor switch contacts are opened and the car gate or door contact is then operative, and the gate must now be closed before the lift can be moved. It will be seen from the above that if the switch fails to open with the weight of a passenger on the floor, a dangerous condition will result, as the occupied lift can then be moved with the car gate or door open. For this reason these switches are not fitted on new lifts, and their use is prohibited in the B.S. Code of Practice, which states that the car door or gate electric switch shall prevent the lift from being started or kept in motion unless the car gates or doors are closed.

Nevertheless, many of these switches still exist and the need for their thorough and periodical maintenance will be clear from what is stated above. The floor must be absolutely free in action and should be raised at every maintenance visit, all dust and dirt cleaned out and the hinges and spindles lubricated.

Car Door or Gate Electric Contact. If this switch closes and opens satisfactorily the car may be moved by the car buttons or car switch if the door is closed but it cannot be moved if the door is open. The cover of the switch should be removed occasionally and if necessary the spindles lightly oiled and the contacts cleaned. The position of this interlock may be such, however, that it may have to be inspected from the top of the car.

Car Switch. It is easy to test the operation of this switch by moving the car up and down and at the various car speeds as indicated on the switch plate. The cover may be taken off, the inside of the switch carefully wiped, working parts lightly oiled, and the contacts cleaned.

Car Push Buttons. Test these by bringing the car to each floor in turn by pressing the appropriate button. It should be possible to stop the motion of the car at any position in the well by operating the stop button. If any button is uncertain in its action, the cover plate should be removed and the contacts cleaned and if necessary a new spring fitted.

Car Safety Gear. If of the wedge clamp type, the removable panel in the floor should be raised and that portion of the drum and safety rope which is visible should be inspected, cleaned, and oiled where necessary.

Emergency Exit. Open the exit to ascertain that, if at the top of the car, it opens outwards and if at the side, inwards. If a top exit, it should be capable of being opened from inside or outside the car and when open should clear any equipment mounted on the top of the car. Where an electric interlock is provided it should be impossible to start the car with the exit open.

Emergency Signal. The bell, buzzer or telephone fitted in the car should be operated and should be clearly audible outside the lift well when the car is midway between adjacent landings. In some buildings this signal is arranged to give a warning in the maintenance engineer's room.

Lighting Fitting. Examine to see that it is securely fastened, that the illumination is adequate, and that the switch operates satisfactorily.

Travelling Conditions. Run the car up and down the well and observe the action of the brake when stopping. The slide of the car should not be excessive nor should the stop be too abrupt. The acceleration and retardation should be noted to find out whether starting and stopping are smooth and reasonably rapid.

(4) INSPECTIONS MADE FROM THE TOP OF THE CAR

To get on top of the car, the crosshead should be brought approximately level with a landing, either by lowering the car with the winding wheel, after switching off and releasing the brake, or by operating the car stop button. In either case a mate is necessary in the car. Then by using the landing gate emergency key or otherwise circumventing the landing lock safety feature, which method should be known to the lift maintenance engineer, the gate may be opened to give access to the car top. The inspector should stand on the car crosshead near the rope fastenings and hold the lifting ropes firmly with one hand when the car is in motion. When in this position it is better, if possible, to arrange for the car to travel in the down direction.

Door Operating Gear. Examine the motor and door operating levers, clean and lubricate if necessary. Watch the operation of the gear when the doors open and close.

Retiring Cam Mechanism. Operate the cam by pulling the connecting chain by hand and note if the cam advances and retires freely. Observe whether it clears the door lock striker arm or the sill trip lever when in the retired position. The operating solenoid or motor and levers should be examined and lubricated.

Governor Rope Release Carrier. Visually examine the rope grip and the springs to see that they are clean and not rusted. This release should be tested by engaging the governor jaws (usually in the machine room) with the governor rope by hand and then lowering the car either by the motor hand wheel or at the slow levelling speed. The shackle should then pull out of its carrier and if the descent of the car is not checked the safety gear will operate.

Car Shoes and Guide Lubricators. Inspect the top pair of shoes for wear of the linings and see that the housings are securely fixed to the car frame. If the shoes are spring loaded, the springs may be tested by rocking the car to and fro sideways. Excessive side play may be cured with most types of shoes by fitting a steel washer between the back of the shoe and its housing. Examine the guide lubricators to ensure that the feed wicks are properly adjusted and that the reservoirs are filled with oil.

Multiplying Pulley. If the roping is 2 to 1, inspect the multiplying pulley for cracks and for adequate lubrication.

Ropes. By moving the car until the top is level with the top of the counterweight the lifting ropes and their fastenings at the car and counterweight may be inspected. Examine the ropes carefully for "needling" and for any irregular shape which may be due to a strand pulling out and carrying no load, this being caused by faulty splicing. If bulldog grips are used, see that a sufficient number is employed and that they are fitted in the correct manner. Any rope equalizing gear may now be inspected and lubricated and the position of the levers noted, particular care being taken to see if a rope has stretched sufficiently to cause the gear to wedge in one of the extreme positions of its movement.

If the car is moved slowly from the top of the well to the bottom, during which it is stopped at intervals of about 5 feet, the lifting ropes above the counterweight may be examined for broken wires, dryness and rust, or excessive lubricant. If dry and showing signs of rust, they should be lightly lubricated, but this should not be overdone or the result, with a traction drive, will be rope slip. At the same time the compensating ropes, governor rope and the final terminal stopping switch rope, tape or wire may be inspected in a similar manner.

Rope Tensions. If rope equalizing gear is not fitted, the rope tensions may be tested with the car about half way between the top of the well and the counterweight. Pull each rope in turn with a spring balance and note the deflections with equal pulls. If any rope can be deflected more than the others, it is not carrying its proper share of the load and adjustment should be made at its screwed support.

Counterweight. When the car is opposite the counterweight, the counterweight shoes and guide lubricators may be inspected in a similar manner to that adopted for the car fittings. Examine the counterweight sections to see that none is displaced in the frame and that the nuts and pins are in their proper positions at the ends of the tie rods. If 2 to 1 roping is employed, see that the pulley at the top of the counterweight is adequately lubricated. In many lifts the counterweight oil buffers are fitted to the bottom of the counterweight and if this is the case examine the oil level in them. Make sure that the counterweight guard at the mid-well position is securely fixed. Inspect the counterweight safety gear as described later for the car gear.

Guides. When moving the car slowly from the top to the bottom of the well, the guide surfaces, joints, clips and brackets should be inspected for wear or looseness and when the car is at the bottom landing any noticeable bend in the guides can be detected by sighting along the face of each car and counter-weight guide in turn.

Landing Gate or Door Locks. These should be examined periodically and if necessary the striker arms adjusted and worn rubbers renewed. The covers should be removed, the inside of the box wiped free of dust, the mechanism lightly oiled and the contacts cleaned. With centre opening swing doors the locks are mounted on the underside of the door top frame and these may be inspected from the landing or the inside of the car.

Slowing and Stopping Switches. If slowing and stopping are performed by switches on the car and cams in the well or vice versa, it should be noted whether the switches and cams are in proper alignment and are securely fastened. The slowing switches should operate sufficiently in advance of the stopping switches to enable the motor to reach its slow levelling speed before the application of the brake. The normal terminal stopping switches are frequently fitted on the car and are operated by cams in the well and these should receive careful attention. After inspection, all these switches should be tested by operating the car at normal speed between the extremes of travel.

Upper Final Terminal Stopping Switch. If this consists of a switch operated by a cam, it may be inspected from the car top. Open the switch by hand and it should then be impossible to start the car. Check the cam and switch for rigidity and alignment. By measuring the distance between the two top terminal switches make sure that the final switch operates as soon as possible after the normal terminal stopping switch without interfering with the operation of the latter.

Travelling Cable. The fixing of this cable at the half-way box in the well may be examined whilst on the car top for any signs of looseness or breakage and the surface of the cable inspected to ascertain whether it is rubbing against anything in the well.

(5) INSPECTIONS MADE FROM THE PIT

The pit will be entered from the bottom floor landing after the car has been raised from this floor about 4 ft. by the winding wheel and the gate lock opened with the emergency key. The car may then be lowered by hand sufficiently to permit the under-car equipment to be inspected.

Bottom Final Terminal Stopping Switch. This may be dealt with in a similar manner to that described for the top switch.

Car Bottom Shoes. Examine these as was done for the top pair.

Car Sling. Inspect the visible fastenings of the car frame and if bolted note if the bolts are tight and also whether the platform is distorted.

Travelling Cable. The lower part of this cable and its fastening to the car should be inspected and the cable should not normally rest upon the pit floor.

Car Safety Gear. If instantaneous safeties are provided, note whether the jaws and the safety block are clear of the guides during normal operation of the car. The governor should be tripped by hand and the empty car lowered by the winding wheel until it rests on the safety jaws. Examine the jaws to see that they are engaged at both sides and note if the operating levers are free and adequately lubricated.

If of the wedge type, examine the jaws for freedom from contact with the guides and turn the drum by hand or pull the safety cable until the jaws touch the guides. Note that there is sufficient cable left on the drum so that it will not be pulled from the drum when the safety operates. See that the levers are adequately lubricated and that there is no excessive slack in the safety rope.

Car and Counterweight Buffers. If these are of the spiral or volute spring pattern, examine to see that they fit vertically and securely in their bases and that the springs are not distorted.

With oil buffers ascertain that they have a sufficient supply of oil by inspecting the gauge provided on each and whether there is any sideplay in the pistons. If the counterweight buffers as well as the car buffers are mounted in the pit, they will all be of the spring return type and may be partly depressed by standing on the top of the plunger. After release, the piston should return to its top position.

Bottom Counterweight Guard. See that this is in position and securely fixed.

Governor Rope Pulley. Inspect during operation to make sure that the pulley frame is free to slide in its guides and that the parts are adequately lubricated.

Compensating Cable Pulley. Examine as for governor rope pulley.

Bottom Clearance. The bottom counterweight clearance should be checked when the car is level with the top landing to ascertain if this has appreciably decreased due to rope stretch. If it is necessary to shorten the ropes because of stretch, care must be taken to ensure afterwards that the bottom counterweight clearance is less than the top car clearance.

ACCEPTANCE TESTS

Inspections and Tests at Works. Purchasers frequently specify that inspection of the various items of equipment shall be made at works to ensure that they conform to the specification before transport to site. These items usually include the motors, generator, brake, gearing, car, controller and other items of control equipment. If works running tests on the motors and generator are not witnessed by the buyer, he often requires copies of the certified test results. A high-voltage dielectric test is made on the controller electrical equipment. This is at ten times the working voltage, with a maximum of 2 000 V, and is carried out between live parts and framework and also between the live parts of adjacent circuits. The test voltage is alternating at 50 cycles per second and is applied for one minute.

Site Tests. Before a new lift is taken over from the manufacturers and put into commission, certain essential tests should be performed to ensure that the lift is satisfactory and conforms to the conditions laid down in the specification. The acceptance tests usually performed are detailed below.

It is desirable that the manufacturers supply and fix in the motor room a framed and glazed wiring diagram of the lift connexions for the use of the maintenance engineer, together with any maintenance recommendations which they wish to make. A plate should be fixed in the car showing the contract load, and in the case of passenger lifts, the maximum number of passengers to be carried, calculated at not less than 150 lb. per passenger. The actual requirements of different purchasers vary in detail, and if the lift has been manufactured and installed in accordance with a specification, it is necessary that those acceptance tests should be performed which will confirm whether the clauses of the specification have been fully met. In some instances the actual tests which will be applied are stated in the specification. Before carrying out any specified tests, the lift should be carefully examined and checked against the general requirements as regards type of equipment, construction of car and doors, type and fixings of guides and shoes, position of safety gear and buffers, methods of guarding, indicators, and other general items. After this has been done, the tests called for may be performed.

Overload Test. This is made to ensure that the equipment is adequately rated. During this test the car is loaded with 10 per cent more than the contract load and the lift is run in both directions with stops at the floors. The starting and running currents, the speeds and the accuracy of levelling are recorded.

Service Temperature Test. A continuous run of 1 hour should be made with a number of starts and stops to reproduce as nearly as practicable the anticipated duty in service. The standard duty cycles for 90 and 180 starts per hour for single-speed polyphase induction motors are stated in Chapter VI and may be used as a guide. It is very difficult in practice to carry out these tests as specified with alternate starts at full load and no load, and it is necessary therefore to simulate these cycles. A suitable test for all motors except squirrel-cage motors is to run the car up from the bottom landing with contract load and stop at each floor. From the top floor a non-stop run is made to the lowest floor and the upward journey with floor stops is then repeated. The time intervals between stops and starts at the floors should be uniform and such as to give about 150 starts in the hour. At the end of this run the temperatures of the armatures and fields of the motors and generator are recorded. The temperature rise should not exceed 55°C, 65°C or 75°C for classes A, E or B insulation respectively. In practice, the usual rises are about 20°C for the motor generator set and 35°C for the hoisting motor.

Efficiency. Manufacturers will frequently guarantee an overall efficiency for their equipment and then it is necessary for this to be checked. The total power input from the mains is measured with a watt meter when the car is loaded with contract load and running at contract speed. The total output is obtained from the product of the out-of-balance load and the contract speed, converted to watts. With variable-voltage geared equipment this is about 40 per cent but with gearless drives a higher efficiency would be obtained. Overall efficiencies exceeding 50 per cent can be obtained with well-chosen machine ratings, good guide erection and efficient shoes.

Energy Consumption. The contract load is required in the car for this test, and an energy meter connected in the supply to the lift, in such a manner that it indicates the sum of the motor and control consumptions. The consumption for a complete up and down journey is very small, and it will be necessary to take the average of, say, five complete journeys to obtain a reliable reading on the meter. When the marked rotating disc of the meter is visible, however, it is often possible to obtain the consumption for a complete journey by counting the number of revolutions made, and applying the ratio stamped on the meter. Excessive consumption may be accounted for by an inaccurate counterweight value.

Full Load Levelling. If the lift is equipped with automatic control, the accuracy of the floor setting gear may now be tested under full load conditions without altering the car loading from that used during the energy consumption test. The car should be called to each floor in turn when travelling in the up direction, and the actual distance that the car stops below each floor measured. Similarly, the stopping distances below each floor on the downward journey must be measured. These distances will be compared with the corresponding measurements taken later when the car is empty.

Irreversibility. Some geared lifts are specified to be self-sustaining and this may be checked by cutting off the control circuit and releasing the brake by hand, when the car, with full load, should remain stationary.

Safety Gear Tests. With the contract load still in the car, the safety gear may now be tested. If the lift operates from a d.c. supply, the excess speed necessary to operate the gear may be obtained by field weakening, but if an a.c. motor is installed the gear may be set to operate at the contract speed or alternatively tripped by hand at the contract speed. On some governors the governor sheave is provided with an extra groove of such a size that it will permit of the governor being driven at a speed corresponding to the safety gear tripping speed when the lift is travelling at its contract speed or special detachable extra governor test weights are supplied for the same purpose. Other governors have special test washers fitted between the spring and its adjusting nut, which when removed reduces the spring tension.

Instantaneous safety gear controlled by a governor should be tested with contract load and at contract speed, the governor being operated by hand. Two tests should be made, however, with wedge clamp or flexible clamp safeties, one with contract load in the car and the other with 150 lb. (equivalent to one person) in the car. The stopping distances obtained should be compared with the specified figures and the guides, car platform, and safety gear should be carefully examined afterwards for signs of permanent distortion. Note if there is sufficient cable left on the safety drum after the gear has operated.

Counterweight safety gear should be tripped by the counterweight governor and the stopping distances noted. In this case, however, the governor tripping speed should exceed that of the car safety governor, but by not more than 10 per cent.

Governor Tripping Speed. During the safety gear tests an inspector with a tachometer should determine the car speed (from the governor or the main sheave) at the instant of tripping. Check the tripping speed with that stated in Chapter XIII. The governor jaws and rope should be examined for any undue wear.

Contract Speed. This should be measured with contract load in the car, with half load and with no load, and should not vary from the contract speed by more than 10 per cent. The most convenient method is by counting the number of revolutions made by the sheave or drum in a known time. A chalk mark on the sheave or drum and a stop watch will facilitate counting and timing, but care must be exercised to ensure that no acceleration or retardation periods are included. If the roping is 2 to 1, the sheave speed is twice the car speed. Alternatively, the speed can be measured by a tachometer applied directly to a rope, immediately below the sheave.

Buffers. The car should be run on to its buffers at contract speed and with contract load in the car to test whether there is any permanent distortion of the car or buffers. The counterweight buffers should be tested similarly.

Size of Car. The floor area of the car should be measured in order to check that the passenger capacity plate is correctly engraved.

Lift Balance. Some of the weights should now be removed until the remaining represent the balance figure specified, say, 50 per cent contract load. It should now be ascertained whether the counterweight is equal to the weight of the car plus 50 per cent contract load. This may be roughly checked by cutting off the supplies and rotating the winding wheel in each direction in turn, and the effort required should be, as nearly as can be judged, the same. A more accurate test consists of wiring an ammeter in the motor supply and taking current readings during the upward and downward journeys. If the lift is properly balanced, these readings should be the same for each direction.

Levelling Empty. If the lift is automatically controlled, the car should be stopped during the upward journey at each floor in turn, and the stopping distances above each floor measured. The "up" levelling gear is correctly adjusted if these "high" distances are approximately equal to the "low" distances obtained during the up full load levelling test. Similar figures must be obtained during the down empty journey, and these distances above the floors should be equal to each other and the same as the "low" stopping figures taken during the down full load levelling test, if the down levelling gear is set correctly.

Car and Counterweight Clearances. The bottom car clearance and the top counterweight clearance may be measured with the car level with the bottom landing. In taking these measurements it may be more convenient to station the car a couple of feet above the bottom landing so that entrance to the pit is made easier. Due allowance must, of course, be made for the distance the car rests above the landing. The top car and bottom counterweight clearances must be obtained from the top of the car and the pit respectively with the car at the top landing. Particular care should be taken to note that the latter clearance is the smaller.

The clearance between the car and landing sills should not exceed $1\frac{1}{4}$ in. and between the car door and the landing door not more than $5\frac{1}{2}$ in.

Car and Landing Doors. The lift should not operate with any door open. The car door delay contact and the retiring lock release cam must be tested. The working of the door operator and of any safety edges or light-ray equipment should be examined.

Controller. The operation of the contactors and interlocks should be examined, also any time lag contacts, and it should be ascertained whether all the requirements laid down in the specification have been met. Insulation tests must be taken on both motor and control circuits at 500 volts d.c. with all wiring in circuit in each case. The insulation resistance to earth should not be less than 0·5 megohms.

The method of earthing the controller and all other electrical equipment should be inspected and tested. An earthing point should be available in the machine room and this will usually be the sheathing or conduit of the incoming supply cable. This earth should be extended to the metal parts of all the lift electrical equipment except, of course, the current-carrying parts and this is usually done either by continuous screwed conduit or separate earth conductors. These latter should be not smaller than 7/0·029 in. conductors. The frames of the motors, generators and rectifiers if not earthed by conduits, should be earthed by 7/0·029 in. or 7/0·064 in. conductors if the name plate ratings do not exceed 50 and 100 amps. respectively and by 19/0·064 in. conductors if the rating exceeds 100 amps. The controller frame should be earthed similarly and a separate earthing lead should be provided in the trailing cable for earthing the electrical equipment in the car. The resistance to earth at any point in the earthing system must not exceed 1 ohm.

Acceleration and Retardation. It can be judged whether these are sufficiently smooth by riding in the car.

Normal Terminal Stopping Switches. Test by letting the car run to each terminal landing in turn, first with no load, and then with contract load in the car. By taking measurements the top and bottom over-travels can be ascertained.

Final Terminal Stopping Switches. The normal slowing switches must be disconnected for this test either by removing the roller from the switch arm or short-circuiting the slowing switches. It is necessary to ensure that these final switches open as close to the normal switches as possible (by measurements) without interfering with the normal stops. If spring buffers are fitted, the final switches must open before the buffers are engaged. With oil buffers the top final terminal switch is tested by arranging for the empty car to strike the switch at half contract speed. The bottom final terminal switch is tested by running the loaded car down so that it strikes the switch at half contract speed. In each case the buffers must be operative.

Oil Buffers. The oil buffers are examined after the above tests have been made to determine if there has been any oil leakage or distortion and to ensure that the buffers return to their normal positions.

Ropes. The size, number, construction and fastenings of the ropes should be carefully checked and recorded.

CHAPTER XIX

LIFT ACCIDENTS

COMPLETE information relating to accidents which have occurred on lifts in this country is not available in any publication and it is doubtful whether such records, useful as they would undoubtedly be, have ever been compiled and preserved. In addition to details of accidents which appear from time to time in the Press, there is the large number, usually less serious, which have occurred and are known only to those closely connected with the persons or lifts concerned. Probably the most comprehensive records of lift accidents are those kept by the Chief Inspector of Factories but even these relate only to accidents that have occurred in factories, and furthermore, which have been reported to his Department. Nevertheless, the information which has been published regarding these is valuable and gives much food for thought.

Lift accidents may be divided broadly into two classes: those due to faulty design or maintenance and those caused by foolishness on the part of users, such as tampering with safety devices. It seems clear that by far the larger number can be attributed to the first cause, and accidents, particularly fatal ones, are now very rare indeed on well designed and adequately maintained lifts. The B.S. Code of Practice for Electric Lifts and the Factories Act prescribe definite standards to be observed in safeguarding lifts, these standards being in accordance with the modern practices of the best lift makers. Although in many cases heavy expenditure has been necessary to alter old type lifts so that they conform to these modern requirements, it is clear, certainly as regards factory lifts, that definite and considerable progress has been made in recent years. The gradual decrease in accidents, as shown by the Chief Inspector of Factories reports, has been due to the observance of the safety requirements of the Act, and in particular the periodical inspections required.

The following are brief descriptions of actual typical accidents which have occurred in factories and which have been notified

engineers was killed and another man injured while engaged on re-roping a 10 cwt. electric goods lift. The car was fitted with serrated steel cam safety gear and this was engaged with the wooden guides in order to support the car. One of the men stood on a step ladder inside the car, with the upper part of his body through the emergency exit. The second man entered the car to give assistance and at that moment the car fell to the bottom of the well. It was found that the safety gear had torn the timber guides and one of the cams had turned over on its shaft.

This illustrates the need for careful design and maintenance of the safety gear and the advisability of using lifting tackle to support the car when re-roping.

(8) *Vol. VII, No. 3.* Two workmen were greasing and oiling a modern passenger lift, but before the work began the man in charge went away to attend to another matter, leaving his mate in the bottom of the 7 feet deep pit. When the charge-hand returned, he signalled to the car attendant for the car to descend to the basement level. At that moment the man in the pit tried to scramble out by climbing up to the landing doors but was struck by the car and received fatal unjuries. He would have been perfectly safe if he had remained in the pit.

This shows the care that must be taken by maintenance men for their own safety and the need to understand thoroughly the principles of their job.

(9) *Vol. VIII, No. 9.* The lift had automatic control and fully interlocked collapsible gates on the landings and car. A man was asked to take the lift down to another floor, but instead of accompanying it he closed the gates, put his arm between the pickets, and operated the car push button. Fortunately he was able to stop the lift before serious injury was done to his arm.

Such accidents are prevented by the use of mid-bar gates and it is also an advantage to place the car buttons out of reach of a person on a landing.

(10) *Vol. IX, No. 10.* A liftman was injured when his car fell from the ground to the basement of a factory. The lift was electrically driven, had worm gear, a *V* sheave, two suspension ropes, and a diverting pulley. The examination after the collapse showed that over a considerable length, the rope wear was serious, and it was also found that the diverting pulley

could not revolve freely on its shaft. The excessive wear was obviously due to the friction of the ropes in the pulley grooves.

This accident would have been prevented if there had been regular and thorough maintenance, which would have revealed the seized pulley and stopped the ropes wearing. The fitting of car safety gear operated on the broken rope principle would have prevented the car from falling.

(11) *Vol. X, No.* 3. The electric lift concerned was of the fully automatic type, with interlocked car and landing gates, a car floor switch, and arranged so that the empty car returned automatically after a brief interval to the ground floor. A workman, whilst painting the inside of the car, was requested by his mate, who was painting the outside of the car from a position on the fifth floor, to move the car a few feet downwards. Having done so, he then opened the car gate and stood on the narrow landing edge sufficiently long for the timing device to bring about the automatic descent of the car. He was trapped and killed.

This reveals the danger attending the use of car floor switches, but the accident would not have occurred if the floor of the car had been temporarily weighted down. The landing gates should be as close as possible to the edge of the landing floor, which would also minimize the risk.

(12) *Vol. X, No.* 4. A workman had crawled into the lift pit to limewash the lower part of the lift well. The understanding between this man and the lift attendant was that the car should be taken to the top landing and left there until all the limewashing within reach was finished. The workman overlooked the fact that as the car moved up the counterweight moved down. He failed to keep clear of the weight, was struck by it and fatally injured.

If a pit counterweight guard had been fitted this accident would not have happened.

(13) *Vol. XII, No.* 1. A maintenance man obtained access to the top of the lift car by removing the emergency exit panel in the roof. After completing the work he left the car, but remembering that he had inadvertently refixed the panel upside down, returned later in the day to turn it over. He got on top of the car through one of the landing gates, but owing to

some misunderstanding with the lift attendant the car was taken up when the maintenance man was in an unsafe position. His back came in contact with one of the guide brackets and he subsequently died from his injuries.

The attendant should have acted entirely under the orders of the man on top of the car. If the exit had been fitted with an interlock switch, however, the accident could not have occurred.

(14) *Vol. XIII, No. 9.* An electric lift was fitted with automatic control, interlocked gates, and a car floor switch. There was, however, a landing space of 10 in. between the car and the landing gates, and a girl leaving the lift stood in this space prior to opening the landing gate. At that moment the lift was called to another landing. The girl was severely injured and fell to the bottom of the well, her injuries proving fatal.

Floor switches should not be used, and the landing gates should be fitted as close as possible to the edge of the well opening.

(15) *Vol. XV, No. 8.* A lightly loaded truck had been taken from the first floor to the basement of a building in the car of a lift serving six floors. A man in the car had pushed the truck half-way out when the car began to ascend. The truck fell out and the car crashed into the top of the well; the man was bruised and his head cut. The cause of the accident was the detachment of the bronze rim of the worm wheel from its centre; one of the bolts had fallen out and the others had sheared. The brake was therefore ineffective and the heavier counterweight took control.

If there had been regular inspections—this worm had not been examined during its ten years life—the defective fastening of the rim would almost certainly have been discovered. It appears also that, for the car to crash at the top of the well, the top car clearance must have been inadequate—a badly designed lift.

(16) *Vol. XVI, No. 20.* The lift was a modern electric one with "self-levelling" gear in which the doors commenced to open when the car reached the levelling zone, which was a few inches from the landing. A man standing on a landing had his foot trapped under the car whilst it was descending slowly

to the landing. On investigation it was found that the apron plate fitted to the front edge of the car was of insufficient depth to fill the maximum possible gap.

It is necessary to ensure that the depth of the apron board is greater than the distance from the landing to the edge of the landing zone.

(17) *Vol. XVI, No.* 21. A boy entered a lift car and while both the car and landing gates were open the car began to descend. The boy was alarmed and tried to scramble out, but he was trapped between the top of the car and the landing. He instantly screamed and the car came to rest. It appears that the lift would not respond to the operation of the car switch and the attendant went to the motor room with another man to find the cause of the trouble. When there he operated one of the contactors by hand, thus setting the car in motion. The probable cause of the original failure to operate was a door not properly closed, as no defect was found in any part of the mechanism.

It is important that only a skilled maintenance engineer should have access to a lift motor room.

(18) *Vol. XVII, No.* 10. The car of a fully automatic lift had been called to the third floor but came to rest below the landing and the gate could not be opened. A girl on the landing, in an endeavour to level the car, reached through the two gates and pressed one of the car buttons. The car began to descend and her arm was sheared between the two gates and badly injured. The car gate was of the mid-bar type but the landing gate of the open picket pattern.

If the landing gate had been of the mid-bar type, or a solid door had been fitted, it would have been impossible to reach the car buttons. Suitably placed car buttons or a safety plate fitted to the gate would have prevented this accident.

(19) *Vol. XVII, No.* 11. A large stand had been placed in the car of a goods lift for conveyance to an upper floor and one of the men found that he could get out of the car only by climbing over the top of the stand. While he was doing so the stand rocked and pressed one of the car buttons. In spite of the fact that the gate was open and was fitted with an interlock, the car began to ascend and the man's leg was caught between the stand and the lintel of the doorway. It was found

afterwards that a spring in the interlock was broken and the contacts thus remained closed when the gate was open.

This illustrates the importance of a thorough examination of the interlocks during the periodical inspections.

(20) *Vol. XVII, No.* 12. A painter's labourer went into a lift motor room and his trousers were caught by the square end of the motor shaft which projected 2 in. beyond the motor casing. Both his legs were cut and bruised.

The square end is necessary to permit winding of the car by hand, but it should be covered by a cap or an open ended shield.

(21) *Vol. XVIII, No.* 10. A man took a drum-driven goods lift up to the upper level and as he was stepping out the car fell and trapped him against the platform, causing fatal injuries. The spur wheel was found to be broken and the motor burnt out. It appears that in the absence of a lower landing limit switch the drum had on some previous occasion continued to revolve after the car had come to rest and the suspension ropes had been taken round the drum in the wrong direction. The men had been in the habit of throwing the car switch to the "down" position when they wanted to go up. When the car was at the upper level, the ropes were consequently at an acute angle to the horizontal instead of being vertical and the rope tension and the torque on the drum and gear shaft were excessive. It was this abnormal pull on the ropes which caused the fracture of the gear wheel.

Limit switches and a slack rope switch are necessary on drum-driven lifts.

(22) *Vol. XX, No.* 14. Two men were taking ironmongery from the ground floor to the basement in an electric lift driven through worm gear and a "V" sheave. They loaded the car without making any estimate of the weight put in, then one man walked down the stairs, leaving his mate to travel down with the goods. When the first man arrived at the basement he looked up and saw his mate trapped between the top of the car and the ground floor. The gate interlocks were subsequently found to be out of order. The accident was undoubtedly due to gross overloading and it was not probable that the gearing was driven "backwards" by the overload. It was not possible to establish the real cause, but the ropes may have pulled through

the sheave grooves and the unexpected movement caused the man to fall over.

A lift car should never be loaded beyond the loading specified in the car. Automatic safe-load indicators which give a visible or an audible warning of overloading may be fitted on lifts.

(23) *Vol. XXI, No.* 7. A man in an automatic lift was taking a load of tea chests from the top floor to the bottom. One of the chests was trapped between the edge of the landing (where there was a space 2 in. wide inside the landing gate) and the top of the car, the car jammed and was brought to rest. At that moment another person pressed the bottom floor button and as it was a drum drive, several feet of the suspension ropes were wound off the drum and hung loose above the car. The projecting chest was knocked back into the car, which immediately fell the few feet of the slack rope. The overhead timber beams, carrying the pulleys, broke and the car fell to the bottom of the well.

The accident would have been prevented if the car gate had been fitted with an electric interlock.

(24) *Vol. XXV, No.* 10. An electrician was on a ladder, wiring conduit near an electric goods lift. The space between the top of the gate and the floor above was not fenced and the man's elbow, projecting into the well, was trapped between the top of the car and an angle iron bracing fitted across the well opening. The man's arm was badly cut.

Well fencing should extend from floor to ceiling on all sides.

(25) *Vol. XXV, No.* 11. After a concert in a large factory, thirty-one people crammed themselves into the car of a lift. The load then was about twice the contract load and the car began to slide down, the ropes slipping through the "V" sheave grooves. When the car struck the bottom several people were injured.

A conspicuous notice showing the maximum number of people to be carried should be placed in every lift, as well as the maximum working load, and the attendant should be given sufficient authority to prevent this from being exceeded.

(26) *Vol. XIII (New Series) No.* 5. Two men had just entered a lift car after having loaded it when, without the controls being actuated, the car moved downwards. One of the men tried to get out of the car and fell awkwardly, breaking

his leg. The lift was not overloaded, but investigation showed that the four suspension ropes which were in good condition were coated with thick oil. This oil caused the ropes to slip on the sheave. The tendency to slip was increased by the presence of a diverting pulley which reduced the angle of contact between the rope and the sheave by 26 per cent. With a traction sheave the drive depends on the grip of the ropes in the sheave grooves, and rope lubrication should be kept to a minimum.

Other factors which affect slipping, e.g. angle and condition of the sheave grooves, did not apply in this case.

(27) *Vol. XIII (New Series) No.* 6. An engineer was examining the suspension rope of a lift in the machine room by allowing it to pass through his gloved hands as the lift was moved slowly. The needled ends of some broken wires caught his gloves and carried his hands into the "nip" between the rope and the groove sheave, where they were trapped and severely injured. Had the man not been wearing gloves it is likely that the accident would not have occurred. The proximity of the sheave increased its severity, but in any case the method of examining a wire rope by allowing it to be moved through the hands is a dangerous one.

(28) *Vol. XV (New Series) No.* 20. A painter, wishing to get a 12-ft. ladder up to the fifth floor of a factory, decided to put it in the lift rather than carry it up the stairway. The trap door in the roof of the car was opened and the ladder was placed in the car with about 5 ft. of it projecting above the roof. As the car ascended, the projecting end of the ladder was caught by the descending counterweight in the well. The end of the ladder in the car struck one of the two men travelling with it, breaking his leg in two places.

The accident illustrates the danger of carrying anything in a lift car which projects through the roof. In this case the ladder struck the counterweight, but it might also have caught on projections in the well, or, if the car had gone to the top floor, it might have struck the top of the lift well. If the trap door had been fitted with an electric interlock, this accident could not have happened.

(29) *Vol. XVII (New Series) No.* 7. A workman entered the ground floor of a warehouse and opened the landing gate of an

electric lift, expecting the car to be at that landing. The car was not there, however, and he fell 15 ft. into the basement, receiving severe injuries. Examination of the gate lock showed that it was in a bad state of repair, which accounted for the injured person being able to open the gate when the car was not at that landing. The locking plate was badly bent, and the internal fittings were loose, so that although the electric part of the lock was in order, the mechanical locking mechanism was ineffective. A lift landing gate lock should not only prevent the car being moved unless the landing gate is closed, but should also prevent the opening of the gate unless the car is at the landing.

The examination of the landing gate locks is a very important part of the statutory six monthly examination of a lift.

(30) *Vol. XVII (New Series) No.* 8. A workman loaded some barrels into an electric lift and then travelled in the car with them to a lower floor. On the way down he moved the car switch to the OFF position and began to tie up his shoelace. He noticed, however, that the car was continuing to descend, and before he could do anything it struck the buffers at the bottom of the well. The impact displaced the barrels and the man received leg injuries.

The safe working load of the lift was 30 cwt., and the weight of the barrels and the man amounted to $31\frac{1}{2}$ cwt. Examination of the brake showed that it was in a bad state of repair. The brake linings were badly worn and cracked, and the brake was obviously in no condition to sustain the car even when carrying its proper load.

Regular and careful maintenance would have prevented this accident.

(31) *No.* 56, *July* 1963. A lift in an old building was fitted with collapsible gates, the spaces between the pickets being $4\frac{1}{2}$ in. Intermediate "toe pickets" had been fitted between the main pickets at the bottoms of the gates to a height of about 2 in. above the floor, thus reducing the spaces here to about 2 in. One of the toe pickets had broken off and a girl travelling upwards in the car had her foot in the broken space. Her foot was crushed between the car floor and the underside of a landing sill. The Hoists Exemption Order 1962, para. 14, requires that all gates on lifts installed after April 1962 shall

have no openings exceeding $2\frac{1}{2}$ in. in width. Efficient maintenance would have prevented this accident.

(32) *No. 56, July* 1963. Four men were taking a large truck to an upper floor in a factory lift, the last man entering, standing in the only available space about 7 in. wide along the front edge of the car floor. In this position he found it impossible to close the car gate so he reached up and held the car gate interlock in the closed position. One of the men set the car in motion and the man on the front edge had his feet badly crushed when they were caught by a projection in the well. This is an example of overcrowding the car and of tampering with one of the safety devices.

(33) *No. 56, July* 1963. A factory service lift had landing openings about 3 ft. above floor level. A man was loading heavy parcels into the car when it suddenly moved downwards and he was trapped between the top of the car and the landing sill and severely injured. The maximum working load was not marked on the lift as required by the Factories Act. It was later ascertained that this was 3 cwt and that the weight of the loaded parcels was 6 cwt, this causing the ropes to slip in the sheave grooves.

(34) *No. 56, July* 1963. A number of accidents have occurred when a lift has moved away from a landing with the gates still open. This is likely to happen when there is poor lighting. A common cause of this type of accident is incorrect wiring of the gate safety interlocks. The most usual arrangement is for the control fuse, the safety interlocks and the motor main contactors to be joined in series across the d.c. control supply with its negative pole earthed. The correct method of wiring is for the fuse and safety interlocks to be wired in this order, directly to the positive pole and the motor contactors to the earthed negative. In these circumstances an earth fault on any of the safety interlocks will blow the fuse and prevent the lift moving. If, however, the safety interlocks are joined direct to the earthed negative, an earth fault on one of the interlocks would still permit the motor to run with the gate open.

The above cases are typical examples of lift accidents, from which the usual causes of such accidents will be evident. The conclusions to be drawn from the information provided in this chapter are that if lifts are constructed and maintained in such a manner that the requirements of the B.S. Code of Practice and the Factories Act are fulfilled and the safety recommendations made in this book are adopted, the possibility of an accident occurring will be very remote.

BIBLIOGRAPHY

1. *Factories Act*, 1961—Part II, Sections 22, 23, 24, 25, 26 and 48. (H.M. Stationary Office).

2. *The Hoists Exemption Order*, 1962 (H.M. Stationery Office).

3. Form 276, March, 1944—*Precautions in the Installation and Working of Hoists and Lifts* (H.M. Stationery Office).

4. Post-war Building Studies No. 9—*Mechanical Installations* (H.M. Stationery Office).

5. *Code of Practice for Electric Lifts and Escalators* (The Building Industries National Council).

6. *Safety Code for Elevators*, American Standard (The American Society of Mechanical Engineers).

7. *Inspection of Elevators, Inspectors' Manual*, American Standard (The American Society of Mechanical Engineers).

8. *Regulations for the Electrical Equipment of Buildings* (Institution of Electrical Engineers).

9. "Modern Electric Lift Practice," by L. S. Atkinson, *Journal of Institution of Electrical Engineers*, 1946.

10. "Electrical Control of Dangerous Machinery and Processes," by W. Fordham Cooper, *Journal of Institution of Electrical Engineers*, 1947.

11. B.S. 205, *Glossary of Terms Used in Electrical Engineering*, Part 7, Section 10 (British Standards Institution).

12. B.S. Handbook No. 4, *Lifting Tackle* (British Standards Institution).

13. B.S. 329, *Wire Ropes for Lifts and Hoists* (British Standards Institution).

14. B.S. 621, *Wire Ropes of Special Construction* (British Standards Institution).

15. B.S. 462, *Bull-dog Grips for Wire Ropes* (British Standards Institution).

16. B.S. 463, *Sockets for Wire Ropes* (British Standards Institution).

17. B.S. 464, *Thimbles for Wire Ropes* (British Standards Institution).

452   ELECTRIC LIFTS

18. B.S. 525, *Fibre Cores for Wire Ropes* (British Standards Institution).

19. B.S. 643, *Capping Metal for Steel Wire Ropes* (British Standards Institution).

20. B.S. 721, *Worm Gearing* (British Standards Institution).

21. B.S. 977, *Braided Travelling Cables for Electric Lifts* (British Standards Institution).

22. "Time-Velocity Characteristics of the High-speed Passenger Elevator," by Bassett Jones, *General Electric Review*, Feb., 1924.

23. "Note on the Probable Number of Stops made by an Elevator," by Bassett Jones, *General Electric Review*, June, 1926.

24. "Some Factors in Roping Traction Elevators," by C. C. Clymer, *General Electric Review*, Nov., 1927.

25. "A Measuring Stick for Elevator Service," by H. B. Cook, *Power*, 7th April, 1931.

26. "Selecting Elevators for an Office Building," by H. B. Cook, *Power*, 15th March, 1932.

27. "Passenger Lifts," by W. A. Dixie, *British Engineering*, Jan., 1949.

28. "Lifts in Ships," *The Shipbuilder and Marine Engine-builder*, Feb., 1952.

29. "Hospital Lifts," by B. P. Hutton, *Hospital and Health Management*, June, 1952.

30. "Lifts," by L. J. Gooch, *Coventry Engineering Society Journal*, Sept.–Oct., 1952.

31. "Modern Trends in Lift Design," by B. P. Hutton, *Electrical Industries*, Oct., 1952.

32. "Electric Lifts in Post-war Housing," by C. G. L. Morley, *Proceedings of the Institution of Electrical Engineers*, Vol. 101, Part II, No. 80, April, 1954.

33. "Standardization of Lift Control Systems," by W. A. Dixie, *G.E.C. Journal*, Vol. XXI, No. 3, July, 1954.

34. "The Application of Lifts for Office Buildings," by W. A. Dixie, *Architectural Design*, March, 1955.

35. B.S. 2655, "Electric Lifts," Parts 1, 2 and 3 (British Standards Institution).

36. B.S. Code of Practice, C.P. 407.101, *Electric Lifts for Passengers, Goods and Service* (British Standards Institution).

37. B.S. Code of Practice, *Fire Precautions in Flats and Maisonettes over 80 ft. in Height* (British Standards Institution).

38. "Installation of Electronically Controlled Passenger Lifts," by P. S. Shilston, *Metropolitan Vickers Gazette*, Nov., 1957.

39. "Transductor-controlled Ward-Leonard Drives for Lifts," by J. Sidler and O. Kolb, *Brown Boverie Review*, Nov., 1957.

40. "Modern Control Systems for Groups of Lifts," by S. T. Hunt, *Electrical Energy*, Jan., 1959.

41. "Speed Control of Lifts," by D. M. C. Dick and D. Embrey, *Electrical Review*, 6 Feb., 1959.

42. "High-speed Lifts of the Centre International Rogier," *The Engineer*, 20 Jan., 1961.

43. "Variable Speed Control Systems," by J. Ben Uri, *Electrotechnology*, March, 1961.

44. "Drive and Control of High-speed Lifts," *Engineer*, 28 July, 1961.

45. *Service Cores in High Flats—The Selection and Planning of Passenger Lifts*, Ministry of Housing and Local Government Design Bulletin, No. 3, Part 2 (H.M. Stationery Office, 1962).

46. "The Costs of Lifts in Multi-storey Flats for Local Authorities," by T. L. Knight and A. E. Duck, *Chartered Surveyor*, Feb., 1962.

47. "A high-performance Elevator Control System," by K. A. Oplinger, L. A. Bobula, A. O. Lund and W. M. Ostrander, *Electrical Engineering*, March, 1962.

48. "Electronic Lift Control," *Engineer*, 17 Aug., 1962.

49. "Ward-Leonard Drives using Silicon-controlled Rectifiers," *Direct Current*, Sept., 1962.

50. "Present State of Ward-Leonard Control Systems," *Direct Current*, July, 1963.

51. "Static Switching applied to Lift Controllers," by V. A. Gault, *British Communications and Electronics*, Sept., 1963.

52. "Automatic Control of Groups af Lifts," by S. T. Hunt and R. J. Bedford, *G.E.C. Journal*, Vol. 31, No 2, 1964.

53. "Control of High-speed Lifts—A Continuous-pattern System," by S. T. Hunt, *G.E.C. Journal*, Vol. 31, No. 3, 1965.

54. "Passenger Lifts in the Post Office Tower, London," by P. E. Marriott, *P.O. Electrical Engineers Journal*, Vol. 58, Part 4, Jan. 1966.

APPENDIX I

Section 22. Hoists and Lifts—General

(1) Every hoist or lift shall be of good mechanical construction, sound material and adequate strength, and shall be properly maintained.

(2) Every hoist or lift shall be thoroughly examined by a competent person at least once in every period of six months and a report of the result of every such examination in the prescribed form and containing the prescribed particulars shall be signed by the person making the examination and shall within twenty-eight days be entered in or attached to the general register.

(3) Where the examination shows that the hoist or lift cannot continue to be used with safety unless certain repairs are carried out immediately or within a specified time, the person making the report shall within twenty-eight days of the completion of the examination send a copy of the report to the inspector for the district.

(4) Every hoistway or liftway shall be efficiently protected by a substantial enclosure fitted with gates, and the enclosure shall be such as to prevent, when the gates are shut, any person falling down the way or coming into contact with any moving part of the hoist or lift.

(5) Any such gate shall, subject to subsection (6) of this section and to section twenty-five of this Act, be fitted with efficient interlocking or other devices to secure that the gate cannot be opened except when the cage or platform is at the landing and that the cage or platform cannot be moved away from the landing until the gate is closed.

(6) If in the case of a hoist or lift constructed or reconstructed before the thirtieth day of July, nineteen hundred and thirty-seven, it is not reasonably practicable to fit it with such devices as are mentioned in subsection (5) of this section, it shall be sufficient if the gate—

    (a) is provided with such arrangements as will secure the objects of that subsection so far as is reasonably practicable, and (*b*) is kept closed and fastened except when the cage or platform is at rest at the landing.

(7) Every hoist or lift and every such enclosure as is mentioned in subsection (4) of this section shall be so constructed as to prevent any part of any person or any goods carried in the hoist or lift from being trapped between any part of the hoist or lift and any fixed structure or between the counterbalance weight and any other moving part of the hoist or lift.

(8) There shall be marked conspicuously on every hoist or lift the maximum working load which it can safely carry, and no load greater than that load shall be carried on any hoist or lift.

Section 23. Hoists and Lifts used for Carrying Persons

(1) The following additional requirements shall apply to hoists and lifts used for carrying persons, whether together with goods or otherwise—

(*a*) efficient automatic devices shall be provided and maintained to prevent the cage or platform overrunning;

(*b*) every cage shall on each side from which access is afforded to a landing be fitted with a gate, and in connection with every such gate efficient devices shall be provided to secure that, when persons or goods are in the cage, the cage cannot be raised or lowered unless the gate is closed, and will come to rest when the gate is opened.

(2) In the case of a hoist or lift constructed or reconstructed before the thirtieth day of July, nineteen hundred and thirty-seven, in connexion with which it is not reasonably practicable to provide such devices as are mentioned in paragraph (*b*) of subsection (1) of this section it shall be sufficient if—

(*a*) such arrangements are provided as will secure the objects of that paragraph so far as is reasonably practicable; and

(*b*) the gate is kept closed and fastened except when the cage is at rest or empty.

(3) In the case of a hoist or lift used as mentioned in sub-section (1) of this section which was constructed or reconstructed after

the twenty-ninth day of July, nineteen hundred and thirty seven, where the platform or cage is suspended by rope or chain, there shall be at least two ropes or chains separately connected with the platform or cage, each rope or chain and its attachments being capable of carrying the whole weight of the platform or cage and its maximum working load, and efficient devices shall be provided and maintained which will support the platform or cage with its maximum working load in the event of a breakage of the ropes or chains or any of their attachments.

Section 24. Teagle Openings and Similar Doorways

(1) Every teagle opening or similar doorway used for hoisting or lowering goods or materials, whether by mechanical power or otherwise, shall be securely fenced and shall be provided with a secure hand-hold on each side.

(2) The fencing shall be properly maintained and shall, except when the hoisting or lowering of goods or materials is being carried on at the opening or doorway, be kept in position.

Section 25. Exceptions and Supplementary Provisions

(1) For the purposes of sections twenty-two and twenty-three of this Act, no lifting machine or appliance shall be deemed to be a hoist or lift unless it has a platform or cage the direction of movement of which is restricted by a guide or guides.

(2) Subsections (3) to (8) of section twenty-two and section twenty-three of this Act shall not apply in the case of a continuous hoist or lift, and in such a case subsection (2) of the said section twenty-two shall have effect as if for the reference to six months there were substituted a reference to twelve months.

(3) Subsections (5) and (6) of the said section twenty-two and the said section twenty-three shall not apply in the case of a hoist or lift not connected with mechanical power; and in such a case—

(a) subsection (2) of the said section twenty-two shall have effect as if for the reference to six months there were substituted a reference to twelve months; and (*b*) any gates to be fitted under subsection (4) of the said section twenty-two shall be kept closed and fastened except when the cage or platform is at rest at the landing.

(4) If it is shown to the satisfaction of the Minister that it would be unreasonable in the special circumstances of the case to enforce any requirement of sections twenty-two to twenty-four of this Act or of subsection (3) of this section in respect of any class or description of hoist, lift, hoistway, liftway or teagle opening or similar doorway, he may by order direct that the requirement shall not apply as respects that class or description.

Section 26. Chains, Ropes and Lifting Tackle

(1) The following provisions shall be complied with as respects every chain, rope or lifting tackle used for the purpose of raising or lowering persons, goods or materials—

(*a*) no chain, rope or lifting tackle shall be used unless it is of good construction, sound material, adequate strength and free from patent defect;

(*b*) subject to subsection (2) of this section, a table showing the safe working loads of every kind and size of chain, rope or lifting tackle in use, and, in the case of a multiple sling, the safe working load at different angles of the legs, shall be posted in the store in which the chains, ropes or lifting tackle are kept, and in prominent positions on the premises, and no chain, rope or lifting tackle not shown in the table shall be used;

(*c*) no chain, rope or lifting tackle shall be used for any load exceeding its safe working load as shown by the table mentioned in paragraph (*b*) of this subsection or marked as mentioned in subsection (2) of this section;

(*d*) all chains, ropes and lifting tackle in use shall be thoroughly examined by a competent person at least once in every period of six months or at such greater intervals as the Minister may prescribe;

(*e*) no chain, rope or lifting tackle, except a fibre rope or fibre rope sling, shall be taken into use in any factory for the first time in that factory unless it has been tested and thoroughly examined by a competent person and a certificate of the test and examination specifying the safe working load and signed by the person making the test and examination has been obtained and is kept available for inspection;

(*f*) every chain and lifting tackle except a rope sling shall, unless of a class or description exempted by certificate of the chief inspector upon the ground that it is made of such material or so constructed that it cannot be subjected to heat treatment without risk of damage or that it has been subjected to some form of heat treatment (other than annealing) approved by him, be annealed at least once in every fourteen months or, in the case of chains or slings of half-inch bar or smaller, or chains used in connection with molten metal or molten slag, in every six months, except that chains and lifting tackle not in regular use need be annealed only when necessary;

(*g*) a register containing the prescribed particulars shall be kept in respect of all such chains, ropes or lifting tackle, except fibre rope slings.

(2) Paragraph (*b*) of subsection (1) of this section shall not apply in relation to any lifting tackle if its safe working load or, in the case of a multiple sling, the safe working load at different angles of the legs is plainly marked upon it.

(3) In this section "lifting tackle" means chain slings, rope slings, rings, hooks, shackles and swivels.

Section 12, Para. (3). Safety (Prime Movers)

Every part of electric generators, motors, and rotary converters, and every flywheel directly connected thereto, shall be securely fenced unless it is in such a position or of such construction as to be as safe to every person employed or working on the premises as it would be if securely fenced.

Section 13. Safety (Transmission Machinery)

(1) Every part of the transmission machinery shall be securely fenced unless it is in such a position or of such construction as to be as safe to every person employed or working on the premises as it would be if securely fenced.

(2) Efficient devices or appliances shall be provided and maintained in every room or place where work is carried on by

which the power can promptly be cut off from the transmission machinery in that room or place.

(3) No driving-belt when not in use shall be allowed to rest or ride upon a revolving shaft which forms part of the transmission machinery.

Section 14. Safety (Other Machinery)

(1) Every dangerous part of any machinery, other than prime movers and transmission machinery, shall be securely fenced unless it is in such a position or of such construction as to be as safe to every person employed or working on the premises as it would be if securely fenced:

(2) In so far as the safety of a dangerous part of any machinery cannot by reason of the nature of the operation be secured by means of a fixed guard, the requirements of subsection (1) shall be deemed to have been complied with if a device is provided which automatically prevents the operator from coming into contact with that part.

Section 17. Construction and Sale of Machinery

(1) In the case of any machine in a factory being a machine intended to be driven by mechanical power—

(*a*) every set-screw, bolt or key on any revolving shaft, spindle, wheel or pinion shall be so sunk, encased or otherwise effectively guarded as to prevent danger; and

(*b*) all spur and other toothed or friction gearing, which does not require frequent adjustment while in motion shall be completely encased unless it is so situated as to be as safe as it would be if completely encased.

(2) Any person who sells or lets on hire or, as agent of the seller or hirer, causes or procures to be sold or let on hire for use in a factory in the United Kingdom any machine intended to be driven by mechanical power which does not comply with the requirements of this section shall be guilty of an offence and liable to a fine not exceeding £200.

Section 48, Para. (4). Safety Provisions in Case of Fire

Every hoistway or liftway inside a building constructed after the end of June 1938 shall be completely enclosed with fire-resisting materials, and all means of access to the hoist or lift shall be fitted with doors of fire-resisting materials: except that any such hoistway or liftway which is not provided with a vent at the top shall at the top be enclosed only by some material easily broken by fire.

The Hoists Exemption Order, 1962

This order exempts certain classes or descriptions of hoists and hoistways from some of the requirements of sections 22, 23 and 25 of the Factories Act 1961. It specifies fourteen different classes of hoist or hoistway and the subsections of the Factories Act 1961 which shall not apply in each case, subject to certain stated conditions or limitations. In the Order the expression "hoist" includes a lift and the expression "hoistway" includes a liftway.

Statutory Rules and Orders, 1908, No. 1312

REGULATION 15.

Every switchboard, having bare conductors normally so exposed that they may be touched, shall, if not located in an area or areas set apart for the purposes thereof, where necessary be suitably fenced or enclosed.

No person except an authorized person, or a person acting under his immediate supervision, shall for the purpose of carrying out his duties have access to any part of an area so set apart.

REGULATION 17.

At the working platform of every switchboard and in every switchboard passageway if there be bare conductors exposed or arranged to be exposed when live so that they may be touched, there shall be a clear and unobstructed passage of ample width and height with a firm and even floor. Adequate means of access, free from danger, shall be provided for every switchboard passageway.

The following provision shall apply to all such switchboard working platforms and passageways, unless the bare conductors, whether overhead or at the sides of the passageways,

are otherwise adequately protected against danger by devices or screens or other suitable means—

(*a*) those constructed for pressures below 650 volts shall have a clear height of not less than 7 ft. and a clear width, measured from bare conductor of not less than 3 ft.

(*c*) bare conductors shall not be exposed on both sides of the switchboard passageway unless either—

(i) the clear width of the passage is in the case of pressures not exceeding 650 volts not less than 4 ft. 6 in. measured between bare conductors, or

(ii) the conductors on one side are so guarded that they cannot be accidentally touched.

REGULATION 21.

Where necessary to prevent danger, adequate precautions shall be taken either by earthing or by other suitable means to prevent any metal other than the conductor from becoming electrically charged.

APPENDIX II

TABLE I

DIAMETERS AND BREAKING STRENGTHS OF 6 × 19 ROPES

1	2	3
Diameter of Rope (in.)	Nominal Breaking Strength	
	Tensile Strength of Wire 70/80 tons per sq. in. (tons)	Tensile Strength of Wire 80/90 tons per sq. in. (tons)
$\frac{1}{4}$	1·5	1·7
$\frac{5}{16}$	2·3	2·7
$\frac{3}{8}$	3·3	3·8
$\frac{7}{16}$	4·6	5·2
$\frac{1}{2}$	6·0	6·8
$\frac{9}{16}$	7·6	8·7
$\frac{5}{8}$	9·4	10·7
$\frac{11}{16}$	11·4	12·9
$\frac{3}{4}$	13·6	15·4
$\frac{13}{16}$	15·9	18·0
$\frac{7}{8}$	18·5	20·9
$\frac{15}{16}$	21·2	24·0

For six-strand Seale ropes of dual tensile strengths, having outer wires of 70/80 tons per sq. in. tensile strength and inner wires of 110/120 tons per sq. in. tensile strength, the approximate breaking strengths are as given in column 3 above.

To find the aggregate breaking strength add $14\frac{1}{4}$ per cent to the figures given in columns 2 and 3.

TABLE II

DIAMETERS AND BREAKING STRENGTHS OF 8 × 19 ROPES

1	2	3
Diameter of Rope (in.)	Nominal Breaking Strength	
	Tensile Strength of Wire 70/80 tons per sq. in. (tons)	Tensile Strength of Wire 80/90 tons per sq. in. (tons)
$\frac{5}{16}$	2·2	2·5
$\frac{3}{8}$	3·1	3·5
$\frac{7}{16}$	4·3	4·8
$\frac{1}{2}$	5·6	6·3
$\frac{9}{16}$	7·0	8·0
$\frac{5}{8}$	8·7	9·8
$\frac{11}{16}$	10·5	11·9
$\frac{3}{4}$	12·5	14·2
$\frac{13}{16}$	14·7	16·6
$\frac{7}{8}$	17·0	19·3
$\frac{15}{16}$	19·5	22·2

For eight-strand Seale ropes of dual tensile strengths, having outer wires of 70/80 tons per sq. in. tensile strength and inner wires of 110/120 tons per sq. in. tensile strength, the approximate breaking strengths are as given in column 3 above.

To find the aggregate breaking strength add 21 per cent to the figures given in columns 2 and 3.

TABLE III

DIAMETERS AND BREAKING STRENGTHS OF FLATTENED-STRAND 6 × 25 ROPES

1	2
Diameter of Rope (in.)	Nominal Breaking Strength
	Tensile Strength of Wire 80/90 tons per sq. in. (tons)
$\frac{3}{8}$	4·0
$\frac{7}{16}$	5·7
$\frac{1}{2}$	7·4
$\frac{9}{16}$	9·1
$\frac{5}{8}$	11·7
$\frac{11}{16}$	14·3
$\frac{3}{4}$	17·0
$\frac{13}{16}$	20·0
$\frac{7}{8}$	23·1
$\frac{15}{16}$	26·0

To find the aggregate breaking strength add $17\frac{1}{2}$ per cent to the figures given in column 2.

APPENDIX III

INTERFERENCE WITH RADIO AND TELEVISION
RECEPTION DUE TO ELECTRIC LIFTS

Sources of Interference. The items of lift equipment which may cause interference with wireless reception consist of the driving motor, which gives a continuous noise in the receiver, and the controller, giving noises of the "click" type. The radiated interference is further increased because the interfering sources are connected to the car flexible cable and to the wiring between gate interlocks and the controller and the supply mains. The latter wiring is frequently in steel conduit, but the flexible cable normally has no metallic protection. Interference may thus reach the wireless receiver by way of the electric mains or by direct radiation from the motor and controller, from wiring in the well, or from the flexible cable acting as a radiating aerial. In addition, the metal enclosure of the well is frequently inadequately earthed, and this gives rise to voltages along the well which may cause serious interference.

The radiated motor interference is less than that due to the control circuits if the motor is properly maintained, the motor noise being negligible if a motor generator set is employed, whilst even for d.c. motors with direct mains supply the noise is generally less than the controller noise except where the motors are not efficiently maintained. Auxiliary motors such as brake or door opening motors are also possible sources of interference.

Controller interference is mainly due to the circuits controlling the driving motor contactor coils and the brake coil, these being most troublesome when supplied direct from the mains. All circuits for remote control, particularly those entering the flexible cable, form systems which give a high frequency radiation, whilst the brake magnet is also a radiating source. A.c. controllers usually cause very little disturbance.

The interference from the sources mentioned above appears also as interfering voltages at the supply terminals, thus causing mains-borne interference.

Methods of Suppression. Two methods are employed to prevent interference at the source—

(*a*) By preventing the generation of the interference.

(*b*) By limiting the effect of the interference and preventing its radiation or its passage to the power supply mains.

The first method can often be adopted on the controller switches by incorporating spark quenching circuits, but in the majority of cases it is necessary to prevent the interference from leaving the source. This is done by screening as far as radiation is concerned, and by the use of filter circuits to prevent the passage of the disturbing radio-frequency currents to the supply mains.

Lift Motors. Most interference is caused by commutator machines and is usually worse if sparking is present at the commutators. In the case of the smaller d.c. motors, capacitors alone often give adequate suppression. Two capacitors (either 2 μF or 4 μF) joined in series across the supply mains are usually sufficient for suppression in the sound broadcast wavebands. The centre point of the capacitors must be connected to the frame of the machine which should be efficiently earthed. A greater degree of suppression is often obtained, however, when the capacitors are connected directly to the brushgear with the leads as short as possible. For the larger d.c. machines two radio-frequency chokes may be necessary in addition to capacitors. The chokes should be about 600 μH, connected one in each main lead to the motor, either on the mains or motor side of the capacitors. If the motor impedance is low and the mains impedance high, the chokes should be fitted on the motor side of the capacitors and vice versa. If suppression is required in the television wavebands the most practicable method is to fit small mica or ceramic capacitors of 300 or 400 $\mu\mu$F between the line terminals and the frame or between the brushes and the frame.

A.c. motors, with the exception of commutator motors, do not usually cause as much interference as d.c. motors. Suppression in the case of a.c. commutator motors is effected by a choke condenser filter similar to that described for d.c. motors. The single-phase repulsion-induction motor is an interfering type, however, and since the armature is normally short-circuited, the fitting of capacitors across the brushes cannot effect a cure. It is therefore necessary to insert capacitors and chokes in the mains leads as for d.c. motors. Induction motors of both the squirrel-cage and slip-ring types do not usually give trouble if they are properly maintained.

Control Circuits. The radiated interference due to the control circuits entering the car flexible cable may be suppressed by chokes of about 6 mH connected in each lead of the flexible cable both at the car and the controller ends. In addition, capacitors of 0·5 μF or 1 μF may be connected from each lead to earth on the flexible cable side of the chokes. Suppression may sometimes be obtained by mounting the chokes at the half-way box in the well instead of in the car, whilst the chokes at the controller end may often be omitted. The former method is generally necessary when the control circuits are energized from the full mains voltage.

Control circuits such as gate interlocks which do not enter the flexible cable are treated in a similar manner, filters being required at the operating switch and the relay coil. If the wiring is in conduit or screened, the chokes may often be omitted from the filters.

The coils of the circuit-breaker and controller contacts should be fitted with capacitors of 1 μF or 2 μF capacitance connected between each end of the coil and earth. It is often advantageous to connect capacitors up to 4 μF across the car switch contacts. Interference due to the brake magnet coil can be suppressed by an arc suppressor of the rectifier type. The metal work of the well should be efficiently earthed.

Any interference passing into the mains supply can be reduced by a choke capacitor filter in the mains.

Suppression Practice. The suppression of interference from lifts usually resolves itself into designing separate suppressor circuits for the sound broadcast bands (medium and low frequency) and the television bands, as the components required for these two bands differ very considerably in their characteristics. At high frequencies, capacitors appear electrically as series tuned circuits, the inductance being that of the foil windings and connecting leads. Capacitors must therefore be chosen that have a low impedance throughout the frequency range to be suppressed. Inductors, however, appear electrically as parallel tuned circuits, the capacitance being that of the inductor windings. Inductors must be chosen therefore, that have a large impedance throughout the frequency range to be suppressed.

Television-band components must be mounted either inside or very close to the machine or source of interference to minimize radiation from the leads connecting the suppressor. Broadcast-band interference being principally mains-borne, the suppressors may be mounted several feet away without their performance being appreciably affected.

The position chosen for the components used and their suppression characteristics are governed partly by the physical construction of the machine and partly by its electrical characteristics in the radio frequency spectrum involved. It is rarely possible to assess the optimum suppressor arrangement by inspection of the lift machine. This leads to a trial-and-error method of assessing the most suitable arrangement of components and their physical location in the machine. This can best be achieved by electrically isolating individual motors or contactors, energizing them and observing the interference caused. Although an interference-measuring set is used for measuring the level of interference, a good deal of preliminary work can be done with a normal broadcast receiver by listening to the effect of fitting the suppressors. For example, in suppressing push-button circuits, all parts of the circuit should be isolated except the supply to one of the car push-buttons and its associated contactor coil. The level of interference is then measured by placing the measuring set at a convenient distance from the lift. An inductor of value 1–2 mH should be placed on each side of the push-button and its effectiveness checked by measuring the residual interference level. This should comply with the requirements of B.S. 800 which specifies limits of the magnitude of radio-noise terminal voltages and radio-noise fields throughout the ranges 200 kc/s to 1605 kc/s and 40 Mc/s to 70 Mc/s. Suppression found to be effective on one push-button will usually be satisfactory on all other push-buttons where the operating conditions are similar.

If inductors are not having any appreciable effect on the measured interference level it indicates that the contacts to which they are fitted are not the prime source of the interference. Careful investigation will usually show that the mechanical shock to the control panel due to the contactor operating

has caused a momentary break in another circuit carrying current via contactor springs. The circuit causing the interference may not normally operate while carrying current, i.e. it may be a safety or interlock circuit. It will, however, require suppression unless contacts can be made completely vibration-proof. The conditions can usually be produced by tapping the panel.

The Post Office Engineering Department, Radio Interference Group, has the responsibility of carrying out tests to locate the source of interference to radio and television receivers and of ascertaining the type of suppressors required to remove the interference. Although the Post Office maintains stocks of interference-suppression components, the cost of any suppressors required must be borne by the owner of the offending equipment. In recent years there has been a great deal of co-operation between Post Office engineers and the lift manufacturers to reduce interference caused by lifts, and the manufacturers are now generally aware of the steps necessary to suppress their own lift equipment.

BIBLIOGRAPHY

A. S. Angwin, Institution of Post Office Electrical Engineers printed paper No. 137.

A. J. Gill and S. Whitehead, "Electrical Interference with Radio Reception," *I.E.E. Journal*, Vol. 83, p. 345.

A. Morris, "Interference of Electrical Plant with the Reception of Radio Broadcasting," *I.E.E. Journal*, Vol. 74, p. 245.

B.S. Code of Practice C.P. 1006, *General Aspects of Radio Interference Suppression* (British Standards Institution).

Post Office Elect. Engrs. J., Vol. 50, Part 4, p. 226.
ibid.   Vol. 51, Part 1, p. 40.
ibid.   Vol. 51, Part 2, p. 115.
ibid.   Vol. 51, Part 3, p. 202.
ibid.   Vol. 52, Part 1, p. 43.

B.S. 2655: Part 1, *Electric Lifts. General Requirements* (British Standards Institution).

APPENDIX IV

NOTES ON WEAR OF WIRE ROPES*

The recommendations regarding working loads given in Appendix II and those relating to minimum diameters of sheaves, pulleys and drums, stated in Chapter IV afford sufficient margin of safety, taking into account bending and reasonable rope life. In the design of new lifts care must be given to securing this reasonable rope life by requiring sheaves and pulleys to be of adequate diameter and to have grooves of correct size and shape, ensuring that sheaves and pulleys are correctly aligned, that the working load is reasonable and, as far as possible, that reverse bends in the roping system are avoided.

In selecting a wire rope, consideration should be given to the diameter of the single wires forming the strand. Wire less than 24 s.w.g. (0·022 in.) allows little wear and 14 s.w.g. (0·080 in.) is about the largest which should normally be used. A range of 22 s.w.g. (0·028 in.) to 16 s.w.g. (0·064 in.) is considered good practice, and on this basis round strand 6/19 ropes should be between $\frac{7}{16}$ in. and $\frac{15}{16}$ in. diameter. Where considerable wear is anticipated it is advisable to install a rope with large diameter outer wires, and this suggests a rope of Seale construction. The construction of a Lang's lay rope, however, exposes a greater length of wire to wear, but Lang's lay has a greater tendency to untwist than ordinary lay ropes. Provided care is taken in handling, this feature is not a disadvantage for lift work, as both ends are secured against rotation. Preformed Lang's lay rope is, of course, not subject to untwisting.

Wire ropes fail chiefly because of abrasive wear of the outer wires or fatigue of the material. The flats on the outer wires can readily be seen, and it is of interest to note that a flat of three-quarters of the diameter of the wire in width reduces the sectional area of the wire by approximately 10 per cent. If there is no internal corrosion or fatigue the inner wires form a

* See B.S. Handbook No. 4—" Lifting Tackle."

reserve of strength which is not affected by the wear on the outer wires, and this reserve is greater the larger the number of layers in the strand.

It is good practice to replace any rope in which a given number of wires in a specified length are broken. The number of broken wires which necessitate the replacement of a rope should bear a relation to the total number of wires in the strand and the proportion of outer to inner wires. The Statutory Rules and Orders under the Docks Regulations, Section VII, clause 20 (*c*), provide—

"No wire rope shall be used in hoisting or lowering if in any length of eight diameters the total number of visible broken wires exceeds 10 per cent of the total number of wires, or the rope shows signs of excessive wear, corrosion, or other defect which, in the opinion of the person who inspects it, renders it unfit for use."

The 1945 Draft Revision of the Building Regulations, Part III, clause 58 (2) provides—

"No wire rope shall be used in raising or lowering or as a means of suspension if in any length of ten diameters the total number of visible broken wires exceeds 5 per cent of the total number of wires in the rope."

In the case of normal six-strand lift ropes, 10 per cent of the total number of wires in eight diameters is equivalent to one wire in twelve in a length equal to one lay, while 5 per cent of the total number of wires in ten diameters is equivalent to one wire in thirty in a length equal to one lay. In the majority of ropes the visible broken wires are outers, but the distribution of the wires is important; there is a considerable difference between broken wires equally distributed between six strands and broken wires all in one or two strands. In the latter case a much smaller number of broken wires may render the rope dangerous.

It will be clear that a greater proportionate number of equally distributed broken outer wires can be tolerated in a 6/37 rope with a total of 222 wires, of which 108 are outers, than in a 6/19 rope with 114 wires, of which 72 are outers. In a 6/37 rope the reserve strength due to the inner wires is 51 per cent, and in a 6/19 rope only 37 per cent.

When inspecting a worn rope the number and distribution of

the broken wires and the amount of visible wear should be taken into account, and an estimate made of the percentage reduction from the original rope strength. If the original Factor of Safety was 12, then a reduction to 83 per cent of the original strength reduces the Factor of Safety to 10, and a reduction to 66 per cent reduces the factor to 8.

In estimating the strength of a worn rope it is convenient to relate the visible wear on the outer wires as equivalent to a reduction in the total number of wires in the rope, subtracting from the total so obtained the number of broken wires in any one lay. The equivalent reduction in the total number of wires can then be calculated as a percentage of the original number of wires. As an example, consider a worn 6/19 rope which has an estimated 20 per cent wear on the outer wires and 5 broken wires in one lay. This rope construction has 42 inner wires and 72 outer wires, making a total of 114 wires. When the rope is new each broken wire results in a percentage loss in strength of 0·877, and five broken wires in a loss of 4·38 per cent. As the wires are now 20 per cent worn, the loss due to these five is $\frac{80}{100}$ of 4·38 per cent = 3·5 per cent. So far as the wear of the wires is concerned, a 20 per cent wear on the outers is equivalent to a loss of $\frac{20}{100} \times 72$ wires = 14·4 wires, which is equivalent to 99·6 good wires remaining. Hence, the remaining percentage rope strength due to 20 per cent wear on the outers is $\frac{99\cdot6}{114} \times 100 = 87\cdot3$, and as the percentage reduction due to broken wires is 3·5, the total remaining percentage strength is therefore 83·8.

During operation, shock loading should, as far as possible, be avoided and gradual acceleration adds materially to the life of a wire rope. Where a rope is under constant load it is a good plan, if possible, to vary the position of the load when periodically at rest, over-night, or at week-ends. The same portion of the rope should not always rest on a sheave if its position can be varied when left standing.

Wire ropes are fully lubricated during manufacture to reduce friction between the wires and to prevent corrosion. In service this lubricant is gradually squeezed out, and it should, as far as possible, be replaced by periodical external applications. Where ropes pass over a sheave drive, however, the sheave grooves and the outer wires of the rope should be kept free of lubricant.

Where rapid wear is evident every effort should be made to ascertain the cause. Both sheaves or pulleys which do not revolve freely and grooves worn so that the ropes "bottom" cause abnormal wear. The method of roping and particularly the number of rope bends considerably affect rope life. The material and construction of the ropes and the relative hardness of the ropes and sheaves are important in minimizing wear. Probably the most important single factor in ensuring a satisfactory rope life is the reduction of the bending stress by employment of sheaves and pulleys of adequate diameters. Considerable wear may result if the ropes bind on the sides when leading on and off the sheave and pulleys. The number of normal stops made and the number of starts and stops due to "inching" by an inefficient attendant are important factors in rope life. High rates of acceleration and retardation which may cause slip will result in rope troubles as also will a badly adjusted brake or unequal loading on individual ropes. Insufficient internal lubrication may cause rusting and breaking of the inside wires.

INDEX

Landing indicator, *see* Indicator
Lang lay, 73, 74, 470
Lay, rope, *see* Ropes
Levelling—
　automatic, 118, 443
　corrective, 6, 286
　floor, 276, 422
　ramp, 287
　speed, 148
　switch, 287, 353, 361
　test, 434, 436
Lift—
　dimensions, 43–8
　layout, 36–42
Lifting joists, *see* Joists
Lighting, 165, 427
"Live" load, 48
Load indicator, 446, 455
Load on structure, 42–9, 65
Lock—
　car gate, 219, 272, 426, 449
　landing gate, 221, 272, 429, 445, 448
　requirements, 223
　wiring, *see* Wiring
Lubrication—
　gearing, *see* Gearing
　guide, *see* Guide

MACHINE room, 31, 42–9, 444
　area, 35, 43–7
　entrance, 35
　finish, 35
　floor, 36
　height, 35, 43–7
　position, 31, 34, 53, 65
　ventilation, 34
Maintenance, 416, 454
Mid-bar gate, *see* Gate
Mileage recorder, 322
Motors—
　a.c., 96, 466
　　commutator, 109, 466
　capacitor, 112
　compound, 89, 124

MARRYAT-SCOTT LIFTS

ESTABLISHED IN LONDON IN 1860

Marryat and Scott Ltd. with its predecessors, associates and subsidiaries, entered the Lift Industry in 1860 and began Electrical Engineering when electricity was hardly more than a new discovery.

The modern lifts now made in the Hounslow, Newcastle, Manchester and Bristol works have been developed progressively from that time and many of the original machines are still in regular use, maintained by our Service organisations with headquarters still in Hatton Garden.

MARRYAT & SCOTT Ltd.
AND ASSOCIATED COMPANIES

WELLINGTON WORKS, HOUNSLOW, MIDDLESEX

Telephone: Hounslow 7799.   Telegrams: Marryat Hounslow

Sales & Service Offices in

BIRMINGHAM, LIVERPOOL, BRISTOL, BRIGHTON, MANCHESTER, LEEDS
NEWCASTLE, GLASGOW, EDINBURGH, DUNDEE, DUBLIN AND OVERSEAS

PLAN WITH AEI—FOR THE RIGHT DRIVE

AEI can provide expert advice to lift-makers, architects and others, for electrical equipment which will 'fit the bill' exactly . . . AEI makes the widest range of equipment . . . Prompt quotations are based on computer analysis of basic needs. Consult AEI at the design stage on the exact requirements of the lift or escalator installation you have in view. This can save you money. Write for new illustrated publication 3861-71.

AEI

Associated Electrical Industries Ltd Motor & Control Gear Division
Industrial Control Systems Department, Rugby, Warwickshire

R06/140

The most
important part of the lift
is hidden away in the machine room.
Let us arrange for you to visit one,
and see for yourself the mechanical
excellence of H & C Lifts.

Hammond & Champness Ltd.

BLACKHORSE LANE, WALTHAMSTOW, E.17
TELEPHONE: LARKSWOOD 1071

H & C
LIFTS

BY APPOINTMENT TO H.M. THE QUEEN
MANUFACTURERS & SUPPLIERS OF PASSENGER LIFTS

THE EXPRESS LIFT CO., LTD.

(CONTROLLED BY THE GENERAL ELECTRIC CO., LTD.)

ABBEY WORKS
NORTHAMPTON

LONDON OFFICE: 9 GREYCOAT STREET, WESTMINSTER, S.W.1

LIFTS AND ESCALATORS

(REGISTERED TRADE MARK)

How to get to the top without even trying

There are several ways of getting to the top. Most involve a considerable expenditure of time and effort. The speediest means of achieving the desired end with the minimum of exertion is, as doubtless you have already appreciated, by lift. And a very important part of any lift is its Control System. The widest range of Lift Control equipment is that made by Dewhurst. Dupar components are largely standardised enabling practically any type of Lift Control System, from a simple push-button to the most advanced electronic multi-bank groups of installations, to be assembled readily and economically for all passenger and service applications.

for dependability

Consult the Dewhurst Technical Advisory Service

DEWHURST & PARTNER LTD
HOUNSLOW · MIDDLESEX

SHAW
LIFTS

for all

PURPOSES

R. J. SHAW & CO., LTD.

Manufacturers of all Types of Lift and Hoist Equipment

ABBOTS ROAD WORKS

GATESHEAD-ON-TYNE   Telephone 71796

WADSWORTH LIFTS

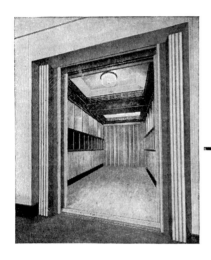

Electronic
V.V. Drive

'Static' variable voltage lift drive, pioneered by Wadsworth, requires no motor-generator set, as it is based on an electronically-controlled static rectifier. Control is considerably more precise; generator installation and maintenance cost are eliminated.

The introduction of electronic control was an advance comparable with the introduction of the variable voltage principle itself more than a quarter of a century before.

Wadsworth 'Static VV' gives maximum passenger comfort, precise levelling, and makes full use of the installed horsepower even for floor-to-floor working and all this is achieved with 'static' reliability. May we tell you more about it?

WM. WADSWORTH & SONS LTD. BOLTON & LONDON
WM. WADSWORTH & SONS (SA) (PTY) LTD., JOHANNESBURG & DURBAN

MCCANDLESS & BARTON